International Series in Operations Research & Management Science

Volume 268

More information about this series at http://www.springer.com/series/6161

Ralf Borndörfer • Torsten Klug
Leonardo Lamorgese • Carlo Mannino
Markus Reuther • Thomas Schlechte
Editors

Handbook of Optimization in the Railway Industry

Springer

Editors
Ralf Borndörfer
Zuse Institute Berlin &
Freie Universität Berlin
Berlin, Germany

Torsten Klug
Department of Optimization
Zuse Institute Berlin
Berlin, Germany

Leonardo Lamorgese
Optrail
Rome, Italy

Carlo Mannino
SINTEF DIGITAL &
University of Oslo
Oslo, Norway

Markus Reuther
LBW Optimization GmbH &
Zuse Institute Berlin
Berlin, Germany

Thomas Schlechte
LBW Optimization GmbH &
Zuse Institute Berlin
Berlin, Germany

ISSN 0884-8289 ISSN 2214-7934 (electronic)
International Series in Operations Research & Management Science
ISBN 978-3-030-10160-2 ISBN 978-3-319-72153-8 (eBook)
https://doi.org/10.1007/978-3-319-72153-8

Printed on acid-free paper

This Springer imprint is published by the registered company Springer International Publishing AG part of Springer Nature.
The registered company address is: Gewerbestrasse 11, 6330 Cham, Switzerland

Preface

The purpose of this book is to promote the use of mathematical optimization and operations research methods in rail transportation. We do so by describing the state of the art in 13 chapters that were contributed by leading scholars. For practitioners, these success stories give an impression of what is currently possible, and they give an inspiration for new applications. For researchers, the chapters provide an up-to-date reference and a broad survey of the field, although we can and do not make any claim of complete coverage.

There is a significant gap between what can be done and has successfully been tested in laboratory experiments, and what has actually been implemented in practice. This is startling considering the historical development of mathematical optimization, in which railway applications always played a central role. Tolstoǐ's early work on the transportation problem dealt with rail transport of salt, cement, and other goods in the Soviet Union (see Толстой [17]). The Max-Flow Min-Cut Theorem by Ford and Fulkerson [4] aimed at the military interdiction (disruption) of the Western Soviet and Eastern European railway systems. The article of Charnes and Miller [2] on the optimization of railway freight train movements is the first reference to a set covering model, whose use in large-scale applications was propelled by the influential F.A.S.T.ER competition of the Italian Operational Research Society AIRO and the Italian railway company Ferrovie dello Stato SpA (see, for instance, Caprara et al. [1]). The recoverable robustness concept of Liebchen et al. [10] was initially developed for timetabling and train platforming problems.

On the other hand, industrial installations appeared only recently. The system DONS (Designer of Network Schedules, see Hooghiemstra et al. [5]) is used by the Dutch infrastructure manager ProRail and the train operator Nederlandse Spoorwegen (NS) to compute countrywide timetables with the CADANS algorithm by Schrijver and Steenbeck [16], and routings through the main stations with the STATIONS algorithm by Zwaneveld [19]. The 2006 timetable constructed by DONS led to an increase in annual profits of 40 million Euros per year (see Kroon et al. [8]). This spectacular success won the prestigious INFORMS Franz Edelman Award for Achievement in Operations Research and the Management Sciences. NS was also a pioneer in crew optimization with the system TURNI (see Kroon and Fischetti [7]). This was followed by Deutsche Bahn as well as the Swedish Statens Järnvägar (see Kharraziha et al. [6]) and by Trenitalia with the system DS-OPT [3]. The optimization kernel RotOR is used by DB Fernverkehr AG in order to optimize the rotations of its ICE high-speed trains (see Reuther [14]). However, integrated and standardized resource planning systems such as ivu.rail (see Scholz [15]) just emerged. Much better established is the use of simulation systems such as OpenTrack [13] or RailSys [18] for capacity and service reliability assessment. Concerning optimal real-time traffic control, a system to dispatch trains was in operation in the Milano underground system in 2007 (see Mannino and Mascis [11]). A semi-automatic dispatching system controlling trains in the Lötschberg base tunnel, which is operated by the Swiss BLS, is in operation since the end of 2007 (see Montigel [12]). Finally, dispatching systems for main lines are in operation in Italy since 2011 (see Lamorgese and Mannino [9]).

There are several reasons for the slow penetration of the railway sector by mathematical optimization methods, including monopolistic structures and problems in the transfer from academia to the industry. The main reason, however, was a lack of algorithms that were capable of dealing with the large, complex, and highly integrated planning challenges that are typical for the railway industry. But this situation has changed.

With this book we intend to give proof that optimization methods are now able to keep some of the long-standing promises. Mathematical optimization can provide better use of capacity, improved efficiency, reduced infrastructure and operating costs, improved reliability, and more punctuality.

These potentials will be demonstrated in a collection of 13 chapters by leading experts in the field, covering almost the entire planning process. The methods described in these chapters are in most cases either already implemented and "up and running" or pilot installations. Indeed, the gap between theory and practice is now mostly in the practical implementation and their engineering.

The structure of the book is visualized in Fig. 1. There are three main clusters of articles, corresponding to the classical stages of the planning process: strategic, tactical, and operational. These clusters are further subdivided into five parts which correspond to the main phases of the railway network planning process: network assessment, capacity planning, timetabling, resource planning, and operational planning. The time-frame ranges from long-term strategic decisions on infrastructure and slots via mid-term tactical planning on the timetable and short-term allocation of resources to real-time operations support. Simulation is at the start to identify the needs and at the end to assess the results, and maybe identify new needs, such that the process starts all over.

The individual planning tasks can differ between passenger ( ) and freight ( ) railways, and therefore most articles have an application focus. There are also different levels of technical maturity presented, ranging from already established industrial standards (e.g., in crew scheduling and train dispatching) to academic proofs of concept. In many chapters the authors tried to explicitly point out the degree of operational readiness that has been achieved.

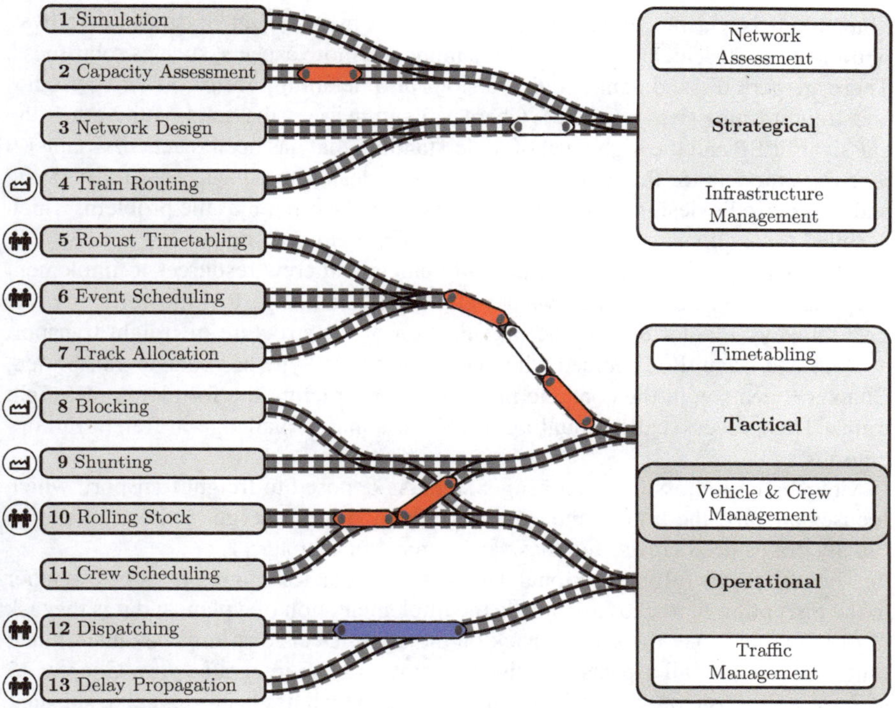

Fig. 1: The book's railway network of contents

We now briefly summarize the contents of each of the five parts of the book.

The first part on network assessment addresses the micro-macro approach in railway optimization. Mathematical models work on a coarsening of the original problem. Data is first aggregated to a macro level, while the solutions are later evaluated for real-world feasibility on a micro model. Simulation is the main tool to do that, and two chapters describe it. Chapter 1 is an overview of the state of the art in simulation methods with a focus on scheduling and train allocation. Simulation is not only useful to generate data for optimization, it can itself use optimization and be a part of an optimization loop.

Chapter 2 addresses the operational feasibility of timetables. Stability is assessed by using algebraic approaches and the outputs can be exploited to identify the need of new infrastructure enhancements.

The second part deals with decisions about the track infrastructure. The solution of these problems is crucial for the success of a railway because they have long-term implications. Chapter 3 is about long-term investments in infrastructure, while Chapter 4 is about the master structure of the freight transportation system.

The third part is devoted to timetabling. The timetable is the interface between the railway and the customers. Its structure is decided on a tactical level with a mid-term time horizon, in both passenger and freight rail. The classical way to approach timetabling is via the PESP model, which is a main object of study in railway optimization. Researchers try to include more and more aspects, such as robustness. There are periodic and nonperiodic versions of timetabling, for pax (more periodic) and freight trains (less periodic). Chapter 5 summarizes the state of the art in the classical PESP and the high level of understanding that has been reached. Chapter 6 discusses models for the robust case in order to deal with delays, both for periodic and non-periodic designs. Chapter 7 is on the freight version of the problem, which is called track allocation, and which is intrinsically non-periodic.

The fourth part is on the allocation of vehicle and crew resources to implement the timetable. Vehicles and crews account for the major part of the variable costs of a railway. Chapter 8 is on the organization of the structure of freight transport at Deutsche Bahn. It is decided what, how, and which type of goods is transported. Chapter 9 deals with the concrete process of constructing the formation of freight trains. This process, called shunting, is indeed a major challenge in freight rail operations.

Chapter 10 addresses the passenger side. As opposed to freight transport, which focuses on cars, the train is the center of attention in passenger rotation planning. Finally, trains need crews, and these are planned in Chapter 11.

The fifth part is fully operational. Unforeseen obstacles ranging from bad weather to the disruption of tracks can impede the implementation of a plan, and it is the task of dispatching to get the service back on the track. Decisions must be taken in real time, dealing with all aspects of railway operation in an integrated way. Chapter 12 addresses the basic problem of train dispatching, which has been subject to substantial mathematical advances. Dispatching systems based on mathematical methods are more and more becoming an industrial standard. Spread of delays is a major problem in passenger railways, whose difficulty stems from the way how trains interact in a railway network. Chapter 13 discusses approaches to tackle and mitigate these effects.

We dedicate this book to the memories of Alberto Caprara and Leo Kroon, who left us far too early. Their pioneering contributions to railway optimization will remain an inspiration for future generations, and keep them alive in our hearts and minds forever.

Berlin, Germany — Thomas Schlechte
Berlin, Germany — Torsten Klug
Berlin, Germany — Markus Reuther
Oslo, Norway — Carlo Mannino
Rome, Italy — Leonardo Lamorgese
Berlin, Germany — Ralf Borndörfer
19 September 2017

References

1. Caprara A, Fischetti M, Toth P (1999) A heuristic method for the set covering problem. Oper Res 47(5):730–743. https://doi.org/10.1287/opre.47.5.730 (cited on page v)
2. Charnes A, Miller MH (1956) A model for the optimal programming of railway freight train movements. Manage Sci 3(1):74–92. https://doi.org/10.1287/mnsc.3.1.74 (cited on page v)
3. DS-OPT (2017) LBW Optimization GmbH. lbw-optimization.com (cited on page vi)
4. Ford LR, Fulkerson DR (1956) Maximal flow through a network. J Can Math 8:399–404. https://doi.org/10.4153/cjm-1956-045-5 (cited on page v)
5. Hooghiemstra JS, Kroon LG, Odijk MA, Salomon, M, Zwaneveld PJ (1999) Decision support systems support the search for win-win solutions in railway network design. Interfaces 29(2):15–32. https://doi.org/10.1287/inte.29.2.15 (cited on page v)
6. Kharraziha H, Ozana M, Spjuth S (2003) Large scale crew rostering. Technical Report CRTR-0305. Carmen Systems (cited on page vi)
7. Kroon LG, Fischetti M (2000) Scheduling train drivers and guards: the Dutch Noord-Oost case. In: Proceedings of the 33rd Hawaii international conference on system sciences, vol 2, pp 20–24. https://doi.org/10.1109/hicss.2000.926668 (cited on page vi)
8. Kroon L, Huisman D, Abbink E, Fioole P-J, Fischetti M, Maróti G, Schrijver A, Steenbeek A, Ybema R (2009) The new Dutch timetable: the OR revolution. Interfaces 39(1):6–17. https://doi.org/10.1287/inte.1080.0409 (cited on page v)
9. Lamorgese L, Mannino C (2015) An exact decomposition approach for the real-time train dispatching problem. Oper Res 63:48–64. https://doi.org/10.1287/opre.2014.1327 (cited on page vi)
10. Liebchen C, Lübbecke M, Möhring R, Stiller S (2009) The concept of recoverable robustness, linear programming recovery, and railway applications. In: Ahuja RK, Möhring RH, Zaroliagis CD (eds) Robust and online large-scale optimization, vol 5868. Lecture notes in computer science. Springer, Berlin, Heidelberg, pp 1–27. ISBN: 978-3-642-05464-8. https://doi.org/10.1007/978-3-642-05465-5_1 (cited on page v)
11. Mannino C, Mascis A (2009) Optimal real-time traffic control in metro stations (English). Oper Res 57:1026–1039. https://doi.org/10.1287/opre.1080.0642 (cited on page vi)
12. Montigel M (2009) Operations control system in the Lötschberg base tunnel. Eur Rail Technol Rev 49(2):42–44. ISSN: 0079-9548 (cited on page vi)
13. OpenTrack (2010) ETH Zürich. opentrack.ch (cited on page vi)
14. Reuther M (2017) Mathematical optimization of rolling stock rotations. PhD thesis. TU Berlin. https://doi.org/10.14279/depositonce-5865 (cited on page vi)
15. Scholz G (2011) IT-Systeme für Verkehrsunternehmen. Deutsch. dpunkt.verlag. ISBN: 978-3-89864-770-0 (cited on page vi)

16. Schrijver A, Steenbeck A (1994) Dienstregelingontwikkeling voor Railned (Timetable construction for Railned). Technical Report C.W.I. Center for Mathematics and Computer Science, Amsterdam (cited on page v)
17. Толстой АН (1930) Методы нахождения наименьшего суммового километража при планировании перевозок в пространстве [Methods of finding the minimal total kilometrage in cargo-transportation planning in space]. Russian. Планирование перевозок, Транспечать НКПС 1:23–55 (cited on page v)
18. Website of RailSys (2017) Rail Management Consultants GmbH. rmcon.de/railsys-en/railsys-suite (cited on page vi)
19. Zwaneveld PJ, Kroon LG, Edwin Romeijn H, Salomon M, Dauzere-Peres S, Van Hoesel SPM, Ambergen HW (1996) Routing trains through railway stations: model formulation and algorithms. Transp Sci 30(3):181–194. https://doi.org/10.1287/trsc.30.3.181 (cited on page v)

Acknowledgments

We would like to sincerely thank all of those who contributed their expertise and efforts to the creation of this book:

Isabel Beckenbach	*Zuse Institute Berlin*
Valentina Cacchiani	*University of Bologna*
Carlos Henrique Cardonha	*IBM Research Brazil*
Twan Dollevoet	*Erasmus University Rotterdam*
Frank Fischer	*Universität Kassel*
Laura Galli	*Università di Pisa*
Boris Grimm	*Zuse Institute Berlin*
Heide Hoppmann	*Zuse Institute Berlin*
Marika Karbstein	*Zuse Institute Berlin*
Christian Liebchen	*Technical University of Applied Sciences Wildau*
Stephen Maher	*Lancaster University*
Güvenç Şahin	*Sabanci University Istanbul*
Stanley Schade	*Zuse Institute Berlin*
Felipe Serrano	*Zuse Institute Berlin*
Steffen Weider	*LBW Optimization GmbH*

Contents

The Editors

Ralf Borndörfer

Torsten Klug

Leonardo Lamorgese

Carlo Mannino

Markus Reuther

Thomas Schlechte

List of Contributors

Erwin Abbink
Netherlands Railways, Process quality and Innovation, HA Utrecht, The Netherlands

Andreas Bärmann
FAU Erlangen-Nürnberg, Lehrstuhl für Wirtschaftsmathematik, Department Mathematik, Erlangen, Germany

Nikola Bešinović
Delft University of Technology, CN Delft, The Netherlands

Markus Bohlin
KTH Royal Institute of Technology, Stockholm, Sweden

Valentina Cacchiani
DEI, University of Bologna, Bologna, Italy

Gabrio Caimi
SBB AG, Bern, Switzerland

Stefano de Fabris
Department of Engineering and Architecture, University of Trieste, Trieste, Italy

Twan Dollevoet
Econometric Institute and ECOPT, Erasmus University Rotterdam, DR Rotterdam, The Netherlands

Frank Fischer
University of Kassel, Kassel, Germany

Armin Fügenschuh
Helmut Schmidt University/University of the Federal Armed Forces Hamburg, Hamburg, Germany

Laura Galli
Dipartimento di Informatica, Università di Pisa, Pisa, Italy

Rob M.P. Goverde
Delft University of Technology, CN Delft, The Netherlands

Ronny Hansmann
Ostfalia – University of Applied Sciences, Salzgitter, Germany

Henning Homfeld
Deutsche Bahn AG, DB Management Consulting, Frankfurt am Main, Germany

Dennis Huisman
Econometric Institute and ECOPT, Erasmus University Rotterdam, DR Rotterdam, The Netherlands

Netherlands Railways, Process quality and Innovation, HA Utrecht, The Netherlands

Marc Johann
Deutsche Bahn AG, DB Analytics, Frankfurt am Main, Germany

Torsten Klug
Department of Optimization, Zuse Institute Berlin, Berlin, Germany

Johanna Törnquist Krasemann
BTH - Blekinge Institute of Technology, Karlskrona, Sweden

Leo Kroon
Rotterdam School of Management, Erasmus University, DR Rotterdam, The Netherlands

Netherlands Railways, Process quality and Innovation, HA Utrecht, The Netherlands

Leonardo Lamorgese
Optrail, Rome, Italy

Frauke Liers
FAU Erlangen-Nürnberg, Lehrstuhl für Wirtschaftsmathematik, Department Mathematik, Erlangen, Germany

Carlo Mannino
SINTEF DIGITAL & University of Oslo, Oslo, Norway

Giorgio Medeossi
Department of Engineering and Architecture, University of Trieste, Trieste, Italy

Dario Pacciarelli
Università degli Studi Roma Tre, Dipartimento di Ingegneria, Rome, Italy

Markus Reuther
LBW Optimization GmbH & Zuse Institute Berlin, Berlin, Germany

Thomas Schlechte
LBW Optimization GmbH & Zuse Institute Berlin, Berlin, Germany

Marie Schmidt
Rotterdam School of Management, Erasmus University Rotterdam, DR Rotterdam, The Netherlands

Anita Schöbel
Institute for Numerical and Applied Mathematics, Göttingen, Germany

Hanno Schülldorf
Deutsche Bahn AG, DB Analytics, Frankfurt am Main, Germany

Anke Stieber
Helmut Schmidt University/University of the Federal Armed Forces Hamburg, Hamburg, Germany

Sebastian Stiller
Institut für Mathematische Optimierung, Technische Universität Braunschweig, Braunschweig, Germany

Paolo Toth
DEI, University of Bologna, Bologna, Italy

Uwe T. Zimmermann
Institute of Mathematical Optimization, TU Braunschweig, TU Braunschweig, Germany

In Memoriam of Alberto Caprara (1968–2012)

written by Valentina Cacchiani and Paolo Toth

On April 21, 2012, our colleague and very dear friend Alberto Caprara unexpectedly and immaturely passed away in a mountaineering accident at "Corno alle Scale" (an Apennine mountain close to Bologna). Born in Ivrea (Italy) on January 9, 1968, Alberto earned his M.S. degree "cum laude" in Electronic Engineering in 1991 from the University of Bologna, and received his Ph.D. degree in "Control System Engineering and Operational Research" in 1996 from the University of Bologna. Alberto spent his academic and scientific career in the Operations Research group of the Department of Electronics, Computer Science and Systems (DEIS), Faculty of Engineering of the University of Bologna. He was an assistant professor from 1996 to 2001 and an associate professor from 2001 to 2005. In 2005 he became a full professor. Alberto's 100 scientific papers (many of which published in prestigious journals as *INFORMS Journal on Computing, Mathematics of Operations Research, Mathematical Programming, Operations Research, Transportation Science*) concerned the definition of mathematical models and the design, implementation, and analysis of exact and approximate algorithms for the solution of Combinatorial Optimization problems, and their application to real-world Transportation, Logistics, Computational Biology, and Railway Optimization problems. His works blended very application-oriented products with academic Operations Research methodologies, and significantly contributed to advance the state of the art in the fields he tackled. His research activity included the following problems: Crew Scheduling and Rostering in railway companies, Set Covering, Train Timetabling, Chvatal-Gomory cuts, Train Platforming, Sorting by Reversals, Knapsack and Bin Packing, Sequence Alignment, Packing of Cycles and Cuts in undirected graphs, Train Unit Assignment, Robust Optimization. Alberto was awarded several prizes, among which:

Special thanks to Juan José Salazar Gonzalez for providing the photo.

- in 1992 the "Francini" prize of AEI (Italian Electrical and Electronic Society) for the best Italian M.S. Dissertation in Electronic Engineering;
- in 1997 the "George B. Dantzig Dissertation Award" of INFORMS for the best Ph.D. Dissertation on Operations Research applications;
- in 2000 the "1999 Best Paper Award" of the *Journal of Combinatorial Optimization*.

Alberto was member of the editorial boards of the journals *Operations Research*, *Mathematical Programming C*, *Operations Research Letters*, and co-editor of *Optima* (the Newsletter of the Mathematical Programming Society) from 2002 to 2010. He delivered several plenary lectures at international conferences and acted as Program Chairman of the Workshop ATMOS 2011 (Algorithmic Approaches for Transportation Modeling, Optimization, and Systems). Alberto was a hard-working person and a dedicated scientist, whose exceptional career touched many of our lives. He is widely recognized for his outstanding professional, ethical, and scientific standards that have inspired all who had contact with him. Alberto always wished to tackle the most difficult problems (both in research and sport) and performed all his activities with enthusiasm and passion. Along his career, Alberto demonstrated his profound academic and human abilities and helped with a lot of professional care younger researchers to grow and become experts in the commonly studied fields.

In Memoriam of Leo G. Kroon (1958–2016)

written by Erwin Abbink and Dennis Huisman

On September 14, 2016, Leo G. Kroon unexpectedly passed away. A gifted mathematician and scientist, Leo Kroon was the world leader in timetable modeling and a scientist who exemplified how to bridge theory and practice. Leo worked as a Professor of Quantitative Logistics at the Rotterdam School of Management (RSM), Erasmus University, and as a Logistics Consultant at Netherlands Railways (NS). He received his master's in mathematics cum laude from the Free University in Amsterdam and obtained his PhD in 1990 at the Erasmus University.

Professor Kroon was foremost interested in railways and worked since 1996 part-time at NS. Leo pioneered in developing a group of practical researchers and student at NS, who worked with "rolled-up sleeves" to solve real-world problems like the labor discussion on variation in crew schedules. Leo had a great ability to present even the most complex problems very clearly. The work of him, his colleagues, and (PhD) students often led to successful implementations in practice at Netherlands Railways in the areas of timetabling, rolling stock scheduling and crew scheduling.

Leo (co-)authored over 100 papers on mathematical optimization models for planning and real-time optimization, in journals like *Operations Research*, *Transportation Science*, *Transportation Research B*, and *Interfaces*. He was also associate editor of the journal *Transportation Science*. Furthermore, Leo has been the promoter of over a dozen PhD students, of which 10 worked on railway-related topics. Twice, he won the PhD Supervisor Award of the national transport research institute TRAIL. Furthermore, five of Leo's PhD students won the RAS student paper competition. Leo had a leading role in many international collaborations: he was the project leader for RSM for the EU-funded projects in railways AMORE, ARRIVAL, and ONTIME. He found great joy in organizing the international 2015 CASPT conference in Rotterdam with a team of his (PhD) students, colleagues, and his dearly beloved wife, Cisca.

Leo and his team received many acknowledgments for their work, including the 2008 INFORMS Franz Edelman Award presented to the NS for model-based contributions to the 2007 NS timetable; the 2008 ERIM Impact Award and best paper award at the 2011 World Congress on Railway Research (theme "Even more trains, even more on time"); and 2013 Strategic University Relationship (SUR) collaboration with IBM. As a result of his work, not only scientists benefit but also millions of passengers in the Netherlands have a better train schedule, resulting in a higher punctuality and a higher customer satisfaction score. Leo will be remembered as a brilliant and erudite researcher who was committed to his students, PhD candidates, and colleagues he worked with. He was honest, humble, and gifted with a subtle sense of humor. He leaves a big void in our community, but will live on in our minds and in his students' future work in optimization for railways.

Chapter 1
Simulation of Rail Operations

Giorgio Medeossi and Stefano de Fabris

Abstract From the long-term design of new infrastructure to the validation of timetables planners need accurate and reliable support tools that allow understanding the effect of their decisions. Microscopic simulation has gradually become widespread, since it allows considering not only the characteristics of infrastructure, signalling and rolling stock, but also human factors. In spite of this success and of the increasing quality of commercial simulation software, the setting up of a simulation requires time and accuracy to ensure that all elements are represented correctly.

1.1 Introduction

Simulation of railway networks has a long tradition, starting many decades ago in railway laboratories, where models in the scale 1:87 were used to reproduce networks and control them using realistic interlocking systems. The growth in the

G. Medeossi (✉) · S. de Fabris
Department of Engineering and Architecture, University of Trieste, Trieste, Italy
e-mail: giorgio.medeossi@units.it; s.defabris@trenolab.com

R. Borndörfer et al. (eds.), *Handbook of Optimization in the Railway Industry*,
International Series in Operations Research & Management Science 268,
https://doi.org/10.1007/978-3-319-72153-8_1

calculating capacity of computers, the creation of graphical user interfaces and the relative simplicity of the basic rules of train traffic have led to the development of simulation tools. Simulation tools were first able to simulate relatively small networks considering all trains in a deterministic way. These tools were mainly used to support infrastructure planning and especially to estimate infrastructure capacity. More recently, the further increase in the performance of computers and the possibility of an automatic import of infrastructure layouts and timetables widened the application spectrum of micro-simulators to large nodes and to more detailed stochastic stability evaluations.

Compared to a deterministic simulation, the stochastic one presents the great advantage of considering also variability in process times, while increasing the precision of the outputs.

Stochastic micro-simulators can reproduce most processes involved in rail traffic and comprehend not only its deterministic aspects, but also human factors. This is particularly relevant in order to simulate traffic under realistic conditions, considering variability at borders, various driving styles and stop times. Obviously, the possibility of obtaining realistic results is strictly related to the quality of the input data used to model stochastic phenomena, which have to be accurately calibrated.

This chapter starts with a brief description of the different types of simulations, followed by its typical application fields. Then a sort of step-by-step guide presents how to set up a simulation model, followed by a brief presentation of the most useful outputs. The chapter is concluded by an explanation of the most important weaknesses of simulations.

1.2 Types of Simulation

A simulation is deterministic if all parameters are defined by the user and do not contain any random components; thus, two runs of a deterministic simulation generate the same outputs. Deterministic models are used to represent real systems which are too complex to be evaluated analytically. In stochastic simulations random components are introduced to better represent one or more phenomena. Since stochastic simulations are used to evaluate the behaviour of a system with some random factors, results of a single simulation run have no statistic relevance; as a consequence, multiple simulations must be performed. Deterministic simulations support timetable planning or the design of new infrastructures, while stochastic models allow timetable robustness or stability analysis.

A simulation is static if it is time-independent. A dynamic simulation model shows how a system evolves over time. Railway simulation models are dynamic, since they are explicitly built to study traffic evolution in a given time interval.

In continuous simulation models, the value of state variables changes continuously in time, therefore it is calculated with analytic continuous resolution of state equations. In discrete-event simulation models, state variables are calculated only when an event occurs, independently from the time-span between two successive events (next-event time). Railway simulators are continuous when the time-step at which the state variables are calculated has a fixed duration, while more simplified models use pre-defined process and event times in a discrete approach.

Macro-simulation models use a simplified infrastructure model to reduce computational time and therefore allow simulations of larger networks. Micro-simulation models offer a description of infrastructure which reproduces the functionality of interlocking, safety and block systems.

In a standard asynchronous simulation, trains with highest priority are simulated first, and conflicts among them solved with a first-come-first-served strategy; the resulting infrastructure occupations are stored. Then the process is repeated for each priority group, more and more saturating the time-windows which are still unused. Therefore, no conflict among trains with different priority is possible: high-priority services are never forced to brake or stop by conflicts with low-priority trains.

On the other hand in synchronous simulations all trains are calculated together, considering the constraints given by the interlocking and signalling system and the behaviour of drivers. As a result, a synchronous simulation reproduces operations as they are in reality, including the consequences of traffic conflicts: a faster train might be forced to brake behind a lower-priority one, and its driver would follow the braking curve as allowed by the *Automatic Train Protection (ATP)* system.

Because of these characteristics, the approach that reproduces railway operations in highest detail is dynamic, synchronous, microscopic and stochastic. These models are also the most flexible and widely-used in commercial simulation software. Thus, we consider only this approach in the following.

1.2.1 Simulation Tools

A number of synchronous micro-simulation models have been presented in the last years. Since all models use the same approach, solving the motion equation of trains which are moving together on a microscopic network with respect to interlocking and safety system functionality, differences between models are mainly in the flexibility to represent different technologies, in the capability to receive automatic inputs or to generate specific outputs.

The EU funded project Optimal Networks for Train Integration Management across Europe (ON-TIME) recently published an "Assessment of State-of-Art of Train Timetabling" [9], which gives a comprehensive overview on the various commercial simulation software.

The most widely-used commercial software at a global level are `OpenTrack` [10], `RailSys` [14], `RAILSIM X` and `RTC` [13]. Although there is an overlap area in their use, each tool has a different basic philosophy, which makes each of them more suitable for specific applications.

`OpenTrack` [10] is a simulation tool developed at the Institute for Transport Planning and Systems (IVT) of the ETH Zurich and now supplied and refined by OpenTrack Railway Technology Ltd. [10] is used by railways, the railway supply industry, consultancies and universities in a significant number of different countries. Additionally [10] features the calculation of power and energy consumption of train services. The key strength of `OpenTrack` [10] compared to the other tools is the extreme flexibility in defining the infrastructure (simply as nodes and edges using the so-called "double-vertex graph") and the interlocking and signalling systems. Instead of directly implementing a set of national systems, the tool features a few standardized signals whose characteristics can be customized by the user to model accurately their functions. As a result, basically all signalling systems worldwide can be represented, and specific signals can be inserted to influence the behaviour of drivers, making it very realistic. Moreover, at present `OpenTrack` [10] appears as the tool with the most complete set of stochastic components to represent drivers' behaviour.

`RailSys` [14] is a comprehensive signal-berth simulation package, which also has modules that facilitate infrastructure management, timetable construction and possession planning. The software database structure allows simple storing of very large models, which is reflected by its use by many infrastructure operators. The model features a timetable construction system and an automatic slot search algorithm, see Hauptmann [3]. Refined simulation techniques to prevent deadlocks, a number of input and output capabilities and the presence of variable process and event times made this tool suitable for large-network stochastic simulations, see Rudolph and Demnitz [11].

`RAILSIM X` developed by SYSTRA, appears as relatively similar to `RailSys` [14], but is extremely focused on the US market. It is formed by several modules, which go far beyond the pure simulation of operations, or simplify some of its most common uses:

- the design and analysis of rail and rapid transit signal systems.
- the analysis of the loads, potentials and regenerative braking receptivity on DC and AC power systems.
- calculating of safe braking distances for "worst case" signal design purposes.
- the development of models, along with the design and modification of signal systems.
- the automotification of signal clearing time (minimum supportable headway) processing for fixed block wayside and/or cab signal/*Automatic Train Control (ATC)* designs.

`RAILSIM X` is used in particular by passenger rail, light rail and metro operators in the US.

Rail Traffic Controller [13] developed by Berkeley Simulation, is the simulation software used by most freight operators in the US to reproduce operations on complex networks, also when formed by long single-track lines that can be used as alternative routes for trains depending on traffic conditions, see `RTC` [13]. Instead of using fixed (or multiple, but pre-selected) itineraries, `RTC` [13] is able to dispatch trains on the network, finding a deadlock-free routing to their destination. This ability has made `RTC` [13] a standard trusted by all stakeholders in the US in particular for estimating the capacity of networks considering a set of delay cases as input.

1.3 Application Fields

Basically, simulation can be used to reproduce rail operations under all conditions. However, it has a few weaknesses that make it more difficult or less accurate to use for certain studies. On the other hand, while a commercial software is a single tool, the way it is used varies based on the goals of the simulation. With a state-of-the-art commercial simulation software it is possible to run:

1. Single train runs, in which a single train is running on the network.
2. Deterministic simulations, in which all trains are running based on their timetable, and no human factor or variability is inserted.
3. Stochastic simulations, in which standard distributions of initial delays, dwell times and/or train performances are inserted. Stochastic simulation have to be run for a statistically relevant number of iterations in order to obtain statistically-relevant data.
4. Advanced stochastic simulations, in which the input distributions are derived from real data. This links the results to the existing data, but at the same time allows comparing real world data with simulation to validate the model, and simulated scenarios among them to forecast reliability.

While (1) is only useful to verify running times, (2)–(4) can be used separately or combined, each to support the specific stages of railway planning.

1.3.1 Simulation to Calculate the Running Times

The most basic use of simulation is to calculate the run times of single trains. Historically, this has been the first use of rail simulation and appears still important when designing a timetable. However, accurate run time calculators are now integrated in most timetable planning tools; thus, microscopic simulators are normally not deployed for this task. On the other hand, compared to run times estimated by macroscopic timetable planning tool (such as `Viriato`), microscopic simulation tools deliver more accurate results in particular in complex stations; however, these differences are normally negligible in timetable planning, in which margins have to be included to consider the human factors, which are significantly higher.

1.3.2 Simulation to Verify a Timetable

A deterministic simulation can be run to verify the feasibility of a timetable on the network. In this case, the entire line or region, or a small portion of it can be simulated. Simulating only the critical area appears sufficient if the planners expect no conflict outside of it; this condition can be reasonably identified on lines by simply checking the timetable graph. A deterministic simulation using train performance reduced to a realistic value (and thus a maximum 95–96%) using the technical dwell times—and not the planned dwell times—can be run as first step. If the timetable appears feasible, a second deterministic test in worse conditions could be run. Different approaches are used by practitioners, with the common goal of creating more stressful conditions for the timetable without setting up a stochastic simulation. The most common ones—often used also in combination—are: reducing the performances of trains by 3–5%, obtaining slower trains and thus testing the capacity of the timetable to work also when drivers are not able to exploit the rolling stock at its maximum; increasing the dwell times by a few seconds at each stop; increasing the route release time; this test leads to longer blocking times and allows identifying conflicts that might occur under normal conditions as an effect of the normal variability of train movements.

1.3.3 Simulation to Estimate Capacity

Estimating the capacity of a line or station is one of the most common tasks of railway planning. As stated in the UIC leaflet on capacity “Capacity as such does not exist”, being a function of “the way the infrastructure is utilized” (see International Union of Railways (UIC) [4]), which is defined at least by the service concept to be operated in it. Synchronous simulation is more demanding, requiring a relatively detailed timetable, which also includes train routings. Once a timetable is created and simulated, the capacity is not yet estimated, since the simulation only reproduces operations with the timetable itself. Different methods are deployed in practice:

1. In an iterative process, the impossibility of adding trains to an existing timetable is shown, normally by displaying the resulting timetable graph, in which the conflicts caused by the additional train are highlighted. Although this method does not lead to an explicit estimation of capacity, but of the residual of the given timetable, it is broadly applied. In fact, the frequency and service pattern of pre-existing passenger trains often lead to a nearly-fixed timetable, which is described and documented by the planner. If this demonstration is considered as valid, evaluating the residual capacity in terms of number of trains that can be added appears correct. Obviously, considering some (or most) services as fixed reduces the total capacity, since the timetable typically contains time that cannot be used by trains.
2. The well-known method described in the already-mentioned UIC 406 leaflet (see International Union of Railways (UIC) [4]) overcomes this limitation by proposing the compression of the timetable followed by an evaluation of the total and residual capacity simply based on percentages of the total time. It is a procedure that can be easily performed in microscopic simulation tools. Some of them (such as `RailSys` [14]) feature a module to automatically estimate the capacity using this method, while in others the slots can be iteratively approached until a conflict is displayed; then the capacity can be estimated. This

approach has the advantage of being widely accepted and easily reproducible. However, the coefficients used to obtain the maximum usable capacity do not consider the effect of technologies that have a beneficial aspect on the usable capacity, such as the continuous ATP systems, nor the impact of the human factors, which vary a lot.

3. To fill this gap it is necessary to use stochastic simulation. It allows obtaining the reliability figures corresponding to a certain timetable, operated under realistic conditions and taking into account the behaviour of drivers, also related to the signaling and ATP systems installed. Medeossi [7] proposed a method that combines the compression with stochastic simulations. In the first step a key part of the real timetable—such as a peak hour—is compressed as explained in the previous paragraph. The compressed hour is the "basic period" of the timetable: it is repeated, ensuring that no buffer time is provided between the last train of a period and the first of the following one and thus obtaining a dense timetable that covers the entire peaks or the entire day. Obviously, the obtained interval has a specific value, which is normally not as simple as 60 or 30 min, and it is not realistic to operate such a timetable. Buffer times can then be gradually added by increasing the headway times among trains by a given amount of time. The results are obtained running a set of stochastic simulations at each increase of buffer time and then calculating the corresponding reliability figures. Since buffer times are directly proportional to the capacity, at each increase of them corresponds a certain capacity: thus, the result is a capacity vs reliability curve that shows how the reliability of the line (or station) decreases at increasing traffic density.

1.3.4 Simulation of Yards

Microscopic simulations, and in particular the most flexible tools such as `OpenTrack` [10] can be used to represent in detail all shunting moves within yards. However, compared to operations in terminals, in yards the times for coupling and decoupling trains and moving switches have to be estimated on field, while the effective speed appear by far more important than the speed limit, the signalling system and the other characteristics that are accurately represented by simulation tools. Moreover, the output statistics of simulation tools are normally not very useful for the goals of these studies. As a result, other tools such as Villon appear more suitable in this field, see Kavička et al. [5].

1.3.5 Simulation to Support the Definition of Infrastructure Improvements

While estimating the capacity appears sometimes more a theoretical than a practical question, the quantitative evaluation of the benefits of infrastructure improvements is a very common task for planners. An improvement may lead to the increase of capacity, reliability and/or train performances (and thus a reduction of running times). To obtain the combined capacity and running time benefits, a similar approach to capacity estimation is used, starting from the construction or adaptation of the timetable. This step is extremely important, since the shorter running times allowed by the speed-up are not exploited if not considered in the timetable. The study continues as a regular capacity estimation: while the effects on capacity can be estimated using deterministic simulations, only the stochastic simulation allows estimating the impact on reliability.

1.3.6 Simulation to Estimate the Robustness of a Timetable

Robustness is the ability of a system to withstand parameter variations or changes in the operational conditions. The estimation of robustness is a typical application of stochastic simulation, in which not only the real variability of process times, but also standard ones can be used as input. Different measures can be used to assess the robustness of a timetable; the most common are:

- the number of conflicts, which has the advantage of being easily estimated and understood. On the other hand, there is no direct proportionality between this number and the effect of the conflicts themselves on operations, which is strongly related to the running time margins inserted in the timetable and the ATP/ATC system installed on the line.
- the propagation of delays, which can be measured as the difference between delays at the beginning and at the end of train journeys. It appears as the explicit measure of robustness as the capacity of the system to recover delays.
- the punctuality or delays. These measures should be used only to compare two scenarios, which are simulated using the same input delays; when a comparison is not possible, the propagation of delays appears preferable.

1.3.7 Simulation to Evaluate the Impact of Maintenance or Construction Works

Track closures or slow speed sections are easily inserted in simulation models. They can then be simulated deterministically when the goal is to decide how to adapt a timetable, and stochastically to estimate the reduction of the reliability due to the works once a timetable is defined. The stochastic simulation of temporary single track sections requires particular attention due to the risk of deadlock that might occur when multiple trains are scheduled on it at the same time. In fact, while simulation easily dispatches conflicts between two trains, the presence of a third one might lead to deadlocks if it is not possible to solve it in two steps. Besides deadlocks, the dispatching criteria used by the simulation tools might give priority to lower-priority trains, leading to an unrealistic distribution of delays.

1.3.8 Simulation to Estimate Ex-ante the Punctuality of a Timetable

This appears as one of the most recent and complex uses of the advanced stochastic simulation, since it requires highest accuracy and at-the-same time appears interesting for the operators in particular on complex or heavily utilized networks. The result is that running these simulations requires time and experience since it is necessary to verify its accuracy. This simulation always requires a careful validation, since it is necessary to verify that the model is able to reproduce operations realistically also under their normal variability. The validation is obtained comparing the results of the simulation of existing operations with the corresponding real operational data. The comparison can be conveniently made using the same performance indicators, measuring points and aggregation of courses that are used for the output. A combination of punctuality and mean delay gives a good synthesis of the quality of operations: the former is the most easily understood by practitioners, but it might be quite sensitive to small changes in the operational conditions: a train with 179 s delay is considered on time, while one with 181 s is late. On the other hand, the mean delay is more stable, but less easily connected to the performance measures used in real operations. It must be noticed that the lower is the punctuality threshold the higher is the residual error; as a result it appears easier to obtain a 1% error in punctuality at 5 s than a 5% error at 1 s. After this validation the results of test scenarios are compared to the reference one, and punctuality is always presented as variation from the reference scenario and not as absolute rates.

1.4 Setting up a Simulation Model

More and more operators own the microscopic model of their network, in several cases already prepared for simulation or even available to the public. However, if the model is not available or validated, or a different tool has to be used, it is necessary to carefully create or check the model before running the simulations. Figure 1.1 shows a typical workflow for a simulation study.

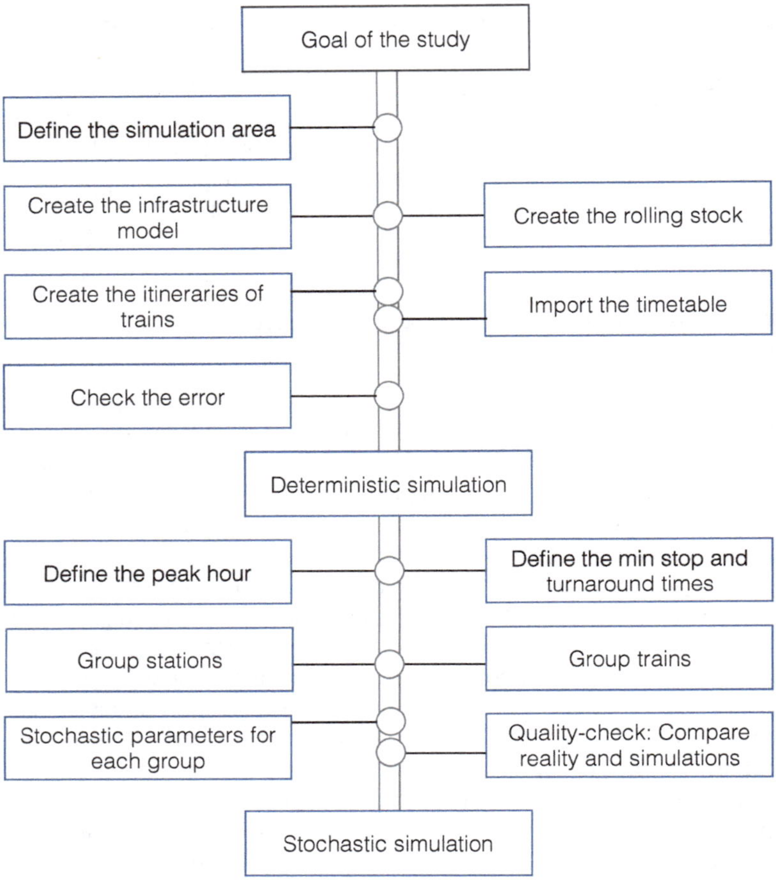

Fig. 1.1: Workflow of a simulation study

1.4.1 Defining the Simulation Area

This simple step appears very important, since choosing properly the simulation area allows including all elements that might have an impact on the outputs by at the same time limiting the time required to set up the model.

No fixed criteria exists for the definition of the simulation area; however there are a few "rules of thumbs":

- On the analysis line(s), consider the entire section between two important stations. This section is further extended to the next station when the first (or last) station is not a terminal one.
- On all lines diverging from the analysis line(s), consider the section to the first station. However, there is no need to simulate a branch line if its trains start/end

at the junction station without any possible conflict with the analysis line(s). The same criteria applies to the other lines converging in the first and last station of the analysis line(s).

- At least the peak hour(s) should be simulated. To allow all planned trains at the beginning of the peak hour to be simulated, the simulation should start at least a complete running time on the line before the peak hour. Similarly, it is important to simulate at least 1 h after the end of the peak. This means that if a line has a running time of 2 h, and the peak hour is 7–9, the simulation time should be at least 5–10.
- Although normally less dense than the morning peak, the evening peak often shows higher delays. Thus, it appears important to simulate it, too.

1.4.2 Creating the Infrastructure Model

To obtain the highest precision, a microscopic simulation model contains all characteristics of the real world that have an influence on train movements and dynamics. On the other hand, since simulation and especially infrastructure modeling are quite time-consuming, parameters which have a very limited influence could be ignored, in order to obtain results more efficiently.

Regarding the line alignment, gradients have to be considered in detail if they are significant for train dynamics (normally $>5\%$). While in many running time calculation software mean gradients for longer sections are used, each gradient has to be modeled, especially when heavy freight trains are considered. The combination of low adhesive weight and wet rails reduces significantly freight train acceleration also if the ramp is just some hundred meters long. Tunnel position and kind (single or double-track, smooth or rough) are also to be inserted into the model, since their additional resistance significantly affects train resistance even at low speed ($V >$ 60 km/h).

On heavy railway lines, a detailed description of curve radii can be strongly simplified considering mean curves on longer sections or adding the curve resistance to the gradients obtaining a total resistance due to line alignment.

The interlocking and block system have to be modelled in high detail, comprehending the complete station layout with all track circuits and all signals with their respective aspects associated to the possible routes. Also the way routes are released has to be modelled, considering overlaps and release groups or other technical parameters, including release times or interdependencies among routes which are not simply identified when creating the track layout.

1.4.3 Characteristics of the Rolling Stock

Inserting the characteristics of rolling stock is significantly less time-consuming, since the required data set is limited, and includes: number, weight, length, position of locomotives and coaches, adhesion weight, tractive effort vs speed curve, and description of braking as function of *Braking Weight Percentage (BWP)* or as deceleration intervals.

The same characteristics should be inserted for all ATP/ATC systems under which the train operates if they influence the braking behaviour of drivers (such as the *European Train Control System (ETCS)* Level 2).

The coefficients of resistance formulas are given as default in the simulation tools and can be customized only when the specific ones are available. This subject is extensively described by Wende [15]; the book also includes the coefficients to be used for several types of locomotives, trailers and multiple units considering both the Davis and Strahl & Southoff formulas. Besides these basic parameters, others have a significant influence but neither often available nor easy to estimate:

- Adhesion: apart from the adhesion weight, the adhesion in expressed as % in motion equation. While recent trains in normal condition can have a 120%, a value used as default in `OpenTrack` [10],
- Acceleration/deceleration delay: the former is the time a train takes to accelerate again after a braking action; it is quite limited on short EMUs higher for loco-hauled passenger trains and even longer on freight trains. The latter is the time following receipt of the signal information before the operator applies the brake and the brake curve can be applied in calculations. It includes a behavioural and a deterministic component, very limited on trains with electric brakes but significant on long freight trains, in which the air pressure takes several seconds to propagate over the entire train. On a long freight train this time can measure nearly 30 s, see Wende [15].
- Correct deceleration on gradients: this parameter allows representing the increase/decrease of braking deceleration on gradients. A value of about 0.01 m/s^2 can be considered as a good approximation. For long freight trains it is necessary to consider that they are not able to keep a constant speed on slopes, and are required to use the so-called "sawtooth method" ("Sägezahnmethode"), see Schweizerische Eisenbahn [12]. In lack of a specific setting in the software it can be simulated calculating the corresponding average speed or using a series of coasting signals.
- Minimum time to hold speed: when the distance between two stops, or a section with higher maximum speed is short, drivers do not reach the maximum possible speed to immediately brake strongly, but keep a more gentle profile. This practice, can be simulated defining the minimum time the train has to keep its maximum speed before braking. The result is that it reaches the maximum speed it can keep for the given time, keep it and then brake. The effect of this parameter is shown in red in Fig. 1.2.
- Coasting: it takes place before braking when the tractive effort is equal to zero and the train is decelerating due to resistance. Some operators promote the use of coasting as an energy-saving measure on early running trains. As demonstrated in Medeossi et al. [8], it is defined as a time interval, in which a train has a deceleration similar to the deceleration caused by train resistance. The duration of coasting depends on the length and speed of the section and the actual

delay of the train and is strongly related to the behaviour of the driver. Due to this variability, unless specifically suggested and defined by the operator or measured on field, it appears quite difficult to accurately model coasting in the existing commercial software. The combined effect of coasting and minimum time to hold speed is shown in green in Fig. 1.2.

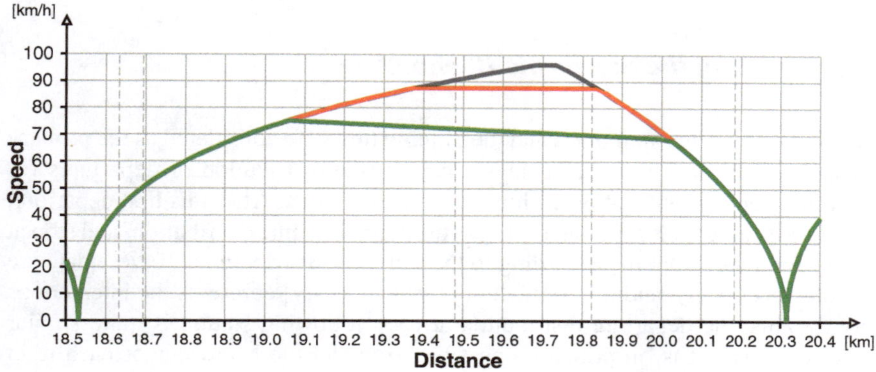

Fig. 1.2: Impact of the "Minimum time to hold speed" (in red) and the coasting (in green) on the speed profile of a train

1.4.4 Running and Checking the Correctness of a Simulation Model

The correctness of a model has to be verified accurately before running the simulation. In the experience of the author even the "official" models of the operators are not completely free from errors. The model can be verified simply based on experience by checking a few aspects of the simulation output of the existing timetable (if available) and lead to a validation report:

- Verify the speed-distance diagrams of at least one example train per line. Are the distances and speed limits correct? Do trains reach their maximum speed? Are acceleration and deceleration similar to their expected rates?
- Run a complete simulation and watch the animation: do all trains use the correct itinerary? Are the stopping positions correct? Are entrance orders in junction stations correct? Are the routes released and blocked as expected?
- On the timetable graph, check the blocking times: do they correspond to the position and aspects of signals?

- On the timetable graph, identify the conflicts and give an explanation to each of them. Are they simply due to the difference between planned and simulated behaviour of trains? (The conflicts due to early-running trains are a typical issue.)
- In a table containing the delays of all trains at all stations check all deviations from the schedule higher than a certain threshold; find them in the graphic timetable and explain them.

1.4.5 Inserting the Stochastic Phenomena

The initial delays, the stop times and the running times are considered as the primary variability sources. The initial delays represent the distribution of departures for each train from its first station within the simulation area. After this first departure, trains move along their route according to their running time distribution, and stop at stations for a variable time according to the stop time distributions. If these data are estimated accurately, and the simulation model correctly represents the interactions among trains, the departure distribution at each following station is quite similar to the real ones. It is, in principle, possible to model the behaviour of each train separately, especially when the distributions are derived from real data using a tool like `TRENO`. On the other hand, besides the difficulty in manually editing real world data when necessary, this would lead to a simulation whose results are extremely linked to the existing timetable. A good compromise can be reached grouping trains and stations with similar characteristics, as:

- Trains with same category and origin and destination (O/D) form one group.
- Stations are grouped based on the number of passengers boarding and alighting. The largest ones are kept separated.
- Peak hours are defined for line sections and direction and train category separately. This allows setting a peak hour in the morning for local services towards a certain regional hub, but not in the opposite direction, nor for the long-distance trains.

As a result, each train group has a distribution of initial delays plus one of dwell time for each station category; each for peak and off-peak hours.

1.4.5.1 Initial Delays

When defining the expected initial delays for a new timetable on the basis of the past data collected at each station, it is important to exclude all major perturbations and all secondary delays (delays due to traffic conflicts). Some researchers use interlocking data and sophisticated mathematical models to identify primary and secondary delays, see Yuan and Hansen [16]. However, these methods require using blocking times collected at block sections of the network, which, currently, are not stored in most countries.

If such accurate methods for filtering out secondary delays cannot be used, a rough filtering process developed by Medeossi et al. [8] can be deployed with the goal of defining a realistic initial delay distribution (removing the effects of conflicts with arriving trains) rather than analysing conflicts in detail based on track occupation. The method uses only train passing data, automatically collected at each timetabling point by the train describers, which identify trains at their first station and then keep track of progress along the route. This simple model is relatively accurate for use in simple stations where complex conflict chains hardly occur. In other cases, for example at large stations, it introduces higher—but in the author's experience, still acceptable—initial delays.

The method consists of two steps. First, major perturbations are simply filtered out by excluding all delays greater than 600 s (5 min). An extensive analysis of major stations in Italy showed that more than 99% of delays due to passengers alighting and boarding or other "normal" phenomena are lower than 5 min. Therefore, this threshold can be used to filter out breakdowns that can be excluded in timetable planning. (The 5-min threshold would need to be reconsidered for application in another context or country.) Second, the conflicts between arriving and departing trains are identified using a mesoscopic model at each relevant station. This process uses data of station arrival and departure times for each train as well as the sequence of signals the train has passed and the corresponding times. All trains with departure delays are checked using these data to determine if a conflicting movement took place. In case of a conflict, the record is simply excluded from the data set.

1.4.5.2 Dwell Times

A precise estimation of stop times appears normally more important than that of initial delays, especially if entire train courses are simulated. At the same time it appears less important on freight lines, as well as on single-track lines, where the stopping times are more influenced by the crossings than by the passengers boarding and alighting. Stop time variability includes two phenomena affecting trains at stations: departure imprecision and dwell time variability. Departure imprecision occurs if a train, which arrived early or punctually at a station, does not depart punctually although all passengers are already on board. Departure imprecision is generally short for regional trains, with mean values between 10 and 20 s, while it is often more than 30 s for long-distance trains. The dwell time depends on the number of passengers boarding or alighting, the presence of late-arriving passengers and on the physical layout of the train-set. The two phenomena must be considered separately to model stop time variability correctly. This process is complicated because, often, the only data available are arrival and departure times. A detailed stop time calculation model was proposed by Longo and Medeossi [6], based on a previous work by Buchmueller et al. [2]. The model can also be used to estimate the effect of an increased demand or different rolling stock, but it requires a set of measures to be collected manually at some stations. However, in most studies it is quite demanding to collect those inputs, so a simpler method was developed by Medeossi et al. [8].

The method is valid for services operated with a frequency greater than every 10 min and uses two assumptions to divide the measured stop time into different parts:

1. For late-running trains, the stop time has no departure imprecision and consists only of the dwell time.

$$D_a > 0 \quad \Rightarrow \quad T_d - T_a = T_s = T_{dw}$$

2. On early-running trains, the stop time includes both dwell time and departure imprecision.

$$D_a < 0 \quad \Rightarrow \quad T_d - T_a = T_s = T_{dw} + T_{dv},$$

where

D_a = Arrival delay
T_a = Arrival time
T_d = Departure time
T_s = Stop time
T_{dw} = Dwell time
T_{dv} = Departure variability.

The same operational data collected at all stations used for defining the initial delays can be used as input to estimate T_d and T_{dw}. Thus, the dwell time T_{dw} is the measured dwell time of late arriving trains, while the departure variability T_{dv} is the departure distribution of the early-arriving ones.

In large stations, where the planned stop time is significantly higher than the minimum dwell time, it is more accurate to separate early- and late-arriving trains at a different threshold to avoid obtaining a dwell time distribution than partially includes waiting for the departure time.

Although this separation appears extremely easy, it is not supported by most commercial simulation tools. In such cases the user is forced to use only one of the two distributions. A quick analysis of real world data allows deciding which to use: when the variability of dwell times is high, the distribution of dwell times should be used, but if the number of passengers boarding and alighting is limited a fixed dwell time plus a distribution of departure times appears more accurate. It must be noticed that this way of considering the dwell times appears not correct for services with frequencies higher than 10 min, in which passengers arrive at stations independently from the expected train departure time. In fact, if a certain train is slightly delayed there will be more passenger waiting for it and, thus, its dwell time will increase, further boosting its delay. This would be represented considering a delay-dependent dwell time function, which is at the moment not implemented in any commercial software.

1.4.5.3 Train Performances

The estimation of the variability of running times to be used in simulations is more complex. It is modeled through one or more performance parameters in the simulation tools. The performance parameters represent the way drivers drive during one or more phases of motion (braking, acceleration, etc.): they are inserted in the motion equation as factors that reduce the maximum performances of the train. During acceleration, the performance parameter reduces the tractive effort of the locomotive; during cruising it reduces the speed limit while during braking it reduces the deceleration. The estimation of performance parameters is by far more complex than that of the initial delay and stop time distribution. It requires to find the performance parameter that splits the motion into five or six phases and reproduces the measured running time at best. A method for the estimation of the set performance parameters was presented by Medeossi et al. [8] and is based on train event recorder or GPS log files; more recently Bešinović et al. [1] presented a method to use track circuit logs instead.

While these algorithms lead to very accurate results, they cannot be used for simulations, because commercial software only uses one performance parameter for the entire motion. Since the best parameter is the one that leads to the same running time between two stations, it could be obtained by simply estimating the distribution that fits at best the running times as recorded in the train describer data.

Train performance parameters are normally between 90 and 98%, with the lower end more common on stopping trains and values quite close to the upper limit for high-speed services. Energy-efficient driving policies have a significant impact on running times, so we suggest analysing a set of GPS or train event recorder logs to evaluate whether or not to model it explicitly, obviously depending on the functionalities provided by the simulation tool.

1.4.6 Incidents

The simulation allow users to examine the impact of disturbances (called "incident") in the infrastructure, rolling stock and schedule systems. Incidents can either be operational failures or operational problems (which allow operations to continue but at a reduced speed or capacity). Examples of the first type of incident include signal failures and broken tracks; examples of the second include slow orders or unplanned train delays. Incidents can be combined into sets of incidents that can be applied during the simulation.

Given the relative detail required to model them, incidents are normally inserted to reproduce specific disturbances and not the normal stochastic behaviour; an exception is represented by a few specific cases in which they are used to represent the systematic interaction of operations with external elements, such as a traffic light on tram operations. Since the set of possible (real) incidents is extremely wide, and the frequency of each extremely low, it is not possible to draw general criteria to define the most significant incidents to be tested on a railway line. Moreover, since most commercial simulation tools are not able to automatically cancel partial or complete train courses, the impact of serious breakdowns can only be simulated by manually and gradually modifying the timetable to reproduce these decisions. The possibility to control dispatching in simulations using an ad-hoc external tool, as possible, for instance in `OpenTrack` [10], opens a broad spectrum of applications, since it is possible (at least in theory) to simulate the correct impact of a much broader set of incidents with very limited manual interventions.

1.4.7 Output

One of the key advantages of microscopic simulations is that it can log the virtual movements of trains on the infrastructure at each integration step. This huge quantity of data can be displayed and aggregated in several ways to obtain the statistics or diagrams.

1.4.7.1 Animation

The animation of trains running on the network is not only an impressive output for the decision-makers; it is also plays a key role for the users of the tool. In fact, it allows the planner to understand and show what happens in the virtual network, and in particular the routes each train takes, the aspect of signals it encounters, the reason of conflicts, etc.

1.4.7.2 Timetable Graph

The timetable graph is a space-time diagram that represents the planned and/or actual movements of trains. A train moving along the line is represented by a line, whose inclination is proportional to the speed of the train. It is the most common technical diagram in railways and the most efficient to view operations along a line, identifying the conflicts and estimating their effect. A qualitative impression of the usage of a line is easily obtained by adding the blocking times to the timetable graph, giving one of the most easily-comprehensible outputs of deterministic capacity estimations. The graphic timetable can also be used in stochastic simulation, to show the effects on operations of an example delay scenario, obviously with no statistical meaning.

1.4.7.3 Diagrams

Diagrams such as speed vs. distance or time, acceleration, power consumption, etc. can be used to show how a train runs on its itinerary, also showing the effect of a different train set or speed profile.

1.4.7.4 Statistics of Occupation

These statistics show the physical occupation or the blocking time of a certain interlocking element or block section. The blocking times represent the time in which the interlocking and signaling system keep the element blocked for a train; they are normally used as an output of deterministic simulation to identify the usage of critical sections in the peak hours or to show the theoretical headway time on it. The occupations represent the time in which an element is physically occupied by a train: they are useful when analyzing the usage of station tracks.

1.4.7.5 Delay Statistics

All outputs listed previously mainly refer to deterministic simulations and do not allow aggregating the results of stochastic simulations. As a result, the most important output of stochastic simulations are the statistics of delays. An advantage of simulations is that reliability of operations can be measured using exactly the same indicators used in real operations. Since they can be measured for all trains and at all stations, the user can decide how to group the trains and where to measure them in order to limit the number of tables and graphs inserted in reports. Punctuality is normally measured as it is in reality, and thus at terminal or important stations, while the mean delay—which is not frequently used in official figures of operators and is more sensitive to slight timetable or infrastructure changes—can also be meaningful if measured at other locations. A classic example is the simulation of small infrastructure improvements whose effect might be negligible if it is measured only at the nearest large station.

1.4.8 Evaluating the Quality of a Simulation Model

Evaluating the quality of the model means comparing it with the reality to obtain a quantitative measure of the differences between simulation and reality. Since the reality is always influenced by stochastic factors, this comparison should be made comparing the real collected data with the stochastic simulation of the same timetable. Obviously, simulations will use the initial delays, as input, dwell times and train performances taken from the same set of collected data.

Once a model is checked for its errors in the deterministic simulation, it must be briefly checked for errors again when running a stochastic one, at least considering a couple of iterations. The goal of this check is to understand whether delays occur and were propagated correctly. In fact, there might be errors in the distributions of dwell times, but also in the settings of the signalling system, whose effect is negligible under deterministic conditions. This check is performed on the graphic timetable, simply pointing out the conflicts and trying to investigate their origin and consequences.

Now it is possible to run a full set of stochastic simulations. The higher the number of iterations is, the less will be aggregated for the analysis of results and the higher is variability of operations on the line. Fifty iterations (corresponding to the working days of about 2 months) appears as a good number in normal conditions.

To allow an effective comparison of reality and simulation it is important to consider a limited, but comprehensive set of indicators. In the experience of the Authors a combination of mean delays and punctuality at 3 or 5 min allows easily identifying the differences that would be relevant in the "official" performance regime, by at the same time mitigating their episodic sensitivity towards small variations with mean delay. Moreover, the mean delay allows appreciating changes also for very low delays. Trains are normally grouped by OD pairs and category, and can be measured in a set of locations. In order to allow the user understanding the differences between model and reality it appears important not to further aggregate them in this phase.

Based on the resulting differences, the user can decide to improve the model and run it again. A simulation model that was checked for errors might still represent the reality very poorly in particular if for some reason the stochastic parameters are not accurate.

The residual error is the accuracy of the model; since no fixed rule exists, it appears very important to set the goal based on the level of detail that is required in the study as a whole. Generally 1–2% punctuality and 20 s mean delay difference appear satisfying. Again, it appears more important to meet this goal or a slightly lower one most measuring stations than on the average value. Similarly, having the same trend in reliability along a line in simulation and reality appears more important than the strict compliance with a given threshold.

Figure 1.3 shows a comparison between simulation and reality considering mean delays (above) and punctuality (below). Apart from the absolute differences and the lack of some measuring points in the real world data, the fact that the two curves follow the same trend proves the quality of the model.

1.5 Weaknesses of Simulation

At present, microscopic simulations are the most accurate way to mimic rail operations, since infrastructure, the interlocking and safety systems as well as train movements are represented in considerable detail. However, even models that are already quality checked at both a deterministic and stochastic level contain some systematic errors. These errors are partially due to the impossibility to represent stochastic phenomena with the parameters of the motion equation and partially to the synchronous nature of simulations. It is paramount for the user to be aware of these limitations to estimate the accuracy he can obtain.

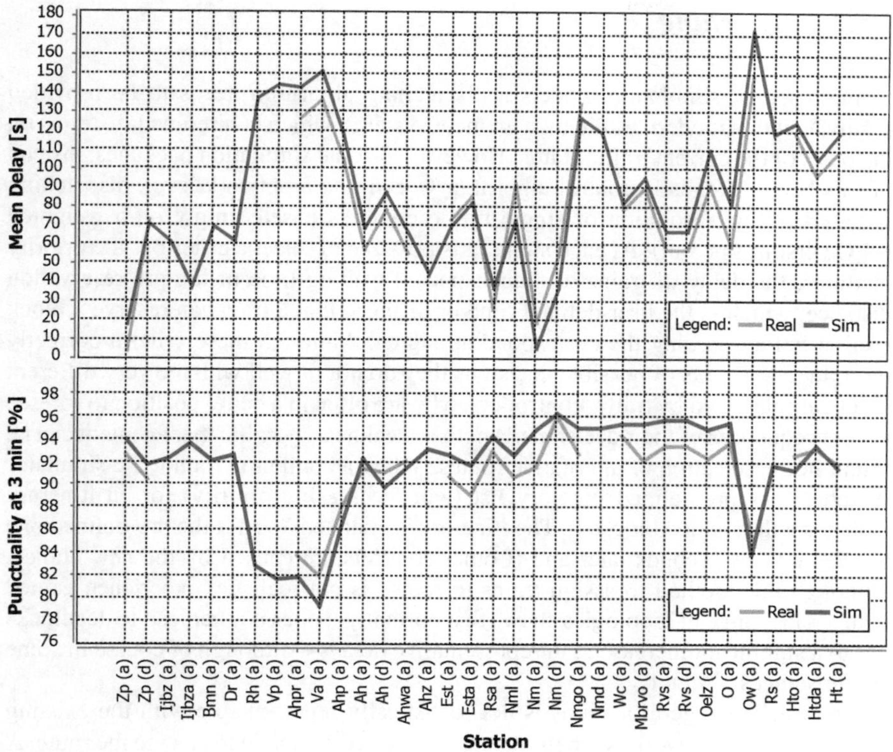

Fig. 1.3: Residual error of simulation as mean delay (above) and punctuality at 3′ (below) for a group of IC trains

1.5.1 Stochastic Nature of Inputs Not Fully Modeled

If the deterministic inputs are set up correctly, the theoretical residual model error is negligible, but the error becomes higher because the stochastic behaviour of some of the inputs is not recognized in the simulation. The traction characteristics of trains, and, hence, the journey times are not perfectly represented using conventional equations as they vary for instance depending on the mechanical condition of the equipment, the voltage being supplied and the rail adhesion.

Moreover, using the described simplified methods to calculate and apply delays and the station dwell time variability does not fully represent the reality; in addition the operational data used as input often contain a measure error, since they are derived from track circuit logs.

Driving variability is an important issue that is hard to fully reflect in simulations, in particular concerning the parts of motion in which the behaviour of drivers is more variable such as coasting or braking. Some models do not even have the capability to take this into account, while others consider them only deterministically.

1.5.2 Dispatching

In microscopic simulation models a dispatching algorithm needs to be provided (equivalent to a modern traffic management system), which can reschedule trains on the basis of the current traffic status. However, in some simulation packages, instead of this being a realistic control algorithm working at a network level, dispatching is based on the resolution of single traffic conflicts based simply on a weighted FIFO (first in-first out). At each signal it is possible to select a criterion: the priority of trains, the delay of trains or their expected waiting time or the pre-reservation distance, which is the user-defined time or distance that a train can reserve a block section before passing the corresponding signal. Although these criteria correctly reproduce wide sets of possible decisions they do not fit well with the very different practical rules used by traffic controllers, who are often in a better position to foresee the consequences of their decisions. Another weakness in traffic management can be found in the selection of alternative routes. When a conflict is found, the simulator searches in a pre-defined itinerary list (with decreasing priority) for an itinerary which is still free and blocks it. The itinerary is changed, even if the pre-defined one is released few seconds later and if other conflicts are caused by the new choice. Although this method to assign trains to routes is not realistic, experience gained with a wide range of topologies shows that normally it does neither lead to deadlocks nor to a significant increase in unrepresentative delays—although of course in some circumstances it might.

The track change at stations is not realistically representable with the existing algorithms, since it works with the above-mentioned principle to change the route. A change of platform forces all passengers to move, often just some minutes before the train's departure. Sometimes this is practicable and sometimes it is not, depending on a number of factors including the passenger numbers and the distance between platforms.

In heavily utilized terminal stations the platform change of one train can cause another change to a second train scheduled to use that platform. Case studies have shown that realistic results in terms of delay per train can often be obtained using the pre-defined platforms, although the real track allocation process is not reproduced.

1.5.3 Effectively Modeling Seriously Delayed Trains

Seriously delayed trains represent a particular challenge for dispatchers, since a de facto new slot has to be inserted, also trying to minimize the impact on other trains. Moreover, if such trains are freight services, overtaking may have to be managed. If they are long-distance passenger services, platforms have to be allocated at terminal stations, and it is necessary to decide whether the train set maintains the planned turnaround (delaying the outbound service) or not.

From the simulation's point of view, the problem appears as a combination of the above issues regarding complex the dispatching on lines and the platform allocation.

With the lookforward rules as currently applied in many simulation packages, it is impossible to ensure that simulations represent such phenomena without generating inappropriately high waiting times for other services.

1.5.4 Modeling Major Disruptions

As mentioned in the Sect. 1.4.6, at present, simulators do not cancel services automatically, often leading to unrealistically high output delays if simulating significant initial delays or disruptions. A further issue with simulating major disruptions is that the reallocation of rolling stock and train crew becomes a critical part of the problem. No microscopic simulations deal with these currently and ignoring it can lead to simulators significantly understating the impact on the performance of particular scenarios.

As major disruptions represent a significant proportion of total train lateness, this seriously limits the usefulness of microscopic simulations to estimate delays ex-ante. Thus, the possibility to control simulations using external tools and the integration of realistic dispatching algorithms are a key area for development of microscopic simulation packages if they are to meet more fully the needs of railway operators and planners.

References

1. Bešinović N, Quaglietta E, Goverde RMP (2013) A simulation based optimization approach for the calibration of dynamic train speed profiles. J Rail Transp Plann Manage 3(4):126–136. ISSN: 2210-9706. https://doi.org/10.1016/j.jrtpm.2013.10.008 (cited on page 17)
2. Buchmueller S, Weidmann U, Nash A (2008) Development of a dwell time calculation model for timetable planning. Comput Railw XI 105–114. https://doi.org/10.2495/CR080511 (cited on page 15)
3. Hauptmann D (2000) Automatische und diskriminierungsfreie Ermittlung von Fahrplantrassen in beliebig großen Netzen spurgeführter Verkehrssysteme, German. PhD thesis, TU Hannover. Eurailpress. ISBN: 978-3-7771-0287-0 (cited on page 4)
4. International Union of Railways (UIC) (2004) Website of UIC, Capacity (UIC code 406). uic.org (cited on page 6)
5. Kavička A, Klima V, Adamko N (2006) Analysis and optimisation of railway nodes using simulation techniques. Comput Railw X Trans Build Environ 88:663–673. https://doi.org/10.2495/CR060651 (cited on page 7)
6. Longo G, Medeossi G (2013) Enhancing timetable planning with stochastic dwell time modelling. Comput Railw XIII Comput Syst Design Oper Railw Other Transit Syst 127:461–471. https://doi.org/10.2495/CR120391 (cited on page 15)

7. Medeossi T (2010) Capacity and reliability on railway networks, a simulative approach. PhD thesis. University of Trieste. HDL: 10077/3675 (cited on page 7)
8. Medeossi G, Longo G, de Fabris S (2011) A method for using stochastic blocking times to improve timetable planning. J Rail Transp Plann Manage 1(1):1–13. ISSN: 2210-9706. https://doi.org/10.1016/j.jrtpm.2011.07.001. Robust modelling of capacity, delays and rescheduling in regional networks (cited on pages 12, 15, 17)
9. ON-TIME Consortium (2013) Website of Task 3.1 assessment of state-of-art of train timetabling. ontime-project.eu (cited on page 3)
10. OpenTrack (2010) ETH Zürich. opentrack.ch (cited on pages 3, 4, 7, 12, 18)
11. Rudolph R, Demnitz J (2003) Simulation of large railway networks. Robustness test for the timetable 2003 in North Rhine Westphalia, Germany. In: Proceedings of world conference on railway research. Springer, Berlin, pp 644–652 (cited on page 4)
12. Schweizerische Eisenbahn (2015) Schweizerische Fahrdienstvorschriften FDV, R300.12 (cited on page 12)
13. Website of Rail Traffic Controller (2017) Berkeley simulation software. berkeleysimulation.com (cited on pages 3, 4)
14. Website of RailSys (2017) Rail management consultants GmbH. rmcon.de/railsys-en/railsys-suite (cited on pages 3, 4, 6)
15. Wende D (2003) Fahrdynamik des Schienenverkehrs. Teubner, Wiesbaden. https://doi.org/10.1007/978-3-322--82961-0 (cited on page 12)
16. Yuan J, Hansen IA (2007) Optimizing capacity utilization of stations by estimating knock-on train delays. Transp Res Part B Methodol 41(2):202–217. ISSN: 0191-2615. https://doi.org/10.1016/j.trb.2006.02.004. Advanced modelling of train operations in stations and networks (cited on page 14)

Chapter 2
Capacity Assessment in Railway Networks

Nikola Bešinović and Rob M. P. Goverde

Abstract Capacity assessment is essential for densely utilized railway networks. To guarantee stable operations, it is necessary to evaluate the capacity occupation and determine possible infrastructure bottlenecks. This requires accurate microscopic models that incorporate detailed infrastructure characteristics, signalling and interlocking logic, train characteristics, and driver behaviour. This chapter presents capacity assessment models based on a novel algebraic approach that builds on accurate running and blocking time computations. The capacity assessment should be undertaken on corridors, station areas, and networks, and as such, support a better understanding of the existing timetable constraints and possible infrastructure investments.

N. Bešinović (✉) · R. M. P. Goverde
Delft University of Technology, Stevinweg 1, 2628 CN Delft, The Netherlands
e-mail: n.besinovic@tudelft.nl; r.m.p.goverde@tudelft.nl

R. Borndörfer et al. (eds.), *Handbook of Optimization in the Railway Industry*, International Series in Operations Research & Management Science 268,
https://doi.org/10.1007/978-3-319-72153-8_2

2.1 Introduction

Passenger and freight railway traffic have increased considerably worldwide over the past two decades, and this trend is expected to continue, see [53]. Many railway networks are already exploited to their maximum capacity and extra measures are needed to satisfy the growing demand. The ON-TIME project has diagnosed multiple capacity issues in several European countries including France, Italy, the Netherlands, Sweden, and the UK, see [37].

The possible implications of a capacity assessment could be constructing new infrastructure, improving the existing one, or using the existing one more efficiently. Upgrading the infrastructure may achieve these objectives, but is very costly and time-consuming. Therefore, more efficient planning of services may be more appropriate. Thus, understanding railway capacity is important to identify the most effective actions.

Various approaches for capacity assessment can be found in the literature and in practice. For example, RMCon [42] and Jensen et al. [25] deployed simulations for this purpose. Schwanhäußer [45, 46] introduced a queuing theory approach for evaluating the capacity. The extensions of this approach are given in [5, 24, 55–57]. Krueger [26] and Lai and Barkan [27] proposed parametric modelling. Analytic approaches based on optimization models for capacity assessment are presented in [6, 7, 34]. However, none of these models consider a timetable with its scheduled arrival and departure times as an input for the capacity assessment.

Based on extensive practical experience, it has been concluded that timetable structures are required to understand the interactions in a dense and complex railway network. Therefore, timetable structures should be used to determine the required infrastructure in terms of numbers of platforms or tracks [36]. Mackie and Preston [32] and Eliasson and Börjesson [11] also stressed the necessity of timetables for estimating the social benefit of railway investment appraisals. In particular, explicit timetable decisions (e.g., train orders and connections) are required assumptions for the analysis. Otherwise, the results will be arbitrary and scenarios will not be comparable.

This chapter describes the main (timetable-based) methods for capacity assessment that are based on *timetable compression*. Particularly, we focus on timetable-based models that consider infrastructure and rolling stock as given and fixed. In addition, the chapter is oriented towards deterministic models for assessing the level of capacity occupation, rather than the maximum (theoretical) capacity. For the latter, we refer to [9]. Section 2.2 introduces the relevant terminology and aspects of railway capacity research. Section 2.3 presents the *compression methods*, the basics of *blocking time theory*, and states the limitations of existing applications. These form the basis for the description of advanced tools for capacity assessment on the different infrastructure levels of corridors (Sect. 2.4), nodes (Sect. 2.5) and networks (Sect. 2.6). Finally, Sect. 2.7 discusses approaches for improving capacity and gives directions for further development.

2.2 Railway Capacity and Blocking Times

In order to discuss railway capacity, it is important to first give some definitions. Railway capacity is highly complex and depends on multiple factors. The *theoretical capacity* of railway lines and station layouts is defined as the maximum number of train paths (i.e., time-distance infrastructure slots) on the infrastructure in a given time window and represents an upper limit for infrastructure capacity. It usually assumes a homogeneous traffic where all trains are identical and optimally spaced throughout the time period, see [51].

The *practical capacity* of railway infrastructure is defined as the maximum number of train paths on the infrastructure in a given time window given the traffic pattern, operational characteristics or timetable structure. Practical capacity thus depends on the mix of train services with different characteristics.

Capacity occupation is defined as the amount of time that the train paths from a given timetable structure in a given time window occupy the infrastructure. Commonly, capacity occupation is expressed in minutes. Moreover, the *capacity occupation rate* (expressed in %) is defined as the ratio of capacity occupation to the given time window. It provides an indication of how a timetable may perform. Other measures for quantifying railway capacity found in the literature like the number of passengers over a given time window and amount of goods over a given time window. Table 2.1 gives an overview of the terminology commonly found in railway capacity research.

Table 2.1: Used terminology in railway capacity research

Term	Synonyms
Theoretical capacity	Design capacity, see [50]
	Absolute capacity, see [7]
	Capacity throughput, see [8, 48]
Practical capacity	Achievable capacity, see [50]
	Effective capacity, see [17]
Capacity occupation	Infrastructure occupation, see [51]
	Occupancy time, see [52]
	Consumed capacity, see [19]
	Capacity utilization, see [15]
	Carrying capacity, see [23]
	Used capacity, see [1]
Capacity occupation rate	Utilization rate, see [28]

Railway capacity depends on various aspects that can be categorized in three groups: infrastructure, rolling stock, and traffic. *Infrastructure* is defined by the railway layout (single-track, double-track, number and length of platform tracks), distance between stations, track speed limits (depending on curves, grades and switches), and the signalling system (block lengths, number of signalling aspects, train protection). For example, Goverde et al. [18] showed the influence of various signalling systems on the capacity occupation. *Rolling stock* characteristics are, among others, train composition (multiple unit or locomotive hauled wagons), length, maximum speed, and traction and braking characteristics. Capacity also depends on *traffic management and operational rules* like dominant train type (passenger, freight or mixed), use of tracks (unidirectional/bidirectional), mix of train services with different characteristics (speed, stopping pattern, frequency), train sequences, dwell times and connections in stations, see [49]. UIC [52] explained that capacity depends on the way the infrastructure is utilized which is represented in the capacity balance of the number of trains, the average speed, the traffic heterogeneity, and stability. A detailed analysis of different aspects affecting capacity can be found in [2, 20, 29, 44, 47], while an empirical comparison of different capacity assessment methods can be found in [43].

Due to the high complexity of capacity assessment, railway infrastructure is often decomposed and assessed independently, see [39]. We distinguish different infrastructure segments such as nodes, line sections (corridors) and networks. A *node* is a track layout with switches and multiple route possibilities. A node may be a small station with only a few platform tracks and limited interlocking areas, but also a big station with higher number of tracks and more complex interlockings, and may serve as a terminal for train lines. In addition, a junction can be considered as a node, which includes only interlocking but does not provide train stopping possibilities. A *line section* is a railway line between two nodes with a fixed number of parallel tracks and no switches. A line section can have one or more parallel tracks and the sequence of trains cannot change. Trains on a line section are usually separated by a block system, where each block can be allocated to at most one train. A *corridor* represents a longer railway line that consists of multiple line sections. Finally, a *network* is an area of various interconnected corridors which are considered at once during the capacity assessment.

2.2.1 Blocking Times

The concept of blocking times (see [39]) is closely related to capacity assessment and the basis for the remainder of this chapter. A *resource* represents a subset of infrastructure elements that can be exclusively allocated to a single train at a given time. In practice, this is a block section or an interlocking route section including one or more switches or crossings. A *train route* defines a set of consecutive resources that can be used by a train to traverse from one point to another (e.g., between two stations). A (time-distance) *train path* extends the train route with the time the route is used.

The *blocking time* of a resource is the time during which the resource is solely dedicated to a single train and cannot be used by any other. The blocking time consists of an approach, running and clearing time, corresponding to the train running time from the approach signal to the point located train length away the signal at the end of the block. In addition, the blocking time includes setup and release times of the route and signals, as well as the driver sight and reaction time before the approach signal. Figure 2.1 illustrates a blocking time computation for a single resource (i.e., block section) of a running train.

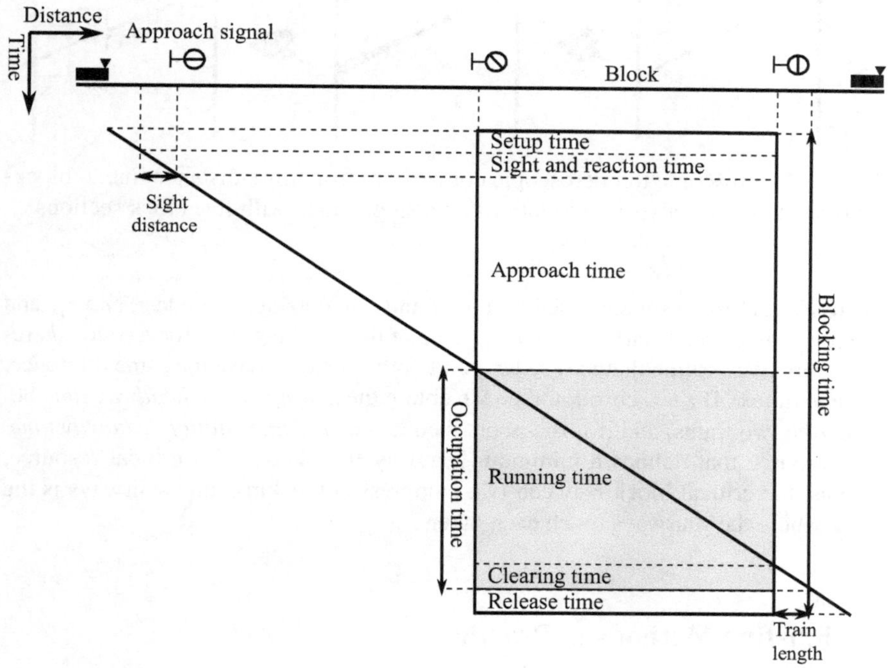

Fig. 2.1: Blocking time for a running train over a block section defined by two signals and the corresponding approach signal

The successive blocking times over a train route form a *blocking time stairway*, which can be computed for all train paths of a given timetable. Generally, a timetable consists of arrival and departure times at nodes, defining *scheduled running time*, which includes running time supplement. For computing blocking times, we need running times over each resource, which are obtained by computing an exact train time-distance speed profile corresponding to a feasible dynamic speed profile for a given scheduled running time. Figure 2.2 illustrates the conversion from timetable departure/arrival times to a train dynamic speed profile and blocking time stairway. The modelling details of the conversion are presented in [4].

Blocking time stairways are applied to compute minimum headways. The minimum headway time h_{ijs} between trains i and j on a corridor or node z is computed as

$$h_{ijz} = \max_{k \in R_{ijz}} (f_{ik} - s_{jk}),$$

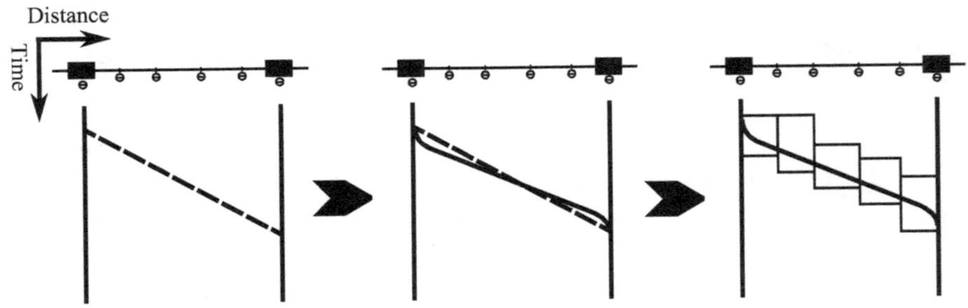

Fig. 2.2: Macroscopic to microscopic conversion: from time-distance line to blocking time stairway between two stations on a single track with five block sections

where R_{ijz} are the resources used by both i and j in corridor or node z, and s_{jk} and f_{ik} are the associated start time and end time of the blocking time for resource k, respectively. We assume that i precedes j and both stairways have the same reference, namely, time 0. If z is a corridor, then we obtain the *minimum line headway time* between the two trains; and if it is a node, then it is a *minimum station headway time*. The resource that defines a minimum headway time is called a critical resource, such as, the critical block between two compressed blocking time stairways is the block where the stairways touch each other.

2.3 Existing Methods in Practice

In Europe, the two most common analytic approaches for capacity assessment are based on the *timetable compression method*. Timetable compression is the process of shifting train paths to each other as much as possible, bringing them to the time distance of minimum headway times. The total time needed for operating such a compressed timetable is the capacity occupation. Here, the *minimum headway time* is the minimum time separation between two train paths that provides conflict-free train runs. The first method has been proposed by the International Railway Association UIC—the UIC 406 capacity method [51]. The second method is the British Capacity Utilization Index (CUI) method, see [13]. Meanwhile, in the US, a timetable compression method has not been applied yet, see [40].

2.3.1 UIC 406 Capacity Method

The UIC 406 capacity method is based on the blocking time theory. Originally, UIC [51] described a method for evaluating capacity of line sections. In the 2nd edition, UIC [52] expanded the approach to the capacity assessment of nodes. The method requires a timetable and a division of the network into line sections and nodes. The original purpose of the UIC 406 capacity method was to measure the capacity occupation of a given timetable, which is achieved by compressing the train blocking time stairways. In addition, the method has been used for assessing practical capacity. This has been done by adding extra trains in the timetable, called *timetable enrichment*.

The UIC 406 capacity method intends to standardize evaluations for obtaining comparable examination results by defining recommended values for the capacity occupation rate of a line section, see [51]. The recommended capacity occupation rates have been proposed only for double tracks and are distinguished between (a) dedicated suburban passenger traffic, dedicated high-speed lines, and mixed traffic lines and (b) peak period and daily period. Suggested capacity occupation rates are 85% and 70% for dedicated suburban traffic (peak and daily period), while they are 75% and 60% for dedicated high-speed lines and mixed traffic lines. UIC [52] proposed some preliminary ranges for nodes, but these still have to be confirmed. It is assumed that these occupation rates would guarantee stable services with respect to small disturbances. These recommendations were based on the practices among European Infrastructure Managers (IMs) at the time, but highly depend on the infrastructure layout, the way it is utilised, and the typical size of delays. Recommended capacity occupation rates are referred to as *saturation rates* (see [2]), while a corridor that reaches these rates is called a *saturated corridor*.

If a corridor is not saturated yet, additional trains may be added. This is done through an iterative process. First, the capacity occupation is computed by timetable compression. If the rate is smaller than the saturation rate, the timetable is enriched by one or more trains. Then, the capacity occupation rate is reassessed. These iterations are repeated until the corridor has been saturated. In addition, enriching can be used to determine a corridors' theoretical capacity. For further details on the enrichment process, see [9, 25].

2.3.2 CUI Method

The CUI is the measure based exclusively upon the headway norms in nodes, given as Timetable Planning Rules, see [35]. Similar to UIC 406, the CUI method builds on a network decomposition into line sections that are evaluated separately by compressing the timetable for each infrastructure segment. A line section for CUI is always determined by two neighbouring nodes, while it may be longer for the UIC 406 method. The method does not consider an exact infrastructure occupation based on blocking times, which makes it less accurate than the UIC 406 method. Thus, we refer to CUI method as to a *normative capacity assessment*. A further comparison between UIC 406 and CUI may be found in [33, 37].

2.3.3 *Open Challenges*

Recently, Lindner [30] evaluated the UIC 406 capacity method. The 2nd edition UIC [52] improved on his observations partially. One of the main remaining limitations of the UIC 406 method is the capacity assessment in nodes. It proposes to decompose a node in switch areas and (platform) track areas, and evaluate each segment independently. More recently, Rotoli [43], gave a descriptive simplified approach for evaluating nodes by using this decomposition and assuming a general node layout. Such a node decomposition may not consider all route dependencies and leads to underestimated capacity occupation. Section 2.5 introduces an analytic model that overcomes this issue.

A second limitation is due to the network decomposition to line sections which causes certain train dependencies to be neglected and result again in an underestimated capacity occupation. Third, the lengths of the decomposed line sections affect the resulting capacity occupation significantly. To overcome these challenges, we propose a network model for capacity assessment that preserves microscopic details of the infrastructure and all train dependencies (Sect. 2.6). Fourth, the given saturation rates represent a rough guideline rather than an exact values to follow. These rates are highly dependent on the infrastructure layout, train characteristics and level of service; and they may vary significantly for different national networks. However, additional research is necessary to achieve better insight.

Armstrong et al. [3] proposed a solution for the limitation of the CUI method, which is mainly applicable on line sections, by an extension for assessing the capacity in nodes. However, due to the coarser level of detail, CUI is a less accurate and rather cumbersome method that is difficult to apply to complex nodes. Following the timetable planning requirements defined by European IMs (see [38]), we encourage using the UIC 406 capacity method for further capacity analyses.

2.4 Capacity Assessment of Corridors

The compression method is quite easy to apply and should allow a natural deployment. However, only the capacity assessment of corridors is straightforward. To that purpose, various analytical and simulation models have been developed. Landex [28] extended the UIC 406 method to single tracks, while Abril et al. [1] applied it on double-track corridors. Čičak et al. [8] proposed an approach for theoretical capacity of single track lines using a normative compression method. Abril et al. [2] and Pouryousef et al. [40] are suggested for further reading on implementations of capacity assessment for corridors in Europe and the USA.

Simulation tools are most commonly used for detailed timetable analyses as described in this chapter. However, only a few of them incorporate a compression method explicitly to evaluate the capacity use of a given timetable, such as `RailSys` [54], and EGTRAIN, see [41].

2.5 Capacity Assessment of Nodes

In this section, we describe the *max-plus automata* model for capacity assessment in nodes and give a numerical example (Sect. 2.5.1). Max-plus automata combine properties of the heaps-of-pieces theory and max-plus algebra, and were introduced by Gaubert and Mairesse [12]. The max-plus algebra is a mathematical technique to model and analyse discrete event dynamic systems (DEDS) such as railway systems. We refer to [15, 22] for more details on max-plus algebra applied to railways.

One of the main advantages of max-plus automata is the ability to explicitly model the infrastructure resources and the blocking times of these resources corresponding to blocking time stairways. This is exactly what is required to compute the capacity occupation of a set of resources by a given set of train paths. Differently from the general max-plus algebra, in the max-plus automata, both the start and end time of each resource by each train are taken into account.

We assume a given timetable with assigned train routes (i.e., a *route plan*) and corresponding blocking time stairways for the trains. In this section, we view a blocking time stairway of a single train as a *piece*. Note that a piece may represent a complete or partial train route through a node. For example, a train route may consist of multiple pieces. Graphically, we may picture a compressed timetable as a *heap* of all blocking time stairways stacked on each other, i.e., a *heap-of-pieces*.

2.5.1 Max-Plus Automata Model

A max-plus algebra is a semiring over $\mathbb{R}_{\max} = \mathbb{R} \cup \{\varepsilon = -\infty\}$, equipped with the two binary operations maximum ($\oplus$) and addition ($\otimes$). For $a, b \in \mathbb{R}_{\max}$ the max-plus operations are defined as

$$a \oplus b = \max(a, b) \quad \text{and} \quad a \otimes b = a + b.$$

The element $\varepsilon = -\infty$ is the neutral element for $\oplus$ and absorbing for $\otimes$. The element $e = 0$ is the neutral element for $\otimes$. Many properties of max-plus algebra are similar to conventional algebra. The scalar max-plus operations are extended to matrices in a standard way. Let $\mathbb{R}_{\max}^{n \times n}$ be the set of $n \times n$ matrices with elements in $\mathbb{R}_{\max}$. Then, for any matrices $A = (a_{ij})$, $B = (b_{ij}) \in \mathbb{R}_{\max}^{n \times n}$ matrix addition $\oplus$ and matrix multiplication $\otimes$ are defined as

$$[A \oplus B]_{ij} = a_{ij} \oplus b_{ij} = \max(a_{ij}, b_{ij}),$$

$$[A \otimes B]_{ij} = \bigoplus_{k=1}^{n} a_{ik} \otimes b_{kj} = \max_{k=1,..,n} (a_{ik} + b_{kj}).$$

A max-plus automaton is a tuple $H = (T, R, M, s, f)$. Here, T is a finite set of tasks that represent all train routes $l \in T$, while R is a finite set of resources that can be block sections or track sections (as defined in Sect. 2.2.1). Also, M is a function that maps a task to the resources it uses. Formally, M is a morphism $T \to \mathbb{R}_{\max}^{R\times R}$ defined uniquely by a finite family of matrices $M(l)$, $l \in T$. We define $s_i(l)$ and $f_i(l)$ as the start and end time of resource i used by task l, respectively. Further, these construct the corresponding R-dimensional row vectors $s(l)$ and $f(l)$. In other words, the task l represents a (partial) train route, while vectors $s(l)$ and $f(l)$ depict the upper and lower contour of the corresponding blocking time stairway. We also assume that each stairway starts at time 0.

The matrix $M(l)$ represents the blocking time stairway, which also equals the capacity occupation, of a task l and is defined as

$$M_{ij}(l) = \begin{cases} e, & \text{for} \quad i = j, i \notin R(l), \\ f_j(l) - s_i(l), & \text{for} \quad i, j \in R(l), \\ \varepsilon, & \text{otherwise.} \end{cases} \tag{2.1}$$

A matrix element $M_{ij}(l)$ gives the time difference between the end time of the resource j and start time of the resource i. If a resource is not used, we assign e, if $i = j$, and ε elsewhere.

We define a route plan w as an ordered sequence of tasks by successive trains $w = l_1 \cdots l_n$, where $l_1, \ldots, l_n \in T$. Then, tasks from the route plan are added one by one to the heap-of-pieces by which the occupation is computed sequentially as

$$M(w) = M(l_1 \cdots l_n) = M(l_1) \otimes \cdots \otimes M(l_n).$$

Thus, matrix $M(w)$ defines the capacity occupation used by all train routes in w compressed together. Moreover, we define $x(e)$ as an empty schedule of length $|R|$. Then an upper contour $x(w)$ of schedule w is given as

$$x(w) = M(w) \otimes x(e).$$

In general, schedule w represents a given train mix (number and types of trains with corresponding routes). For practical reasons, the first train may be added as an additional train at the end of the sequence. The start time of this final train is the end point of the capacity occupation. In case of a periodic timetable, adding this first train from the next period is required, as it determines the earliest possible time to schedule the next period, which completes a full cycle. This will guarantee a necessary separation between the last train of the current period and the first of the next one. To do so, let a be the first task in a schedule of tasks w. Then the capacity occupation $\mu(w)$ of a schedule w is computed as

$$\mu(w) = \min_{i \in R(a)} \left(x_i(wa) - (f_i(a) - s_i(a)) \right), \tag{2.2}$$

where wa is the schedule for one period w with an additional train route a that belongs to the next period. We use an added train route a to determine the earliest possible start of the next period. Here, $x(wa)$ represents the capacity occupation including repeated train a. However, as mentioned, the actual occupation is defined until the start time of a, so we subtract the occupation time of a from $x(wa)$, that is, the difference $f(a)-s(a)$. Next, the capacity occupation $\mu(w)$ is defined between the start time of each element of the first train in w and a. Accepting that w starts at 0, then μ can take the minimum value of the vector $x(wa)-(f(a)-s(a))$. So, (2.2) computes the occupation μ of a node for a given route plan w that specifies an ordered sequence of blocking time stairways $l \in T$. Note that the model complexity depends on the route choices and not on the station layout complexity, so the set R can be limited to the set of used resources in the given route plan.

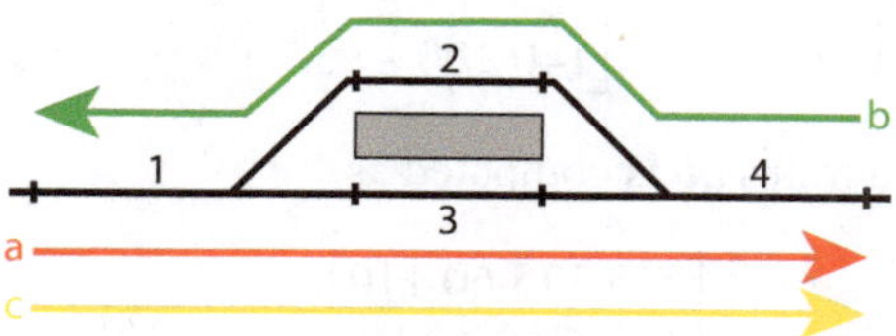

Fig. 2.3: Example 1: simple node infrastructure with trains a, b and c

Consider the following example for computing the capacity occupation of the node presented in Fig. 2.3. Consider three trains a, b, and c, timetable $w_1 = abc$ and resources $r = 1,..,4$. Trains a and c use resources $\{1,3,4\}$, while b uses $\{4,2,1\}$. Note that the order of resources defines the direction of each train. The train blocking times are given in the blocking stairways (in seconds) as follows:

Route r	$s(r)$ [s]	$f(r)$ [s]
a	$[0, \varepsilon, 25, 40]$	$[40, \varepsilon, 60, 75]$
b	$[80, 25, \varepsilon, 0]$	$[140, 100, \varepsilon, 35]$
c	$[0, \varepsilon, 35, 100]$	$[50, \varepsilon, 120, 160]$

Figure 2.4 shows individual train routes, i.e., pieces, of a, b and c. Each piece is physically connected in reality. However, it does not have to be connected in a two-dimensional plot since the horizontal axis reports all resources in the node (Fig. 2.3)

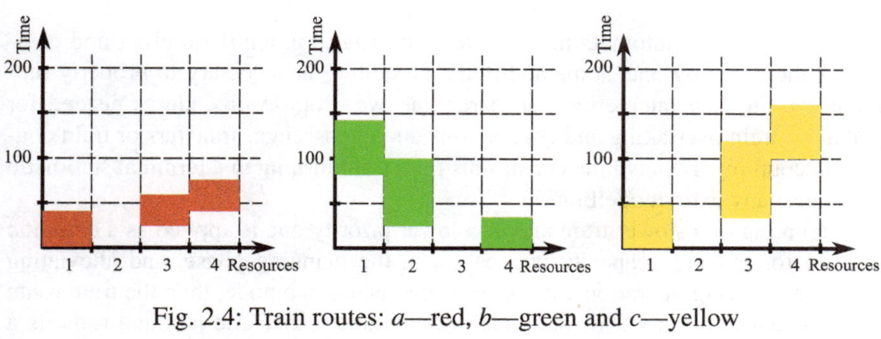

Fig. 2.4: Train routes: a—red, b—green and c—yellow

and more, these resources can be ordered randomly. So, if a train route does not use a resource, a gap in a piece may be observed. For example, resource 2 is not used by train route *a*.

The corresponding matrices *M* for train routes *a*, *b* and *c* are defined by applying (2.1) as follows

$$M(a) = \begin{bmatrix} 40 & \varepsilon & 60 & 75 \\ \varepsilon & e & \varepsilon & \varepsilon \\ 15 & \varepsilon & 35 & 50 \\ 0 & \varepsilon & 20 & 55 \end{bmatrix}, \quad M(b) = \begin{bmatrix} 60 & 20 & \varepsilon & -45 \\ 115 & 75 & \varepsilon & 10 \\ \varepsilon & \varepsilon & e & \varepsilon \\ 140 & 100 & \varepsilon & 35 \end{bmatrix}, \quad M(c) = \begin{bmatrix} 50 & \varepsilon & 120 & 160 \\ \varepsilon & e & \varepsilon & \varepsilon \\ 15 & \varepsilon & 85 & 125 \\ -50 & \varepsilon & 20 & 60 \end{bmatrix}.$$

The matrix *M* for a trains *ab* is computed as

$$M(ab) = M(a) \otimes M(b) = \begin{bmatrix} 215 & 175 & 60 & 110 \\ 115 & 75 & \varepsilon & 10 \\ 190 & 150 & 35 & 85 \\ 175 & 135 & 20 & 70 \end{bmatrix}.$$

Matrix $M(ab)$ defines the capacity occupation of *ab*, representing that *a* is immediately followed by *b*. Similarly, train route *c* is added to the route plan as $M(abc) = M(ab) \otimes M(c)$. The upper contour of the route plan *abca* is then computed as $x(abca) = M(abca) \otimes x(e) = (375, 175, 395, 410)^T$. And the capacity occupation for the route plan *abc* is then computed using (2.2) as

$$\mu(abc) = \min(x(abca) - (f(a) - s(a))) = \min\left(\begin{bmatrix} 375 \\ 175 \\ 395 \\ 410 \end{bmatrix} - \begin{bmatrix} 40 \\ \varepsilon \\ 35 \\ 35 \end{bmatrix}\right) = 335[s],$$

where the minimum is taken over the vector entries. Figure 2.5 shows the final result.

2.5.2 Satisfying Additional Timetable Constraints

Since the max-plus automata model takes train routes in temporal order and compresses them one by one, some additional modelling is necessary to properly represent certain train interactions. In particular, we propose procedures needed for modelling train overtaking and connections due to passenger transfers or train coupling/decoupling. Meanwhile, constraints for a train turning in a terminal station do not request any extra modelling.

Overtaking of a slower train and/or a lower priority one is applied as a common measure for reducing capacity occupation in the planning phase, and alleviating train delays during operations. If a train is overtaken in a node, then the train route is partitioned in an inbound route and an outbound route. The inbound route is a train route from the node entry point to the platform track, while the outbound route

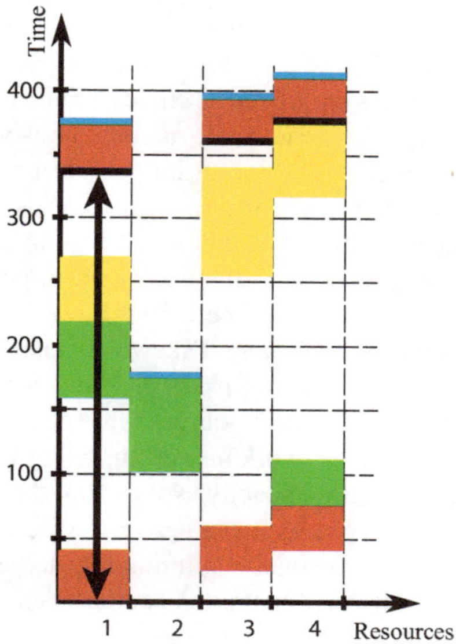

Fig. 2.5 Capacity occupation for a route plan $w_1 = abc$. The upper contour $x(abca)$ is showed by the blue line. The capacity occupation $\mu(w)$ is presented with a double arrow

runs from the platform track to the exit from the node. Coupling or decoupling of trains can be treated similarly.

A timetable often includes constraints that represent traffic requirements such as passenger transfers, which are not necessarily related to the infrastructure limitations. In other words, connecting trains often use a dedicated infrastructure. In order to maintain the timetable dependency in the max-plus automata model, additional modelling is necessary to keep the two trains together. To do so, train routes of these trains are modelled as a single task.

2.6 Capacity Assessment in Networks

Capacity assessment of railway networks is not a general practice yet. CAPRES is a railway network capacity assessment tool based on saturation of a periodic timetable with extra train paths, see [31]. PETER is an analytical tool for evaluating the capacity occupation rate and stability of a periodic timetable on the network level based on max-plus algebra, see [15, 16]. These models are based on a macroscopic network description and were originally developed for normative headway times, like the CUI. On the other hand, KABAN is a microscopic capacity assessment tool built on a detailed modelling of infrastructure, periodic timetable and train routes, which also applies max-plus algebra to compute the capacity occupation, see [10]. However, due to the high level of details, KABAN is limited only to small-sized networks. This section focuses on the general max-plus algebra modelling such as used in PETER. For similar approaches, see also [21, 22]. Note that instead of using normative headway times, we also explain how to deploy the results of capacity assessments of corridors and nodes as input to the network capacity assessment. This provides improved accuracy similar to UIC 406 over the CUI method and allows evaluating large-scale national networks.

When considering large-scale networks, the microscopic detail of the capacity assessment of corridors and nodes is aggregated into a macroscopic model that connects all the corridors together at the nodes. For the capacity assessment at corridor level, the train paths were split in parts and tackled separately over the successive line sections. At the network level, the successive train paths must again be considered as a whole. Likewise, existing interactions in nodes between various train paths over successive or crossing corridors must be regarded at the network level as well. Since the resources were already taken into account at the corridor and node level, the network model can be formulated using only time constraints. Instead, for a normative capacity assessment, the events can be any arrival, departure or passing-through events in the network which are connected by minimum running and dwell times or normative minimum headway times. Moreover, on the network level other operational constraints can be taken into account, such as passenger transfers and rolling stock connections.

In general, the *network model* consists of event times at nodes and precedence constraints between them, which represent the interconnection structure of the various trains. Before an event time may occur, it must satisfy all precedence constraints which take the form

$$x_i \geq a_{ij} + x_j,$$

where x_i and x_j are two event times and $a_{ij} \geq 0$ is the minimum time duration from event time x_j to event time x_i. This precedence constraint is very general and can be used to define a directed acyclic graph (DAG) with the event times as the nodes and the minimum time durations as the arc weights between the nodes. The minimum time durations may correspond to minimum line or station headway times, or to scheduled activities between events such as the aggregated running time between nodes or a minimum dwell or transfer time in a node (see Fig. 2.6). A critical path between two nodes in a graph represents the longest path between the two nodes, that is, the path with the highest sum of weights. A critical path algorithm over the DAG then finds the earliest occurrence times of all the events in the network which correspond to a compressed timetable with all precedence constraints respected. Finding a critical path in DAG can be done by any shortest path algorithm after negating the weights.

For periodic timetables, it is more convenient to consider periodic event times and assess the network capacity occupation in a basic timetable period. For this, an event i represents a triple $i = (E_i, L_i, S_i)$, where E_i is the event type (arrival or departure), L_i is the associated train line and S_i the station. Denote by $x_i(k)$ the event time of a periodic event i in timetable period k. So, the event time of an event i in the first period is $x_i(1)$, in the second period it is $x_i(2)$, and so on. If the events occur on time with a scheduled cycle time T, then $x_i(k+1) = x_i(k) + T$. Now the

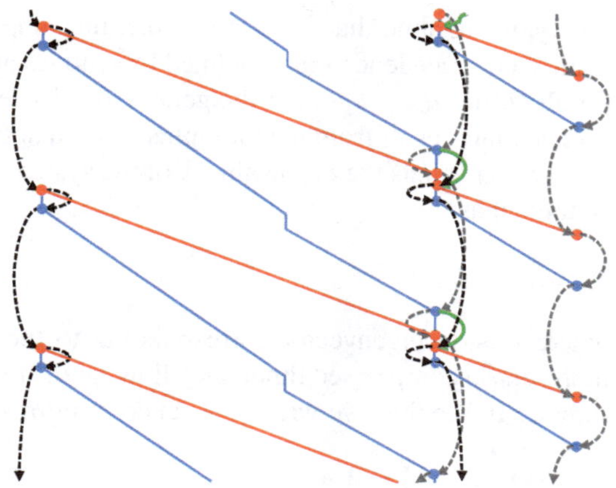

Fig. 2.6 Modelling timetable constraints in a network including event times (dots), runs, stops and transfers (solid arcs), and minimum headway times (dashed arcs)

precedence constraints can be written as

$$x_i(k) \geq a_{ij} + x_j(k - m_{ij}),$$

for all predecessor events j of i, where a_{ij} is the same for each period and m_{ij} is a non-negative integer indicating the period shift between the two events. For example, event j is scheduled m_{ij} periods before event i. Mostly, $m_{ij} \in \{0,1\}$, corresponding to two events that are scheduled in the same period ($m_{ij} = 0$) or in successive periods ($m_{ij} \geq 1$), so that the time separation crosses a period boundary. Any scheduled activity that covers more than one period can be split in parts with dummy events, so in the sequel we assume $m_{ij} \in \{0,1\}$.

The earliest occurrence of an event time is now obtained by

$$x_i(k) = \max_j \left(a_{ij} + x_j(k - m_{ij})\right), \tag{2.3}$$

where j ranges over the predecessors of i. This can be formulated conveniently in max-plus algebra. Let $x(k) = \left(x_1(k), \ldots, x_n(k)\right)'$, and collect the minimum activity and headway times in two matrices A_0 and A_1, with $[A_{m_{ij}}]_{ij} = a_{ij}$ and fill the empty entries by $\varepsilon = -\infty$ indicating that there is no direct precedence relation from event j to i. If there are parallel arcs between the same events with the same period shift, then only the maximum arc weight has to be added to the matrix. The recursive equation (2.3) can now be written for all events together as $x(k) = A_0 \otimes x(k) \oplus A_1 \otimes x(k-1)$, where k ranges over the successive timetable periods. It is a straightforward result from max-plus algebra theory that any max-plus system can be reformulated as a purely first-order system of the form

$$x(k) = A \otimes x(k-1),$$

where $A = A_0^* \otimes A_1$, with the Kleene star operator $A^* = A^0 \oplus A^1 \oplus \ldots \oplus A^{n-1}$ and the powers are understood in the max-plus algebra, e.g., $A^2 = A \otimes A$, see [22]. For simplicity, we assume that A is irreducible, meaning that it corresponds to a strongly-connected precedence graph defined by n nodes and an arc (j,i) with arc weight a_{ij} for all entries $a_{ij} \neq -\infty$. For the general results see [15].

The main result from the max-plus algebra approach is that the network capacity occupation equals the eigenvalue λ of the system matrix A. The eigenvalue problem is defined as

$$A \otimes v = \lambda \otimes v, \tag{2.4}$$

where v is an eigenvector corresponding to the eigenvalue λ. The eigenvector v represents a compressed timetable allowing the railway system to operate with cycle time λ. To see this, we write (2.4) in conventional form as

$$\max_j \left(a_{ij} + v_j\right) = \lambda + v_i. \tag{2.5}$$

Considering v as a timetable vector in some period, then the left-hand side gives the earliest occurrence time for event i in the next period and the right-hand side says that this occurrence time is exactly λ after the previous event time v_i. If $G(A)$ is strongly connected, then the eigenvalue λ is unique (see [14, 22]), and so (2.5) holds for each v_i with the same λ. Since the a_{ij} are the minimum activity and headway times, v is the compressed timetable, and λ is the network capacity occupation.

A *critical circuit* is a circuit in the precedence graph with the maximum ratio of total arc weight to the number of arcs in the circuit, which equals λ. To obtain a stable timetable that can cope with delays, the timetable must be operated with a period length $T > \lambda$. The events on the critical circuit also identify the critical activities and headways in the network, similar to the critical blocks in the capacity assessment of corridors. Figure 2.7 shows a large network where the critical circuit is the traffic over a partial single-track line. Efficient algorithms are available for solving the max-plus eigenvalue problem, and in particular graph algorithms based on the critical circuit, see [14, 22].

2.7 Conclusions and Future Developments

Railway capacity research plays an important role in railway planning and operations. In this chapter, we gave an overview of methods for railway capacity assessment, with the focus on deterministic timetable-based models. We first presented common methods based on timetable compression, UIC 406 and CUI. The CUI is a normative method, while the UIC 406 model considers a higher level of detail that allows more accurate estimation of capacity occupation. We also described the existing and advanced models for assessing different infrastructure segments independently like corridors and nodes, but also whole networks.

The benefit of capacity assessment is manifold. First, evaluating existing or new timetables to determine capacity occupation will provide insight into the ex-

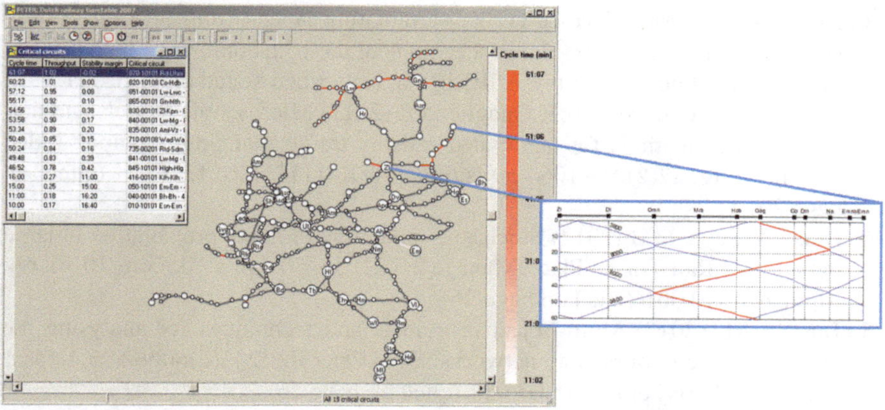

Fig. 2.7: Critical circuit in a large network (PETER)

pected level of service. Second, the infrastructure bottlenecks can be determined in a network. Third, capacity assessment may suggest possible improvements in traffic management like using alternative train routes that are more efficient. Fourth, proposing the most attractive and beneficial infrastructure projects based on capacity assessment is particularly valuable to infrastructure managers and governments in allocating available funds most efficiently. Fifth, the impact of scheduled construction and maintenance works on traffic can be estimated.

The future development of capacity assessment models should stay in line with the existing compression method, the UIC 406. To make it a standard evaluation tool and apply it internationally, additional research on capacity saturation rates and required levels of service for punctuality and regularity is essential. The network models should gain more attention, as only these are able to incorporate all interactions occurring in a timetable. In addition, it is necessary to maintain the high level of accuracy by transforming data from micro to macro models.

References

1. Abril M, Salido MA, Barber F, Ingolotti L, Lova A, Tormos MP (2005) A heuristic technique for the capacity assessment of periodic trains. In: Proceedings of the 2005 conference on artificial intelligence research and development. IOS Press, Amsterdam, pp 339–346. ISBN: 1-58603-560-6 (cited on pages 27, 32)
2. Abril M, Barber F, Ingolotti L, Salido MA, Tormos MP, Lova A (2008) An assessment of railway capacity. Transp Res Part E Logist Transp Rev 44(5):774–806. https://doi.org/10.1016/j.tre.2007.04.001 (cited on pages 28, 31, 32)

3. Armstrong J, Preston J, Hood I (2015) Evaluating capacity utilisation and its upper limits at railway nodes. In: Proceedings of the 13th conference on advanced systems in public transport (CASPT2015), Rotterdam (cited on page 32)
4. Bešinović N, Goverde RMP, Quaglietta E (2017) Microscopic models and network transformations for automated railway traffic planning. Comput Aided Civ Inf Eng 32(2):89–106. https://doi.org/10.1111/mice.12207 (cited on page 29)
5. Büker T, Seybold B (2012) Stochastic modelling of delay propagation in large networks. J Rail Transp Plann Manage 2(1–2):34–50. https://doi.org/10.1016/j.jrtpm.2012.10.001 (cited on page 26)
6. Burdett RL (2015) Multi-objective models and techniques for analysing the absolute capacity of railway networks. Eur J Oper Res 245(2):489–505. https://doi.org/10.1016/j.ejor.2015.03.020 (cited on page 26)
7. Burdett RL Kozan E (2006) Techniques for absolute capacity determination in railways. Transp Res B Methodol 40(8):616–632. https://doi.org/10.1016/j.trb.2005.09.004 (cited on pages 26, 27)
8. Čičak M, Mlinarić TJ, Abramović B (2012) Methods for determining throughput capacity of railway lines using coefficients of elimination. PROMET-Traffic Transp 16(2):63–69. https://doi.org/10.7307/ptt.v16i2.575 (cited on pages 27, 32)
9. Delorme X, Gandibleux X, Rodriguez J (2009) Stability evaluation of a railway timetable at station level. Eur J Oper Res 195(3):780–790. https://doi.org/10.1016/j.ejor.2007.06.062 (cited on pages 26, 31)
10. Ekman J (2011) Kaban-a tool for analysis of railway capacity. In: Computational methods and experimental measurements XV, pp 693–702. WIT Press, Ashurst. https://doi.org/10.2495/cmem110611 (cited on page 37)
11. Eliasson J, Börjesson M (2014) On timetable assumptions in railway investment appraisal. Transp Policy 36:118–126. https://doi.org/10.1016/j.tranpol.2014.08.008 (cited on page 26)
12. Gaubert S, Mairesse J (1999) Modeling and analysis of timed Petri nets using heaps of pieces. IEEE Trans Autom Control 44:683–697. https://doi.org/10.1109/9.754807 (cited on page 33)
13. Gibson S, Cooper G, Ball B (2002) Developments in transport policy: the evolution of capacity charges on the UK rail network. J Transp Econ Policy 36(2):341–354. https://jstor.org/stable/20053906 (cited on page 30)
14. Goverde RMP (2005) Punctuality of railway operations and timetable stability analysis. PhD thesis. Delft University of Technology. UUID: a40ae4f1-1732-4bf3-bbf5-fdb8dfd635e7 (cited on page 40)
15. Goverde RMP (2007) Railway timetable stability analysis using max-plus system theory. Transp Res B Methodol 41(2):179–201. https://doi.org/10.1016/j.trb.2006.02.003 (cited on pages 27, 33, 37, 40)
16. Goverde RMP (2010) A delay propagation algorithm for large-scale railway traffic networks. Transp Res Part C Emerg Technol 18(3):269–287. https://10.1016/j.trc.2010.01.002 (cited on page 37)

17. Goverde RMP, Hansen IA (2013) Performance indicators for railway timetables. In: 2013 IEEE international conference on intelligent rail transportation (ICIRT). IEEE, New York, pp 301–306. https://doi.org/10.1109/icirt.2013.6696312 (cited on page 27)
18. Goverde RMP, Corman F, D'Ariano A (2013) Railway line capacity consumption of different railway signalling systems under scheduled and disturbed conditions. J Rail Transp Plann Manage 3(3):78–94. https://doi.org/10.1016/j.jrtpm.2013.12.001 (cited on page 27)
19. Hansen IA, Pachl J (2014) Railway timetabling & operations: analysis, modelling, optimisation, simulation, performance evaluation. Eurailpress, Hamburg. ISBN: 978-3-7771-0462-1 (cited on page 27)
20. Harrod S (2009) Capacity factors of a mixed speed railway network. Transp Res Part E Logist Transp Rev 45(5):830–841. https://doi.org/10.1016/j.tre.2009.03.004 (cited on page 28)
21. Heidergott B, de Vries R (2001) Towards a (max,+) control theory for public transportation networks. Discrete Event Dyn Syst 11(4):371–398. https://doi.org/10.1023/A:1011225209640 (cited on page 37)
22. Heidergott B, Jan Olsder G, van der Woude J (2014) Max Plus at work: modeling and analysis of synchronized systems: a course on Max-Plus algebra and its applications. Princeton University Press, Princeton. ISBN: 978-0-691-11763-8 (cited on pages 33, 37, 39, 40)
23. Hu J, Li H, Meng L, Xu X (2013) Modeling capacity of urban rail transit network based on bi-level programming. In: 2013 joint rail conference. American Society of Mechanical Engineers. https://doi.org/10.1115/jrc2013-2429 (cited on page 27)
24. Huisman T, Boucherie RJ, van Dijk NM (2002) A solvable queueing network model for railway networks and its validation and applications for the Netherlands. Eur J Oper Res 142(1):30–51. https://doi.org/10.1016/s0377-2217(01)00269-7 (cited on page 26)
25. Jensen LW, Landex A, Nielsen AO, Kroon LG, Schmidt M (2017) Strategic assessment of capacity consumption in railway networks: framework and model. Transp Res Part C Emerg Technol 74:126–149. ISSN: 0968-090X. https://doi.org/10.1016/j.trc.2016.10.013 (cited on pages 26, 31)
26. Krueger H (1999) Parametric modeling in rail capacity planning. In Simulation conference proceedings, 1999 Winter. vol 2. IEEE. Phoenix, AZ, pp 1194–1200. https://10.1109/wsc.1999.816840 (cited on page 26)
27. Lai Y-C, Barkan C (2009) Enhanced parametric railway capacity evaluation tool. Transp Res Rec J Transp Res Board 2117:33–40. https://doi.org/10.3141/2117-05 (cited on page 26)
28. Landex A (2009) Evaluation of railway networks with single track operation using the UIC 406 capacity method. Netw Spat Econ 9(1):7–23. https://doi.org/10.1007/s11067-008-9090-7 (cited on pages 27, 32)
29. Lindfeldt A (2015) Railway capacity analysis: methods for simulation and evaluation of timetables, delays and infrastructure. PhD thesis, KTH Royal Institute of Technology. ISBN: 978-91-87353-65-9 (cited on page 28)

30. Lindner T (2011) Applicability of the analytical UIC Code 406 compression method for evaluating line and station capacity. J Rail Transp Plann Manage 1(1):49–57. https://doi.org/10.1016/j.jrtpm.2011.09.002 (cited on page 32)
31. Lucchini L, Rivier R, Emery D (2000) CAPRES network capacity assessment for Swiss North-South rail freight traffic. In: Computers in railways VII. WIT Press, Ashurst, pp 221–230. https://doi.org/10.2495/CR000211 (cited on page 37)
32. Mackie P, Preston J (1998) Twenty-one sources of error and bias in transport project appraisal. Transp Policy 5(1):1–7. https://doi.org/10.1016/s0967-070x(98)00004-3 (cited on page 26)
33. Melody KS (2012) Railway track capacity: measuring and managing. PhD thesis. University of Southampton, Faculty of Engineering and the Environment. https://eprints.soton.ac.uk/348816 (cited on page 31)
34. Mussone L, Calvo RW (2013) An analytical approach to calculate the capacity of a railway system. Eur J Oper Res 228(1):11–23. https://doi.org/10.1016/j.ejor.2012.12.027 (cited on page 26)
35. Network Rail (2015) Timetable planning rules. Oct 2015. https://networkrail.co.uk/aspx/3741.aspx (cited on page 31)
36. Odijk MA, Romeijn HE, van Maaren H (2006) Generation of classes of robust periodic railway timetables. Comput Oper Res 33(8):2283–2299. https://doi.org/10.1016/j.cor.2005.02.004 (cited on page 26)
37. ON-TIME (2012) Methods for capacity assessment in Europe. Technical report ONTWP03-I-UDB-009-03 (cited on pages 26, 31)
38. ON-TIME (2014) Benchmark analysis, test and integration of timetable tools. Technical report ONT-WP03-D-TUT-037-02 (cited on page 32)
39. Pachl J (2014) Railway operation and control, 3rd edn. VTD Rail Publishing, Mountlake Terrace. ISBN: 978-0-9719915-6-9 (cited on page 28)
40. Pouryousef H, Lautala P, White T (2015) Railroad capacity tools and methodologies in the US and Europe. J Modern Transp 23(1):30–42. https://doi.org/10.1007/s40534-015-0069-z (cited on pages 30, 32)
41. Quaglietta E (2014) A simulation-based approach for the optimal design of signalling block layout in railway networks. Simul Model Pract Theory 46:4–24. https://doi.org/10.1016/j.simpat.2013.11.006 (cited on page 32)
42. RMCon (2012) RailSys 8 Classic, Railsys 8 Enterprise, network wide timetable and infrastructure management. Technical report. Rail management consultants GmbH (RMCon), Hanover (cited on page 26)
43. Rotoli F, Cawood EN, Soria A (2016) Capacity assessment of railway infrastructure: tools, methodologies and policy relevance in the EU context. Technical report. EUR 27835 EN, Joint Research Center, European Union. https://doi.org/10.2791/037759 (cited on pages 28, 32)
44. Schmidt MJR (2014) The effect of infrastructure changes on railway operations. PhD thesis. University of Louisville. https://doi.org/10.18297/etd/1769 (cited on page 28)

45. Schwanhäußer W (1978) Die Ermittlung der Leistungsfähigkeit von großen Fahrstraßenknoten und von Teilen des Eisenbahnnetzes. AET 33:7–18. ISSN: 0066-6300 (cited on page 26)
46. Schwanhäußer W (1994) The status of German railway operations management in research and practice. Transp Res A Policy Pract 28(6):495–500. https://doi.org/10.1016/0965-8564(94)90047-7 (cited on page 26)
47. Shih M-C, Dick C, Sogin S, Barkan C (2014) Comparison of capacity expansion strategies for single-track railway lines with sparse sidings. Transp Res Rec J Transp Res Board 2448:53–61. https://doi.org/10.3141/2448-07 (cited on page 28)
48. Sogin S, Lai Y-C, Dick CT, Barkan C (2013) Comparison of capacity of single and double-track rail lines. Transp Res Rec J Transp Res Board 2374:111–118. https://doi.org/10.3141/2374-13 (cited on page 27)
49. Strategic Rail Authority (2014) Capacity utilisation policy: network utilisation strategy. SRA, London (cited on page 28)
50. TRB (2013) Transit capacity and quality of service manual, 3rd edn. Transit Cooperative Research Program (TCRP) Report 165. Transportation Research Board, Washington. https://doi.org/10.17226/24766 (cited on page 27)
51. UIC (2004) Code 406: capacity, 1st edn. International Union of Railways, Paris (cited on pages 27, 30, 31)
52. UIC (2013) Code 406: capacity, 2nd edn. International Union of Railways, Paris (cited on pages 27, 28, 30, 31, 32)
53. UNECE (2015) Number of railway passengers by country, passengers and time. Sept 2015. https://w3.unece.org/PXWeb2015/pxweb/en/STAT/STAT_40-TRTRANS_03-TRRAIL (cited on page 26)
54. Website of RailSys (2017) Rail management consultants GmbH. June 2017. https://rmcon.de/railsys-en/railsys-suite (cited on page 32)
55. Weik N, Niebel N, Nießen N (2016) Capacity analysis of railway lines in Germany – a rigorous discussion of the queueing based approach. J Rail Transp Plann Manage 6(2):99–115. https://doi.org/10.1016/j.jrtpm.2016.06.001 (cited on page 26)
56. Wendler E (2007) The scheduled waiting time on railway lines. Transp Res B Methodol 41(2):148–158. https://doi.org/10.1016/j.trb.2006.02.009 (Not cited.)
57. Yuan J, Hansen IA (2007) Optimizing capacity utilization of stations by estimating knock-on train delays. Transp Res B Methodol 41(2):202–217. https://doi.org/10.1016/j.trb.2006.02.004 (cited on page 26)

Chapter 3
Aggregation Methods for Railway Network Design Based on Lifted Benders Cuts

Andreas Bärmann and Frauke Liers

Abstract Rail freight traffic in Germany has experienced significant growth rates over the last decade, and recent forecasts expect this tendency to continue over the next 20 years due to the increases in national and international trade. Internal predictions of Deutsche Bahn AG, the most important German railway enterprise, assume a mean increase of 2% per year for rail freight traffic until 2030. At this pace, the German railway network in its current state would reach its capacity limit way before this date. As investments into the network infrastructure bear a very high price tag, it is crucial to use the available budget in the most efficient manner. Furthermore, the large size of the networks under consideration warrants the development of efficient algorithms to handle the complex network design problems arising for real-world data. This led us to the development of network aggregation procedures which allow for much shorter solution times by compressing the network information. More exactly, our framework works by clustering the nodes of the underlying graph to components and solving the network design problem over this aggregated

A. Bärmann (✉) · F. Liers
FAU Erlangen-Nürnberg, Lehrstuhl für Wirtschaftsmathematik, Department Mathematik, Cauerstraße 11, 91058 Erlangen, Germany
e-mail: Andreas.Baermann@math.uni-erlangen.de; Frauke.Liers@math.uni-erlangen.de

R. Borndörfer et al. (eds.), *Handbook of Optimization in the Railway Industry*,
International Series in Operations Research & Management Science 268,
https://doi.org/10.1007/978-3-319-72153-8_3

graph. This kind of aggregation may either be used as a quick heuristic, or it can be extended to an exact method, e.g. by iterative refinement of the clustering, The latter results in a cutting plane algorithm, which also introduces new variables with each refinement. This idea developed in Bärmann et al. (Math Program Comput 7(2):189–217, 2015) is extended in this chapter such that it is able to incorporate the costs for routing flow through the network via lifted Benders optimality cuts. Altogether, our algorithm can be seen as a novel kind of Benders decomposition which allows to shift variables from the subproblem to the master problem in the process. Computations on several benchmark sets demonstrate the effectiveness of the approach.

3.1 Introduction

Rail freight traffic in Germany has seen high growth rates between 2001 and 2011, only shortly interrupted by the economic crisis in 2009, as can be seen from Fig. 3.1. The total amount of goods transported on the German railway network has risen

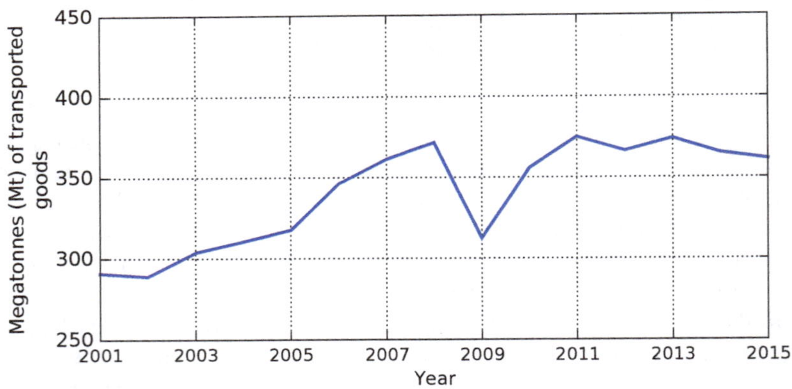

Fig. 3.1: Development of rail freight traffic in Germany between 2001 and 2015 in mega-tonnes (Mt) of transported goods per year; source: [17]

from 291 Mt of transported goods to 375 Mt in these years, which means a total increase by 29% or an average increase of 2.6% per year in this time. Experts in the field speak of a mere "renaissance" of rail freight traffic, making the decline in earlier decades almost forgotten. On the one hand, this development is explainable by an overall increase in freight traffic, which is due to the role of Germany as an exporting and also as a freight transit country. Furthermore, ecological considerations gave rise to political incentives aiming at the shift from road transport to rail transport. In the years since 2011, however, there has mainly been a sideways trend in the transportation figures. This can be attributed to the fact that line capacities in the

German rail freight network are already exhausted in many regions. Along several of its main corridors, congestions and resulting delays are part of daily business; in extreme cases, transportation orders have to be rejected entirely due to insufficient capacities. In order to improve this situation, and to allow for the growth of rail freight traffic to continue over the years to come, it will be necessary to invest decisively in higher track capacities, using different available measures. This is the only way to ensure that a significant part of the projected increase in freight traffic overall can be transported along the railway network instead of contributing to road traffic.

Our industry partner Deutsche Bahn AG has put much effort into a reliable forecast for the demand potential in rail freight transportation until 2030. The data presented here is the current basis for infrastructure planning at Deutsche Bahn AG, and we therefore use it for our computational experiments in this chapter. Figure 3.2 shows the growth in demand between 2010, the base year of our study and the target year 2030 according to the internal demand forecast by GSV, which is their department for traffic prognosis, simulation and optimization. Note that the figure is a joint presentation for freight traffic and long-distance passenger traffic.

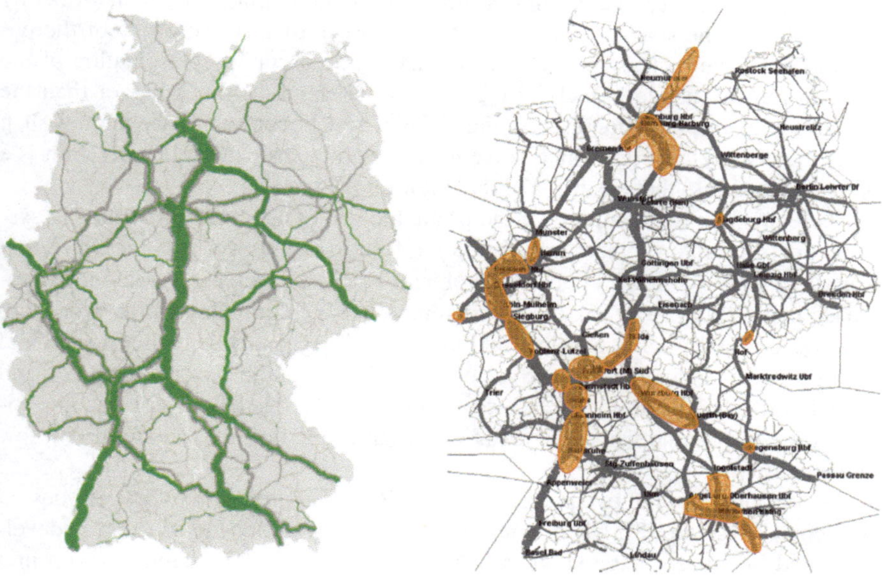

(a) (b)

Fig. 3.2: Visualization of the GSV forecast for the growth in transportation demand until the year 2030; source: [9], see also [3]. (**a**) Joint growth in freight traffic and long-distance passenger traffic on the main transportation corridors between 2010 and 2030. (**b**) Expected bottlenecks in the German railway network in 2030 without the creation of new capacities on the links

The projected demand increases between 2010 and 2030 can be seen in Fig. 3.2a. The links are coloured in grey, a thicker line represents a higher utilization of the link in the base year 2010. The expected utilization of the links in the target year 2030 is superimposed in green, with a thicker line representing a higher growth in demand. It is clearly visible that there are certain corridors on which GSV expects a high increase in transportation demand. Figure 3.2b shows the locations in the railway network where this increase is expected to lead to bottlenecks if network capacities are not increased. In order to avoid a decreasing punctuality and an increase of denied route reservations, it will be necessary to construct new links and to upgrade existing links.

From a mathematical point of view, the expansion of the railway network leads to a large-scale multi-commodity network design problem on a graph with 1600 nodes and 3600 origin-destination pairs. Together with the requirement to accompany the choice of links to upgrade with a detailed schedule for the implementation of the upgrades, the task becomes very challenging. This was the motivation to develop a decomposition approach called multiple-knapsack decomposition in [2]. It works by an efficient decomposition along the timescale, and the above paper reports very favorable results in an extensive case study conducted on the above data. In the present chapter, we develop a different solution approach which works by aggregation of the spacial dimension of the network to shrink the size of the optimization problem. This reflects the fact that a microscopic representation of the German railway network leads to a graph that is about 10 times bigger than the macroscopic representation used in the above study. In contrast to techniques such as Lagrangean relaxation, which solve local subproblems exactly, aggregation is a coarsening process that ensures a global view on the problem.

There are several examples for the profitable use of aggregation in railway systems. In [15], a micro-macro transformation for railway networks is presented that is used to solve the track allocation problem. An extensive case study on this approach is given in [5]. In [6], the authors introduce a column generation procedure with a coarsened pricing-problem and successfully apply their method to optimal rolling-stock planning in railway systems. In [7], the authors propose a two-level approach for the generation of railway timetables which is based on a decomposition of the railway network into areas with high traffic density and areas with low traffic density.

We are only aware of two previous applications of aggregation to network design problems. These are [8], where a multi-level tabu search for the problem is developed, and [4], where iterative aggregation is applied to the planning horizon in a network design problem with time-dependent demand. Further literature on the use of aggregation in mixed-integer programming can be found in [1].

In this chapter, we extend the idea of [1] for an iterative aggregation scheme for network design problems. It works by clustering the nodes in the graph to components in order to reduce the size of the network design problem and thus the necessary computation time. Via adaptive refinement of the clustering, this approach can be made an exact algorithm for the problem. The original version of this algorithm is for the case of network design without routing costs, i.e. it only considers the costs for upgrading the links but not those for routing flow in the network. However, the consideration of routing costs is an important requirement, not only in the problem context considered here, to come to efficient routing patterns for the given demand. We show how to integrate routing costs into the existing framework by introducing additional cutting planes. These will take the form of lifted Benders optimality cuts.

Aggregation techniques in combination with Benders decomposition have mostly been considered so far as a means to shrink the scenario tree for stochastic programming problems. Publications like [10, 13, 18] may serve as a reference here. In [14], the authors investigate a Benders decomposition scheme with an aggregated mixed-integer linear subproblem, which is used to determine an optimal expansion of power generation capacities in an electricity supply system. In [16], a Zipkin-like node aggregation scheme is proposed in the context of network design problems. Under quite restrictive aggregation rules (cf. [19]), it is possible to solve the problem via a Benders approach and to establish error bounds on the obtained solution. The idea for a general purpose Benders algorithm for aggregated formulations can be found in [11]. The authors propose a Benders master problem that is an aggregated version of the original problem with additional variables modelling the coupling between the coarse and the fine decision space. The original problem functions as the Benders subproblem in this approach. Branching on the original problem variables ensures an exact algorithm. However, computational results are not available yet.

Our approach is differs from the above in that we will integrate the routing costs of the arcs into the basic aggregation scheme by incorporating an additional subproblem. Its task is to calculate the component-wise routing cost of feasible solutions and to pass this information to the master problem via a lifted Benders optimality cut. The main difference to the algorithm in [11] is that the subdivision between the master problem and the subproblem does not stay fixed in our approach. Our algorithm can be described as a novel kind of Benders decomposition which allows to shift variables from the subproblem to the master problem in the process—an idea that can be generalized beyond the application to network design. The algorithm in its present form is able to solve the network design problem in its single-commodity, single-period form. An extension to more general network design problems is a worthwhile goal for future research.

The rest of this chapter is organized as follows. In Sect. 3.2, we review the aggregation scheme developed in [1]. Then we derive our extension of this approach to include routing costs in Sect. 3.3. Our computational results in Sect. 3.4 show the effectiveness of the method for instances derived from the railway network design problem described above. In Sect. 3.5, we give our conclusions on the presented algorithm and outline directions for further research.

3.2 Basic Aggregation Scheme

In this section, we give a short review of the aggregation algorithm for network design problems from [1]. We state the algorithm as well as our extension of it for a very basic setting, namely a canonical single-commodity variant of the problem. On the one hand, this problem is already NP-complete, and on the other hand, the algorithmic framework developed for this basic problem can be generalized to broader classes of network design problems. Results presented in [1] demonstrate that a straightforward extension to multi-commodity problems is possible.

We consider a symmetric directed graph $G = (V, A)$ with a set of nodes V and a set of arcs A. Each arc $a \in A$ possesses initial arc capacities $c_a \geq 0$. They may be increased by installing a module with an additional capacity of C_a at a price of k_a per unit which is available in integral multiples. This is modelled by variable $y_a \in Z_+$. Let δ_v^- be the set of arcs entering node v and δ_v^+ be the set of arcs leaving node v. The flow on arc a shall be represented by variable x_a, and it incurs a routing cost f_a per unit on this arc. The aim is to determine a feasible routing of a specified demand vector $d \in \mathbb{R}^{|V|}$ that minimizes the total cost of routing and capacity expansion. A mixed-integer programming (MIP) formulation is given by

$$\begin{aligned} \min \; & \sum_{a \in A} f_a x_a + \sum_{a \in A} k_a y_a \\ \text{subject to } \; & \sum_{a \in \delta_v^+} x_a - \sum_{a \in \delta_v^-} x_a = d_v && (\forall v \in V) \\ & x_a \leq c_a + C_a y_a && (\forall a \in A) \\ & x_a \geq 0 && (\forall a \in A) \\ & y_a \in \mathbb{Z}^+ && (\forall a \in A). \end{aligned} \tag{3.1}$$

The first set of constraints is referred to as the *flow conservation constraints*, while the second set are the *capacity constraints*. For the rest of this section, we assume $f = 0$, i.e. we consider the case without routing costs. The extension to non-zero f will be part of Sect. 3.3. The main algorithmic idea of [1] is a decomposition of the problem which—from a bird's eye view—can be stated as follows:

1. Partition the node set of the graph into components, i.e. choose an initial aggregation.
2. Master problem: Solve the network expansion problem over the aggregated graph.
3. Subproblem: Check the feasibility of the network upgrade w.r.t. the original graph.
4. (a) In case of feasibility: Terminate and return the feasible network expansion.
   (b) In case of infeasibility: Refine the partition and go to Step 2.

In the next two subsections, we give the definition of the master and the subproblem as well as an explanation of their interplay.

3.2.1 The Aggregation Master Problem

The term *aggregation* as it is used in the following refers to clustering the nodes of a directed graph $G = (V, A)$ into subsets. The *aggregated graph* $\mathcal{G}_\varphi = (\mathcal{V}_\varphi, \mathcal{A}_\varphi)$

is defined by a surjective *clustering function* $\varphi\colon V \to \{1,\dots,k\}$, $k \in \mathbb{N}$. Its node set $\mathscr{V}_\varphi = \{V_1,\dots,V_k\}$ is a partition of V into k *components* with $V_i := \varphi^{-1}(i)$ for $i = \{1,\dots,k\}$. Its arc set $\mathscr{A}_\varphi$ contains a directed arc from $V_i \in \mathscr{V}_\varphi$ to $V_j \in \mathscr{V}_\varphi$ for each arc $(u,v) \in A$ with $i = \varphi(u) \neq \varphi(v) = j$, i.e. u and v belong to different components. Note that G as well as $\mathscr{G}_\varphi$ are allowed to contain multiple arcs between the same two nodes.

The master problem of the aggregation algorithm then is Problem (3.1) applied to some aggregation $\mathscr{G}_\varphi$ of G, which is done as follows. The aggregated demand vector d_φ is defined via $d_\varphi(V_i) = \sum_{v\in V_i} d_v$ for all $i = 1,\dots,k$, i.e. the demand of a component is the net demand of the nodes it contains. The capacity $c_\varphi(a)$ and the installable module of an arc $a \in \mathscr{A}_\varphi$ are those of the corresponding original arc. In order to simplify the notation, we identify a component $V_i \in \mathscr{V}_\varphi$ with its index i, especially defining $d_i := d_\varphi(V_i)$, and identify each arc $a \in \mathscr{A}_\varphi$ with the corresponding original one in A. The master problem with respect to $\mathscr{G}_\varphi$ can then be stated as

$$
\begin{aligned}
\min \quad & \sum_{a\in\mathscr{A}_\varphi} k_a y_a \\
\text{subject to} \quad & \sum_{a\in\delta_i^+} x_a - \sum_{a\in\delta_i^-} x_a = d_i && (\forall i \in \mathscr{V}_\varphi) \\
& x_a \le c_a + C_a y_a && (\forall a \in \mathscr{A}_\varphi) \\
& x_a \ge 0 && (\forall a \in \mathscr{A}_\varphi) \\
& y_a \in \mathbb{Z}^+ && (\forall a \in \mathscr{A}_\varphi).
\end{aligned}
\tag{3.2}
$$

Note that the flow conservation constraint for component $i \in \mathscr{V}_\varphi$ arises as the sum of the original flow conservation constraints belonging to the nodes $v \in V_i$. Furthermore, the capacity constraint for some arc $a \in A$ of the original problem (3.1) is part of the aggregated problem (3.2) if and only if $a \in \mathscr{A}_\varphi$ holds. This applies similarly to the summands in the objective function. Therefore, the master problem is a relaxation of the original network expansion problem, and its optimal value w.r.t. an arbitrary clustering function φ provides a lower bound for the optimal value of the latter.

By construction of the master problem, a feasible solution naturally translates to a (not necessarily feasible) solution for the original problem by performing two steps: First, all upgrade decisions (y-variables) corresponding to arcs within any of the components are set to zero. Second, the induced flows (x-variables) within the components are computed by solving a maximum-flow problem. This *extendibility test* is the purpose of the subproblem, which is derived next.

3.2.2 Definition of the Subproblem and Graph Disaggregation

The solution of the master problem (3.2) induces new demands within the aggregated components. The purpose of the subproblem is to validate whether these demands can be routed without additional capacity upgrades inside the components. This validation decomposes into separate subproblems for each component. An example of this situation is depicted in Fig. 3.3.

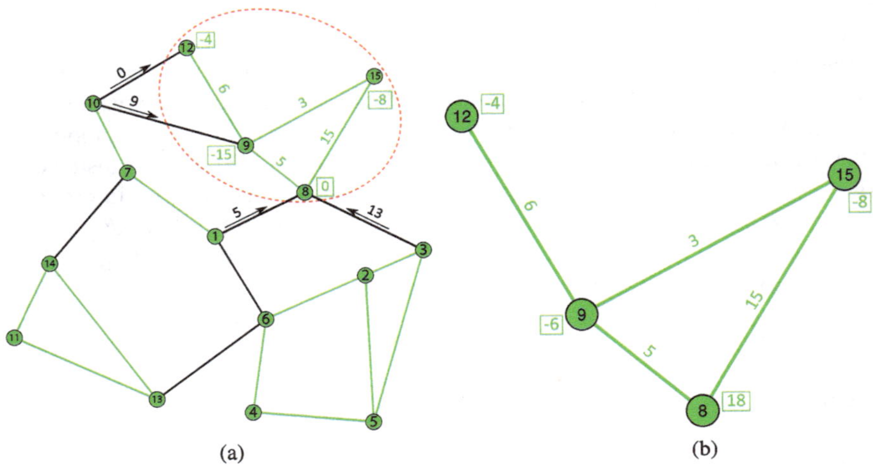

Fig. 3.3: Illustration of the subproblem for some component of $\mathscr{G}_\varphi$. (**a**) Induced demands for component $\{8, 9, 12, 15\}$. (**b**) The associated feasibility subproblem

Figure 3.3a shows part of the solution of the master problem which induces a new demand vector within some component of the aggregated graph. The corresponding subproblem for this component is depicted in Fig. 3.3b.

Checking the feasibility of a network expansion involves the solution of a maximum-flow problem within each component as follows. Let $H_i = (V_i, A_i)$ be the subgraph of $G = (V, A)$ induced by component V_i of the partition of V according to φ. The nodes V_i of H_i have an original demand of d_v, $v \in V_i$. The optimal flows of the master problem induce new demands within H_i as each flow x_a on an arc $a = (u, v) \in \mathscr{A}_\varphi$ with $u \in V_i$ and $v \in V_j$ changes the demand of u to $\bar{d_u} := d_u - x_a$ and that of v to $\bar{d_v} := d_v + x_a$. By introducing artificial nodes s as super source and t as super sink, the check whether a feasible flow exists can be formulated as the following single-source maximum-flow problem:

$$
\begin{aligned}
&\max \textstyle\sum_{v \in V_i^+} z_v \\
&\text{subject to } \sum_{a \in \delta_v^+} x_a - \sum_{a \in \delta_v^-} x_a = \begin{cases} -z_v, & \text{if } v \in V_i^+ \\ z_v, & \text{if } v \in V_i^- \\ 0, & \text{otherwise} \end{cases} && (\forall v \in V_i) \\
&\qquad x_a \le c_a && (\forall a \in \mathscr{A}_\varphi) \\
&\qquad z_v \le |\bar{d_v}| && (\forall v \in V_i^+ \cup V_i^-) \\
&\qquad x_a \ge 0 && (\forall a \in \mathscr{A}_\varphi) \\
&\qquad z_v \ge 0 && (\forall v \in V_i^+ \cup V_i^-).
\end{aligned}
\tag{3.3}
$$

Here, $V_i^+, V_i^- \subseteq V_i$ are the nodes with positive (resp. negative) induced demand $\bar{d}_v$. Variable z_v for $v \in V_i^+$ then models the flow from the super source s to node v, while z_v for $v \in V_i^-$ is the flow from node v to the super sink t. If the maximum s-t-flow attains a value of $\sum_{v \in V_i^+} \bar{d}_v$, the induced demands within component V_i are feasible, otherwise they are infeasible.

A feasible component V_i requires no further examination in the current iteration. An infeasible subproblem can occur for two reasons. It either suggests that the initial capacities within the associated component are not sufficient to route the demands induced by the master problem solution. In this case, it was not justified to neglect the capacity limitations within the component. Or the algorithm might not be able to prove that all upgrade decisions are already optimal. Whenever an infeasible subproblem is encountered, the partition is refined in order to consider additional arcs in the master problem. This arc set is chosen as a minimum s-t-cut that limits the flow.

Updating the master problem (3.1) is done by disaggregating an infeasible component along this minimum cut. Without loss of generality, let V_k be an infeasible component and let $V_k^1, \ldots, V_k^l$ be the components into which V_k disaggregates. We define a new clustering function $\psi\colon V \to \{1, \ldots, k+l-1\}$ with $\psi(v) = \varphi(v)$ if $v \notin V_k$ and $\psi(v) = k+i$ if $v \in V_k^i \subset V_k$. In the next iteration of the algorithm, the master problem is solved for the resulting aggregated graph $\mathcal{G}_\psi$. The algorithm terminates as soon as all subproblems are feasible.

3.2.3 Properties and Implementations of the Aggregation Scheme

In [1], it was shown that the above algorithm always yields the optimal solution to the original problem within a finite number of iterations.

Theorem 3.1 ([1]). *For $f = 0$ and $k \geq 0$ in Problem* (3.1)*, the aggregation scheme always terminates after a finite number of iterations and returns an optimal solution to the network expansion problem for the original graph.*

The referenced paper also proposes three different implementations of the approach. The first one is a sequential version (named SAGG), in which master problem and subproblem are solved in an alternating fashion. In each iteration, the network expansion master problem is solved to optimality, beginning with a trivial aggregation of the graph to a single node. In case of feasibility of all subproblems, the algorithm terminates. It has the obvious drawback that only very limited information is retained when moving from one iteration to the next. To overcome this problem, the authors of [1] propose the integration of the aggregation scheme into a branch-and-bound framework (named IAGG). For each solution candidate found at a node in the branch-and-bound tree, the subproblem is solved in order to check for feasibility. A feasible solution may be accepted as the new incumbent, while each infeasible solution is cut off via the cutting planes representing the disaggregation of the infeasible components. They also consider a hybrid of the two approaches (named HAGG), which begins by solving a few iterations sequentially before switching to branch-and-bound. They also present several more enhancements to make the algorithm more efficient.

Finally, Bärmann et al. [1] observe a very close relation of their algorithm to Benders decomposition.

Theorem 3.2 ([1]). *Let φ be a clustering function according to a given network graph G. For disaggregation of G along a minimal cut, the primal constraints introduced to the master problem* (3.1) *in the proposed aggregation scheme strictly imply the Benders feasibility cut obtained from the corresponding subproblem* (3.3).

This result is the basis for the algorithm developed in the next section. It shows that the Benders feasibility cuts can be replaced by the primal cuts of the aggregation scheme to obtain tighter relaxations. Doing so implies passing from a static subdivision between master problem and subproblem, as is the case in traditional Benders decomposition, to a dynamic subdivision. This dynamic element will allow us to derive lifted Benders optimality cuts to incorporate the routing costs into the master problem. Altogether, this approach can be described as a hybrid between aggregation and Benders decomposition.

3.3 Integration of Routing Costs via Lifted Benders Cuts

The aggregation scheme presented in Sect. 3.2 is an exact method for the solution of network design problem (3.1) when there are no routing costs, i.e. $f = 0$. For nonzero f, it is straightforward to consider the routing costs of the arcs in the aggregated graph in the objective function of the master problem. However, this neglects the routing costs arising within the components of the aggregation, which means a loss of information. The scheme still delivers feasible solutions in this case, but they need not be optimal any more. Therefore, we now derive an extension of the aggregation scheme from [1] which projects the routing costs within each component $V_i \in \mathscr{V}_\varphi$ onto the arcs of the master problem graph $\mathscr{A}_\varphi$. This can be achieved by introducing an additional subproblem which evaluates the costs for a feasible routing within a component and reformulates them as a cutting plane to be added to the master problem.

A suitable adaptation of the aggregation master problem (3.1) for a given clustering function φ is as follows:

$$\begin{aligned} \min \ & \sum_{a \in \mathscr{A}_\varphi} f_a x_a + \sum_{a \in \mathscr{A}_\varphi} k_a y_a + \Phi(x,y) & \\ \text{subject to} \ & \sum_{a \in \delta_i^+} x_a - \sum_{a \in \delta_i^-} x_a = d_i & (\forall i \in \mathscr{V}_\varphi) \\ & x_a \leq c_a + C_a y_a & (\forall a \in \mathscr{A}_\varphi) \\ & x_a \geq 0 & (\forall a \in \mathscr{A}_\varphi) \\ & y_a \in \mathbb{Z}_+ & (\forall a \in \mathscr{A}_\varphi). \end{aligned} \tag{3.4}$$

This new master problem directly includes the routing costs of the arcs in $\mathscr{A}_\varphi$ in the objective function. The routing costs within the components of the aggregated graph are incorporated via a piecewise-linear cost function Φ. This function Φ is not incorporated directly, but is iteratively underestimated via Benders cutting planes which project the term $\sum_{a\in A\setminus\mathscr{A}_\varphi} f_a x_a$ onto variables x_a for $a \in \mathscr{A}_\varphi$ and y_a for $a \in A$. This procedure is described in the following.

Let $(\bar{x}, \bar{y})$ be a solution to the master problem (3.4) which can be extended to a feasible solution to the original problem (3.1) in the sense of Sect. 3.2.2, i.e. we are able to find a feasible routing within the components. In this case, the routing costs arising within the components are given as the optimal value of the following minimum-cost flow problem:

$$\begin{aligned} \min \sum_{a\in A\setminus\mathscr{A}_\varphi} & f_a x_a \\ \text{subject to} \sum_{a\in\delta_v^+\setminus\mathscr{A}_\varphi} x_a - \sum_{a\in\delta_v^-\setminus\mathscr{A}_\varphi} x_a &= \bar{d}_v \ (\forall v \in V) \\ x_a &\leq \bar{c}_a \ (\forall a \in A \setminus \mathscr{A}_\varphi) \\ x_a &\geq 0 \ (\forall a \in A \setminus \mathscr{A}_\varphi), \end{aligned} \tag{3.5}$$

where $\bar{d}_v := d_v - \sum_{a\in\delta_v^+\cap\mathscr{A}_\varphi} \bar{x}_a + \sum_{a\in\delta_v^-\cap\mathscr{A}_\varphi} \bar{x}_a$ for $v \in V$ are the induced node demands, and where $\bar{c}_a := c_a + C_a\bar{y}_a$ are the arc capacities within the components.

From the dual to Problem (3.5), we can derive Benders optimality cut for the current solution $(\bar{x}, \bar{y})$. It is given by:

$$\begin{aligned} \min \sum_{v\in V} \bar{d}_v \alpha_v - \sum_{a\in A\setminus\mathscr{A}_\varphi} & \bar{c}_a \beta_a \\ \text{subject to } \alpha_v - \alpha_w - \beta_a &\leq f_a \ (\forall a = (v,w) \in A \setminus \mathscr{A}_\varphi) \\ \beta_a &\geq 0 \ \ (\forall a \in A \setminus \mathscr{A}_\varphi) \end{aligned} \tag{3.6}$$

with dual variables α for the flow conservation constraints and β for the capacity constraints. Whenever the current underestimation of the component-wise routing costs $\Phi(\bar{x}, \bar{y})$ within the master problem is lower than the optimal value of Subproblem (3.6), we can introduce a Benders optimality cut to update function Φ. By back-substitution for $\bar{d}$ and $\bar{c}$, this cut can be stated as:

$$\begin{aligned} &\sum_{v\in V} \bar{\alpha}_v d_v - \sum_{a\in A\setminus\mathscr{A}_\varphi} \bar{\beta}_a c_a \\ &\quad + \sum_{a=(v,w)\in\mathscr{A}_\varphi} (\bar{\alpha}_w - \bar{\alpha}_v) x_a - \sum_{a\in A\setminus\mathscr{A}_\varphi} (C_a \bar{\beta}_a) y_a \leq \Phi(x,y), \end{aligned} \tag{3.7}$$

given a dual optimal solution $(\bar{\alpha}, \bar{\beta})$. The interpretation of this cutting plane is that a given suboptimal solution with respect to the original problem can be improved by rerouting the flow in the master problem graph or by creating new capacities within the components.

The procedure starts with a coarse, possibly trivial underestimation of Φ. Adding a violated optimality cut of the form (3.7) to the master problem then yields a better estimate for the routing costs within the components. The main idea is to iterate this until either no more violated optimality cuts exist or until the induced flow within any of the components becomes infeasible. In the first case, we know that we have found an optimal solution, as the master problem is a relaxation of the original problem. In the latter case, one could use a Benders feasibility cut to cut off the infeasible solution. However, we already know that they are dominated by the cutting planes produced by the aggregation scheme. Therefore, we perform a disaggregation step which updates the distribution of the arcs between master problem and subproblem and restart the method.

As [1] show, the aggregation scheme is most effective when it is integrated into a branch-and-bound procedure. The important consideration in the presence of routing costs is to guarantee the global validity of the optimality cuts. As long as no disaggregation is performed after starting the branch-and-bound phase of the algorithms, this is no problem as the method then behaves like ordinary Benders decomposition. After a disaggregation step, however, the cutting plane has to be modified as the graph corresponding to the master problem grows while the components shrink. The arcs moving from inside the components to the master graph are no longer valued in Benders dual subproblem (3.6). In the extreme case of complete disaggregation, this subproblem is empty with an optimal value of zero.

The solution is a suitable lifting of the optimality cuts obtained from Subproblem (3.6). We add the primal routing cost of all arcs that have passed to the master problem after the root node to the left side of (3.7). If φ is the clustering function corresponding to the initial aggregation and ψ is the clustering function describing the current state of aggregation at a node within the branch-and-bound tree, the new optimality cut reads as follows:

$$\begin{aligned}\sum_{v\in V} \bar{\alpha}_v d_v - \sum_{a\in A\setminus\mathscr{A}_\psi} \bar{\beta}_a c_a + \sum_{a\in\mathscr{A}_\psi\setminus\mathscr{A}_\varphi} f_a x_a \\ + \sum_{a=(v,w)\in\mathscr{A}_\psi} (\bar{\alpha}_w - \bar{\alpha}_v) x_a - \sum_{a\in A\setminus\mathscr{A}_\psi} (C_a\bar{\beta}_a) y_a \leq \Phi(x,y).\end{aligned} \tag{3.8}$$

A solution produced within the branch-and-bound procedure can then be accepted as feasible if no violated cutting plane of the form (3.8) exists. This can be checked via the optimality cuts already added to the master problem. Let $(\bar{x},\bar{y})$ be the current solution candidate, let $\bar{\Phi}$ be the estimate of the routing costs within the components according to the initial clustering φ and let $\Phi(\bar{x},\bar{y})$ be the actual value of the routing costs within the components according to the current clustering ψ. If

$$\bar{\Phi} - \sum_{a\in\mathscr{A}_\psi\setminus\mathscr{A}_\varphi} f_a \bar{x}_a = \Phi(\bar{x},\bar{y})$$

holds, the cost estimate is correct, and the solution may be accepted. Otherwise, we add the cutting plane from Eq. (3.8) and reject the solution. Altogether, this leads to a hybrid algorithm between the original aggregation scheme and Benders decomposition. Note that a trivial but powerful heuristic within the branch-and-bound process is to "repair" a solution that has been cut off by (3.8) by providing the actual value of Φ for the routing costs within the components. If its correct cost is still better than that of the best solution found so far, it can be retained and becomes the new incumbent.

The following theorem states the correctness of the proposed method, for which it is possible to give a very general proof. It shows that the method can be generalized to a Benders decomposition that allows shifting variables from the subproblem to the master problem in the process—an approach that might be beneficial in broader problem contexts as well.

Theorem 3.3 (Correctness of the Aggregation Scheme with Routing Costs). *For $k \geq 0$ and f such that G does not contain negative cycles, the sequential, the integrated and the hybrid version of the algorithm are all finite and return an optimal solution to the network expansion problem* (3.1) *for the original graph G.*

Proof. The correctness of cutting plane (3.8) can be proved in a very general setting. Consider the following optimization problem with variables $x \in R^n$ and $y \in R^m$ for $n \geq 0$ and $m \geq 1$ as well as vectors b and c and matrices A and B of suitable dimensions:

$$\min c^T x + d^T y$$

$$\text{subject to } Ax + By \leq b$$

$$x \geq 0$$

$$y \geq 0.$$

Now, assume that the x-variables are projected out of the problem. Let $\bar{y}$ be a choice of variables y such that a feasible solution $(\bar{x}, \bar{y})$ exists for this problem. In this case, Benders optimality cut for an optimal dual solution $\bar{\alpha}$ according to $\bar{y}$ is of the form

$$\Phi(x,y) \geq -\bar{\alpha}^T b + (B^T \bar{\alpha})^T y.$$

We now consider the situation that part of the x-variables are shifted from the subproblem to the master problem. Without loss of generality, we assume that this is done for the first p variables $x_1 \in \mathbb{R}^p$, while the remaining vector of variables $x_2 \in \mathbb{R}^{m-p}$ remains in the subproblem. Let $A = (A_1, A_2)$ and $c = (c_1, c_2)$ be the corresponding subdivision of the constraint matrix and the objective function. If x_1 had been part of the master problem from the start, the corresponding Benders optimality cut would have taken the form

$$\Phi(x,y) \geq -\bar{\alpha}^T b + (B^T \bar{\alpha})^T y + (A_1^T \bar{\alpha})^T x_1.$$

Now, we would like to show that the addition of its lifted version

$$\Phi(x,y) \geq -\bar{\alpha}^T b + (B^T \bar{\alpha})^T y + (A_1^T \bar{\alpha})^T x_1 + c_1^T x_1$$

to the master problem leads to a correct estimation of the subproblem cost after moving x_1 from the subproblem to the master problem. This is exactly what happens in the aggregation scheme when the graph is disaggregated in the course of the branch-and-bound process. Let $\bar{\Phi}$ be the current estimate for the subproblem cost according to the initial division of the variables, let $\bar{x}_2$ be a feasible completion to the partial solution $(\bar{x}_1, \bar{y})$, and let $\Phi(\bar{x}, \bar{y})$ be the corresponding value of the shrunk subproblem. If

$$\bar{\Phi} - c_1^T \bar{x}_1 = \Phi(\bar{x}, \bar{y})$$

holds, we know that $\bar{\Phi}$ is the exact value for the cost of the initial subproblem incurred by solution $\bar{x}$ in the initial master problem. Thus, $\bar{\Phi} = c_1^T \bar{x}_1 + \Phi(\bar{x}, \bar{y})$ is the correct objective value of the y-variables within the original problem. In the case

$$\bar{\Phi} - c_1^T \bar{x}_1 < \Phi(\bar{x}, \bar{y}),$$

we have underestimated the cost of solution $(\bar{x}, \bar{y})$ and thus add the lifted optimality cut to correct the estimation.

It remains to show that all the optimality cuts that have been added to the master problem before reintroducing variables x_1 remain valid. In other words, we never overestimate the objective value of a solution. This can be seen as follows. The optimality cut that would have been calculated for $\bar{y}$ without reintroducing variables x_1 dominates all the cutting planes that have been added so far in the point $\bar{y}$. Thus, the corresponding cost estimate $\hat{\Phi}$ is at least as high as $\bar{\Phi}$. Let $\hat{x} = (\hat{x}_1, \hat{x}_2)$ be the optimal completion calculated by the subproblem according to the initial division of the variables. Then we have

$$\bar{\Phi} \leq \hat{\Phi} = c_1^T \hat{x}_1 + c_2^T \hat{x}_2 \leq c_1^T \bar{x}_1 + c_2^T \bar{x}_2.$$

This leads to

$$\bar{\Phi} - c_1^T \bar{x}_1 \leq c_2^T \bar{x}_2 = \Phi(\bar{x}, \bar{y}),$$

which proves the claim.

It is now easy to see the correctness of the proposed aggregation scheme. It is finite because disaggregation can only occur until reaching the original graph and because there is only a finite number of extreme points from which non-dominated optimality cuts can be derived. The optimality of the returned solution follows from three facts. Firstly, the non-negativity of k ensures the boundedness of the master problem. Secondly, the feasible completability of the master problem solution is always ensured by disaggregation of infeasible components. Thirdly, the absence of negative circles in G with respect to f ensures that both Benders primal and dual subproblem are always feasible. This completes the proof.

The above reasoning does not only show the correctness of the proposed aggregation scheme with lifted optimality cuts, but also introduces the idea of a generalized Benders decomposition. In the traditional Benders decomposition, the splitting of the variables between master and subproblem remains fixed. Following the outline in the proof of Theorem 3.3, it is possible to convert Benders decomposition from a pure row generation scheme into a row-and-column-generation scheme. It is an interesting line for future research to assess the potential of such a method in general.

3.4 Computational Results

We present computational experiments for the above aggregation scheme on a set of benchmark instances which are derived from a real-world railway network expansion problem. Note that the instances we use are based on the real-world data for the railway network design investigated in [2], where the problem is solved via decomposition along the time axis. As the aggregation scheme in its current form is still only capable to solve single-period, single-commodity network design problem, we have used the original data to create test instances that match this (simpler) problem, and which exhibit a similar capacity usage on the links of the network. The idea of the chapter is to lay the algorithmic foundation to solve network design problems on much larger graphs than currently possible—a microscopic representation of the German railway network is about ten times bigger than the macroscopic representation for which the problem can be currently solved. A detailed case study for the expansion of the German railway network on the original data can be found in the paper cited above.

We begin by describing the instances used to test our aggregation scheme as well as the computational setup before we present its results.

3.4.1 Test Instances

The test instances originate from a joint project with our industry partner Deutsche Bahn AG on the expansion of the German railway network. They consist of an instance describing the problem on the Germany-wide network as well as a collection of subnetworks of it. The characteristics of these instances are described in Table 3.1. As can be seen, the instances vary in size from 62 nodes for the smallest network up to 1620 nodes for the complete German network. The seven subnetworks are composed of one to three German federal states. Together they cover the complete Germany instance with a small overlap at the borders.

The original project consisted in determining an optimal expansion of the given railway network to accommodate the demand pattern fore-casted for 2030. For the use in this chapter, the instances were adapted to fit the single-commodity problem setting considered here by balancing the actual multi-commodity demands, i.e. by calculating the net demand of each node.

Table 3.1: Layout of the railway networks under consideration

Network	Involved federal states	#Nodes	#Arcs
SH-HA	Schleswig-Holstein, Hamburg	62	200
B-BB-MV	Berlin, Brandenburg, Mecklenburg-Vorpommern	185	592
HE-RP-SL	Hessen, Rheinland-Pfalz, Saarland	214	720
NRW	Nordrhein-Westfalen	251	814
BA-BW	Bayern, Baden-Württemberg	246	778
NS-BR	Niedersachsen, Bremen	369	1180
TH-S-SA	Thüringen, Sachsen, Sachsen-Anhalt	389	1288
Deutschland	The entire German instance	1620	5162

The initial capacities of each instance were then scaled by a constant factor in order to obtain different percentages of initial demand satisfaction l, which was done by solving an auxiliary network flow problem. Consequently, the parameter l indicates which portion of the demand can be routed given the initial state of the network. Varying these initial capacities has a significant impact on the solution time and the solvability in general and is therefore an important parameter for the forthcoming analysis.

To assess the influence of the routing costs on the performance of the aggregation algorithms, we introduce a parameter σ to describe their magnitude. The routing cost f_a of an arc $a \in A$ is then given by $f_a = \sigma k_a$, i.e. it is chosen to be proportional to its expansion cost. This is motivated by the instances from railway network expansion, where both the expansion cost and the routing cost are proportional to the length of the corresponding arc. Parameter σ takes values within the set $\{0, 10^{-5}, 10^{-4}, 10^{-3}, 10^{-2}\}$ to which we refer as the cases of *no*, *small*, *medium*, *high*, and *very high* routing costs respectively.

Finally, the instances have been filtered in order to exclude "too easy" and "too hard" instances. The upper bound is given by the original time limit of 10 h, while the lower bound was chosen as 100 s to account for the overhead of the aggregation scheme that does not pay off on trivial instances. This is done in both cases excluding and including routing costs to enable a unified presentation and to account for the fact that the aggregation schemes entail a larger overhead in the latter case and start to pay off on somewhat more difficult instances.

3.4.2 Computational Setup

For the case without routing costs, we use the implementations SAGG, IAGG and HAGG of the aggregation scheme from [1], see Sect. 3.2.3. For the case including routing costs, we have implemented our extension of the aggregation in two different variants. The first one, denoted by IAGGB, clusters the nodes of the original graph into a single component and directly proceeds to the branch-and-bound phase. The second version, called HAGGB, disaggregates the graph in a sequential fashion until the first feasible solution is found before switching to branch-and-bound. They are generalizations of the algorithms IAGG and HAGG respectively. All the aggregation schemes have been implemented via the C++-API of the MIP solver `Gurobi` Gurobi Optimization, Inc. [12]. For the aggregation schemes operating within a branch-and-bound procedure, it was necessary to adjust Gurobi's parameter settings, which involves a more aggressive cutting plane generation, a focus on improving the bound and downscaling the frequency of the heuristics. Additionally, dual reductions had to be disabled in order to guarantee correctness as these algorithms use *lazy cuts*. The different aggregation methods are compared to Gurobi's default branch-and-bound algorithm under standard parameter settings. These reference solution times are denoted by MIP.

All computational experiments were carried through on a queuing cluster of Intel Xeon™ E5-2690 v2 3.00 GHz computers with 24 MB cache and 128 GB RAM, using version 5.6.3 of `Gurobi` Gurobi Optimization, Inc. [12]. Each job was run on four cores and with a time limit of 10 h.

3.4.3 Results Without Routing Costs

We begin our experiments with a comparison of IAGG, SAGG and HAGG for the case without routing costs. The results are presented as a performance diagram in Fig. 3.4.

We see that methods IAGG and HAGG both solve the same percentage of instances fastest. However, IAGG is able to solve all instances under consideration within the given time limit while HAGG still solves about 90% of them. Implementation SAGG is clearly not competitive.

The overall best of the three algorithms, IAGG, is now compared to MIP on the same set of instances. Figure 3.5 shows the corresponding performance diagram. It shows that IAGG clearly outperforms MIP as it solves more than 90% of the instances fastest. Furthermore, it is able to solve all the instances, while MIP only solves somewhat less than 60% of them. A detailed summary of the computation times is given in Table 3.2. It confirms that IAGG solves almost all the instances faster than MIP, sometimes significantly faster, while it is not decisively slower on the few others.

3.4.4 Results with Routing Costs

In the following, we present our results for the extended aggregation scheme proposed in this chapter. We compare the two implementations IAGGB and HAGGB on the same railway network instances, this time including routing costs as described in Sect. 3.4.1. We did not implement a sequential variant corresponding to SAGG as the high number of necessary optimality cuts is likely to render this approach

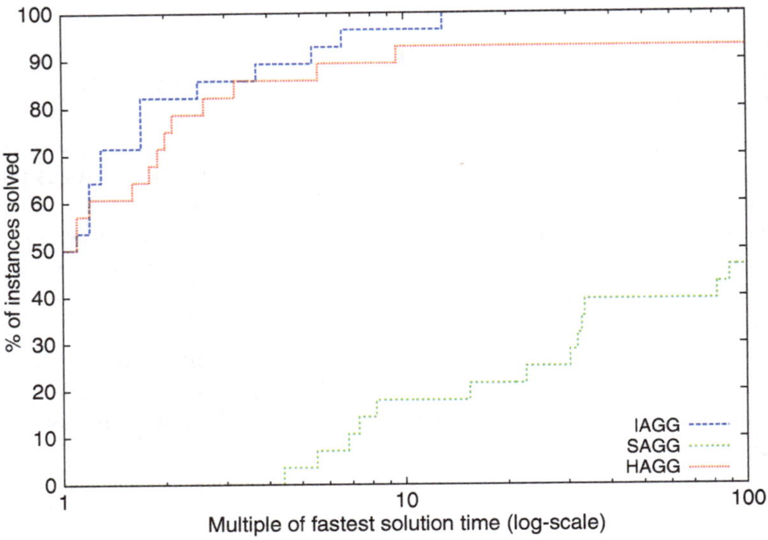

Fig. 3.4: Performance profile comparing the three basic aggregation schemes on railway instances

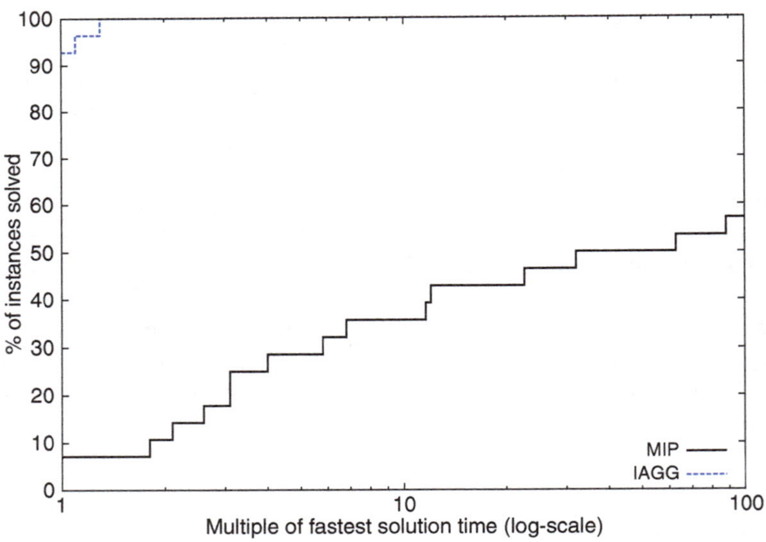

Fig. 3.5: Performance profile comparing MIP and IAGG on railway instances

Table 3.2: Solution times [s] for MIP and IAGG on the railway instances with initial demand satisfaction of l percent – best-performing method for each instance in bold

Instance	l	MIP	IAGG
B-BB-MV	0.5	1353.06	**199.33**
B-BB-MV	0.55	**209.06**	218.97
BA-BW	0.3	∞	**8234.56**
BA-BW	0.4	1231.68	**19.63**
BA-BW	0.45	∞	**13.89**
BA-BW	0.5	15.06	**8.48**
Deutschland	0.7	∞	**6898.27**
Deutschland	0.75	∞	**65.08**
HE-RP-SL	0.4	∞	**5742.33**
HE-RP-SL	0.45	∞	**2786.14**
HE-RP-SL	0.5	2336.17	**73.03**
HE-RP-SL	0.55	2909.57	**33.08**
NRW	0.6	∞	**8378.55**
NRW	0.65	5730.42	**497.96**

Instance	l	MIP	IAGG
NRW	0.7	140.04	**24.31**
NRW	0.75	224.81	**18.79**
NS-BR	0.5	12,003.63	**3035.58**
NS-BR	0.55	∞	**24,310.02**
NS-BR	0.6	**1404.99**	1826.26
NS-BR	0.65	75.39	**24.73**
TH-S-SA	0.4	∞	**29,243.07**
TH-S-SA	0.45	∞	**4176.81**
TH-S-SA	0.5	∞	**210.66**
TH-S-SA	0.55	∞	**1167.42**
TH-S-SA	0.6	1842.68	**81.69**
TH-S-SA	0.65	133.27	**52.72**
TH-S-SA	0.7	40.28	**20.12**
TH-S-SA	0.75	39.30	**13.06**

even more uncompetitive. Figure 3.6 shows the performance diagrams of the two aggregation schemes under consideration for different choices of the routing cost parameter.

We see that HAGGB performs consistently better from medium-scale routing costs on. For small routing costs, IAGGB might be preferred as it solves more instances in total, which shows that this case is closely related to the case without routing costs. Note that about 7% of these computations were not finished due to numerical difficulties within `Gurobi` Gurobi Optimization, Inc. [12] which might have been induced by the Benders cutting planes. When we compare the victorious aggregation scheme HAGGB to MIP in the following, the corresponding instances are treated as unsolved for HAGGB.

Figure 3.7 shows four performance diagrams comparing HAGGB and MIP for the different choices of the routing cost parameter.

We see that HAGGB is able to solve a majority of the instances fastest in all the cases with clear advantages for small, medium, and high routing costs. We also recognize the same tendency as for the scale-free networks with 100 nodes which lets MIP catch up for higher routing costs, although it is somewhat concealed by the numerical problems affecting HAGG especially in Fig. 3.7a. The decreasing difficulty of the instances for higher routing costs can be seen from the increasing percentage of instances solved by MIP within the time limit, and MIP increases the share of instances that it solves fastest.

A complete summary of the computation times of MIP and HAGGB is given in Table 3.3, where instances causing numerical difficulties are marked by "–". Note that the varying number of instances for each magnitude of the routing costs is due to our filter criterion for overly easy or hard instances. It supports the observations from the performance diagrams by underlining the strong performance of HAGGB on most instances. The aggregation scheme is largely favoured up to high routing costs and still competitive for very high routing costs.

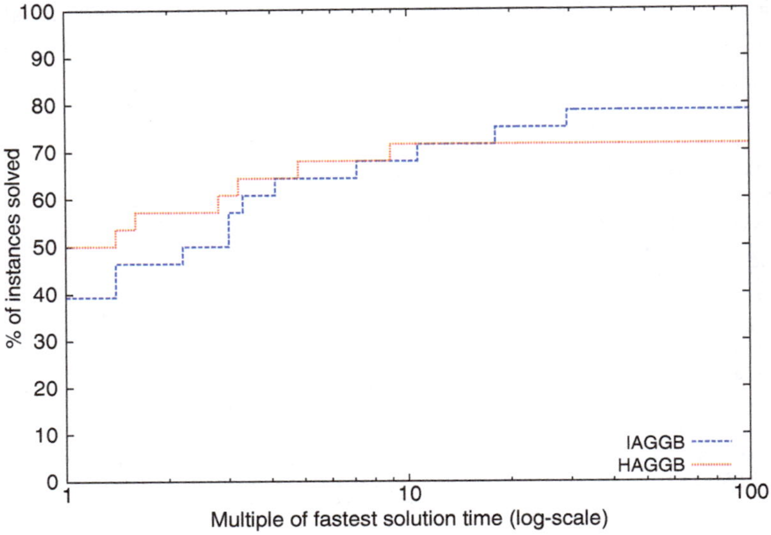

100
90
80
70
60
50
40
30
20
10
0
% of instances solved
1
10
100
Multiple of fastest solution time (log-scale)
IAGGB
HAGGB

(a)

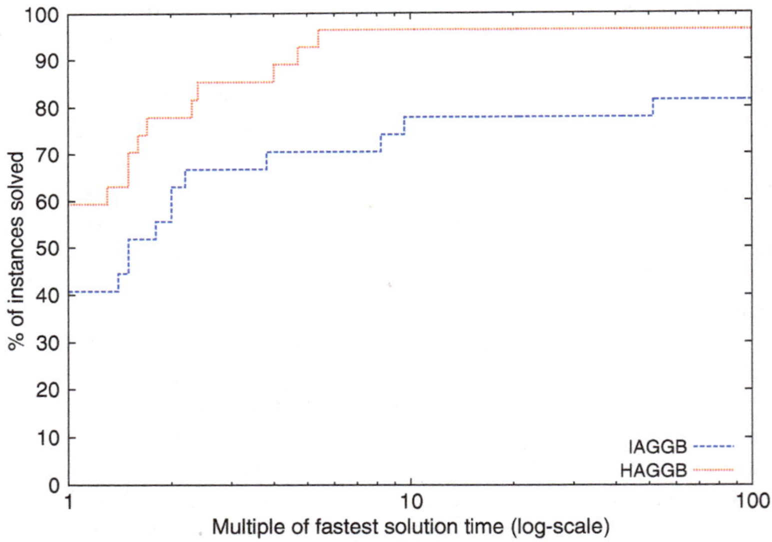

100
90
80
70
60
50
40
30
20
10
0
% of instances solved
1
10
100
Multiple of fastest solution time (log-scale)
IAGGB
HAGGB

(b)

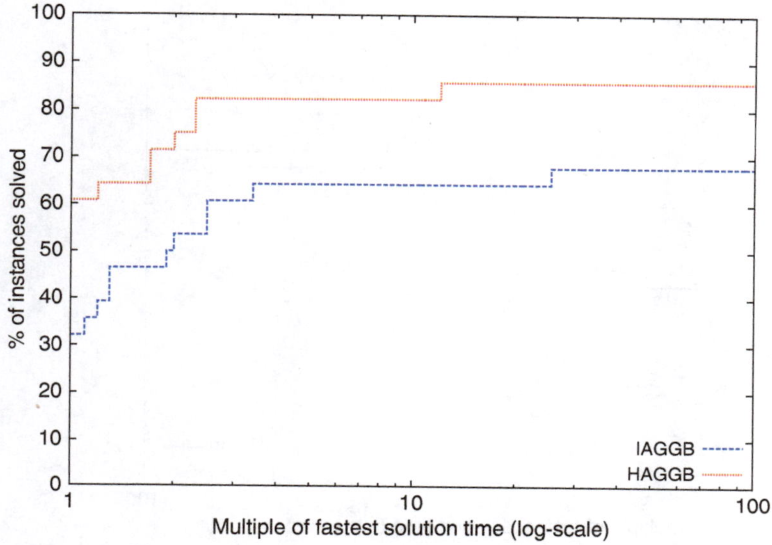

(c)

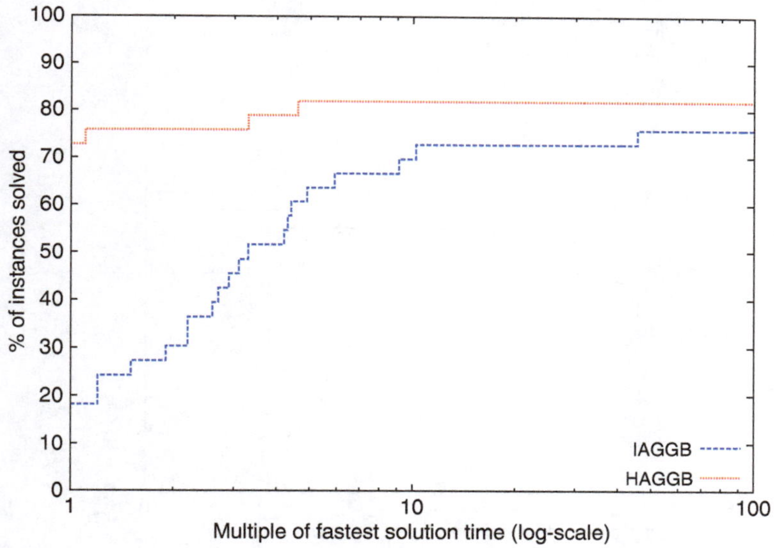

(d)

Fig. 3.6: Performance profiles comparing methods IAGGB and HAGGB on the railway instances for different sizes of the routing costs. (**a**) Small routing costs. (**b**) Medium routing costs. (**c**) High routing costs. (**d**) Very high routing costs

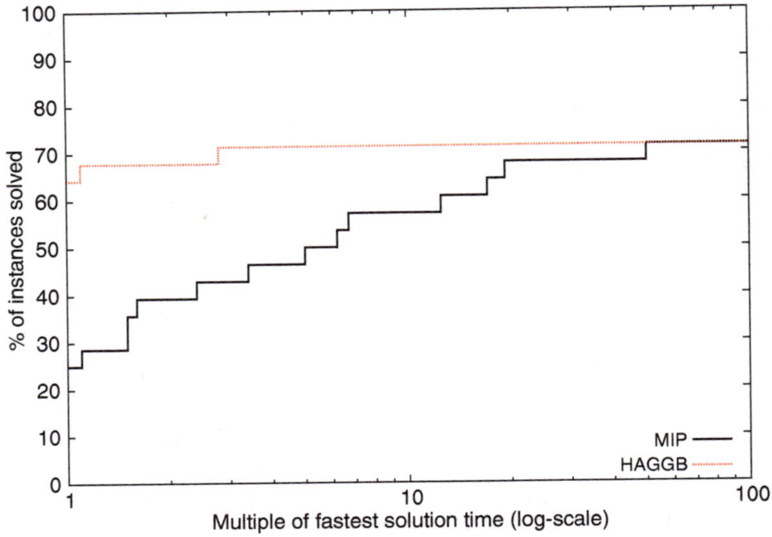
100
90
80
70
60
50
40
30
20
10
0
% of instances solved
1
10
100
Multiple of fastest solution time (log-scale)
MIP
HAGGB

(a)

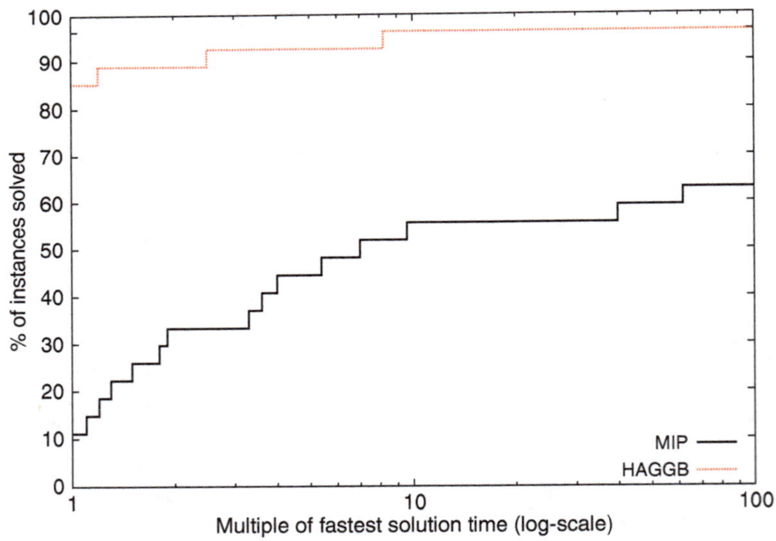
100
90
80
70
60
50
40
30
20
10
0
% of instances solved
1
10
100
Multiple of fastest solution time (log-scale)
MIP
HAGGB

(b)

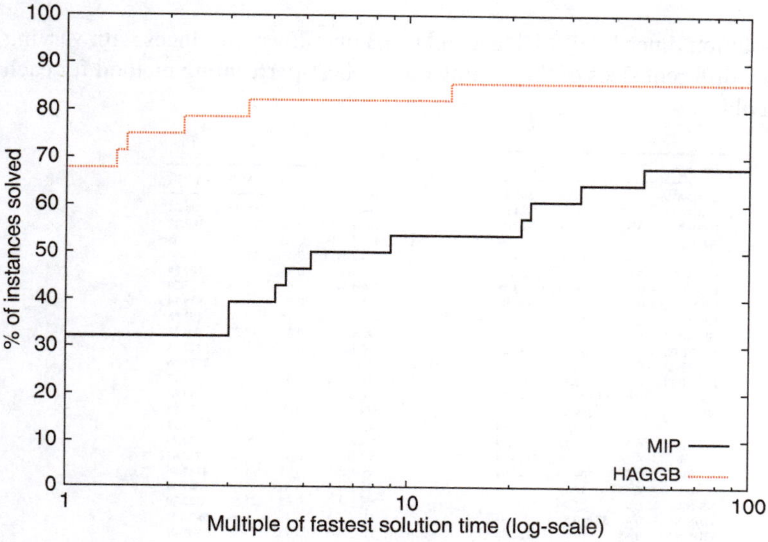

(c)

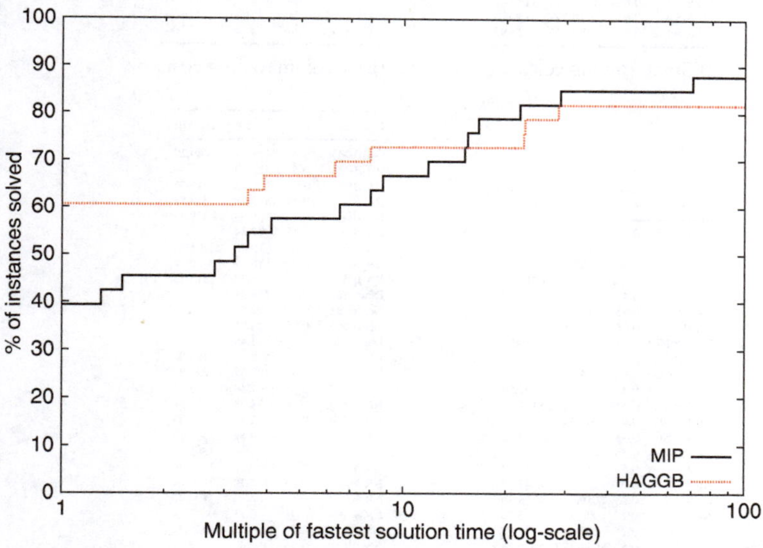

(d)

Fig. 3.7: Performance profiles comparing methods MIP and HAGGB on the railway instances for different sizes of the routing costs. (**a**) Small routing costs. (**b**) Medium routing costs. (**c**) High routing costs. (**d**) Very high routing costs

Table 3.3: Solution times [s] of MIP and HAGGB on railway instances with varying values of l for different sizes of the routing costs – best-performing method for each instance in bold

Instance	l	MIP	HAGGB
B-BB-MV	0.5	731.89	**507.45**
B-BB-MV	0.55	198.20	**128.31**
B-BB-MV	0.65	**3.71**	10.08
BA-BW	0.35	∞	∞
BA-BW	0.4	742.08	**14.83**
BA-BW	0.45	∞	**51.96**
BA-BW	0.55	143.55	**7.46**
Deutschland	0.75	∞	**330.56**
HE-RP-SL	0.4	∞	∞
HE-RP-SL	0.45	∞	**5107.47**
HE-RP-SL	0.5	566.04	**33.23**
HE-RP-SL	0.55	775.01	**324.97**
HE-RP-SL	0.6	**5.75**	–
NRW	0.65	12,329.86	**2471.76**
NRW	0.7	**169.42**	–
NRW	0.75	139.36	**41.54**
NS-BR	0.5	5441.22	**436.49**
NS-BR	0.55	∞	**5396.39**
NS-BR	0.6	2709.92	**440.91**
NS-BR	0.65	**16.18**	–
NS-BR	0.7	**1.06**	–
TH-S-SA	0.4	**31,210.83**	–
TH-S-SA	0.45	∞	**21,752.85**
TH-S-SA	0.5	4650.49	**701.91**
TH-S-SA	0.55	∞	–
TH-S-SA	0.6	1712.90	**1142.46**
TH-S-SA	0.65	**118.79**	130.10
TH-S-SA	0.8	23.13	**23.07**

(a) Small routing costs

Instance	l	MIP	HAGGB
B-BB-MV	0.5	1368.97	**768.31**
B-BB-MV	0.55	**170.00**	189.91
B-BB-MV	0.65	**3.48**	8.68
BA-BW	0.4	4017.62	**18.50**
BA-BW	0.45	∞	**109.89**
BA-BW	0.5	16.69	**5.20**
Deutschland	0.7	∞	**29,928.14**
Deutschland	0.75	∞	**252.88**
Deutschland	0.8	**12.83**	104.11
HE-RP-SL	0.4	∞	**722.55**
HE-RP-SL	0.45	∞	**2723.46**
HE-RP-SL	0.5	802.73	**20.17**
HE-RP-SL	0.55	2729.76	**44.23**
NRW	0.65	2417.67	**1322.88**
NRW	0.7	345.11	**49.77**
NRW	0.75	120.20	**30.49**
NS-BR	0.5	∞	**3715.65**
NS-BR	0.55	4701.26	**4137.90**
NS-BR	0.6	3117.81	**327.65**
NS-BR	0.65	121.52	**22.70**
TH-S-SA	0.4	∞	**9072.72**
TH-S-SA	0.45	∞	**18,781.02**
TH-S-SA	0.5	2929.61	**2321.07**
TH-S-SA	0.55	∞	∞
TH-S-SA	0.6	2220.04	**627.63**
TH-S-SA	0.65	176.76	**118.90**
TH-S-SA	0.8	17.52	**17.42**

(b) Medium routing costs

Instance	l	MIP	HAGGB
B-BB-MV	0.5	1565.10	**521.98**
B-BB-MV	0.55	**84.64**	120.58
B-BB-MV	0.65	**2.44**	8.27
BA-BW	0.3	∞	**14,012.36**
BA-BW	0.4	310.06	**13.57**
BA-BW	0.55	246.18	**7.69**
BA-BW	0.75	**0.51**	6.79
Deutschland	0.75	∞	**1323.57**
Deutschland	0.8	**9.15**	–
Deutschland	0.85	**4.50**	–
HE-RP-SL	0.4	∞	**791.89**
HE-RP-SL	0.45	∞	**5197.57**
HE-RP-SL	0.5	101.34	**25.20**
HE-RP-SL	0.55	1911.69	**38.96**
HE-RP-SL	0.75	**0.63**	–
NRW	0.5	∞	**16,212.22**
NRW	0.6	∞	**28,241.39**
NRW	0.65	**2019.42**	4258.36
NRW	0.7	1000.37	**193.74**
NRW	0.75	105.39	**11.95**
NS-BR	0.5	8044.32	**375.19**
NS-BR	0.55	∞	**4925.51**
NS-BR	0.6	917.58	**212.88**
TH-S-SA	0.4	∞	**7607.41**
TH-S-SA	0.45	**8199.26**	11,189.70
TH-S-SA	0.5	∞	**426.97**
TH-S-SA	0.55	**6519.13**	∞
TH-S-SA	0.6	772.52	**257.78**

(c) High routing costs

Instance	l	MIP	HAGGB
B-BB-MV	0.45	7428.37	**449.29**
B-BB-MV	0.5	319.19	**117.30**
BA-BW	0.4	101.13	**15.57**
BA-BW	0.45	4313.88	**368.54**
BA-BW	0.5	∞	**24,318.80**
BA-BW	0.55	**87.34**	–
BA-BW	0.6	**79.90**	2263.22
BA-BW	0.7	116.79	**29.06**
BA-BW	0.75	**0.59**	13.45
Deutschland	0.75	**3248.31**	–
Deutschland	0.8	**92.13**	–
Deutschland	0.85	**24.95**	–
Deutschland	0.9	**1.48**	–
HE-RP-SL	0.4	∞	**161.92**
HE-RP-SL	0.45	3493.37	**231.90**
HE-RP-SL	0.5	1161.44	**53.22**
HE-RP-SL	0.55	78.38	**24.58**
HE-RP-SL	0.65	**1.18**	9.43
HE-RP-SL	0.75	**0.82**	5.13
NRW	0.5	∞	**1565.48**
NRW	0.6	1599.58	**1068.06**
NRW	0.65	**565.79**	2165.09
NRW	0.7	**56.02**	194.42
NRW	0.9	**0.21**	–
NS-BR	0.5	1803.06	**528.92**
NS-BR	0.55	493.40	**32.17**
NS-BR	0.6	257.26	**29.58**
NS-BR	0.85	**0.13**	2.85
TH-S-SA	0.4	∞	**2861.21**
TH-S-SA	0.45	251.13	**206.33**
TH-S-SA	0.5	3796.55	**132.21**
TH-S-SA	0.55	4524.58	**64.43**
TH-S-SA	0.6	628.44	**79.17**

(d) Very high routing costs

3.5 Conclusion

Altogether, we have shown once more that aggregation is a powerful means to reduce the solution times of network design problems without sacrificing the optimality of the obtained result. Moreover, we saw that the basic idea from [1] can be extended to different problem settings occurring in real-world problem instances. In this case, this entails the incorporation of routing costs, which we did using lifted Benders optimality cuts. The computational results show that the resulting algorithm can help to make traditional branch-and-bound-based algorithm much more efficient. The method works very well on single-commodity network design problems, which can be seen from the drastic reductions in solution time on many instances. A direction for future research is the further generalization of the approach to multi-commodity network design. An efficient extension might be achieved via an additional aggregation of the demand vector and finding a trade-off between the use of Benders feasibility cuts, which add fewer information, and our aggregation cuts, which add more information, but also let the linear system grow faster.

Acknowledgements We thank our co-authors from [1], Alexander Martin, Maximilian Merkert, Christoph Thurner and Dieter Weninger, for the interesting discussions we had on the topic. Furthermore, we gratefully acknowledge the computing resources provided by the group of Michael Jünger in Cologne. In particular, we thank Thomas Lange for his technical support. Last but not least, we thank the BMBF for its financial support under research grant 05M10WEC.

References

1. Bärmann A, Liers F, Martin A, Merkert M, Thurner C, Weninger D (2015) Solving network design problems via iterative aggregation. Math Program Comput 7(2):189–217. https://doi.org/10.1007/s12532-015-0079-1 (cited on pages 50, 51, 52, 55, 56, 58, 62, 71)
2. Bärmann A, Martin A, Schülldorf H (2017) A decomposition method for multiperiod railway network expansion – with a case study for Germany. Transp Sci. https://doi.org/10.1287/trsc.2017.0747 (cited on pages 50, 61)
3. Beck MJ (2013) Kapazitätsmanagement und Netzentwicklung: Erfahrungen mit Kompromissen zum Fahrplan 2018. Presentation by DB Netz AG. Available at: http://www.deutschland-takt.de/deutschlandtakt/index.php?option=com_docman&Itemid=64&task=doc_download&gid=84 (cited on page 49)
4. Boland N, Ernst A, Kalinowski T, Rocha de Paula M, Savelsbergh M, Singh G (2013) Time aggregation for network design to meet time-constrained demand. In: Piantadosi J, Anderssen RS, Boland J (eds) Proceedings of the 20th international congress on modelling and simulation (MODSIM2013), pp 3281–3287 (cited on page 50)

5. Borndörfer R, Erol B, Graffagnino T, Schlechte T, Swarat E (2014) Optimizing the Simplon railway corridor. Ann Oper Res 218(1):93–106. https://doi.org/10.1007/s10479-012-1260-9 (cited on page 50)
6. Borndörfer R, Reuther M, Schlechte T (2014) A coarse-to-fine approach to the railway rolling stock rotation problem. In: Proceedings of the 14th workshop on algorithmic approaches for transportation modelling, optimization, and systems (ATMOS' 14), vol 42, pp 79–91. https://doi.org/10.4230/OASIcs.ATMOS.2014.79 (cited on page 50)
7. Caimi GC (2009) Algorithmic decision support for train scheduling in a large and highly utilised railway network. PhD thesis, Eidgenössische Technische Hochschule Zürich. https://doi.org/10.3929/ethz-a-005947637 (cited on page 50)
8. Crainic TG, Li Y, Toulouse M (2006) A first multilevel cooperative algorithm for capacitated multicommodity network design. Comput Oper Res 33(9):2602–2622. https://doi.org/10.1016/j.cor.2005.07.015 (cited on page 50)
9. DB Netz AG (2013) Netzkonzeption 2030: Zielnetz der DB Netz AG für die Schieneninfrastruktur im Jahr 2030. Information Booklet (cited on page 49)
10. Dempster MAH, Thompson RT (1998) Parallelization and aggregation of nested benders decomposition. Ann Oper Res 81:163–187. https://doi.org/10.2139/ssrn.37765 (cited on page 51)
11. Gamst M, Spoorendonk S (2013) An exact approach for aggregated formulations. Technical report, DTU Management Engineering Report 3.2013. Technical University of Denmark (cited on page 51)
12. Gurobi Optimization, Inc. (2017) Website of Gurobi. June 2017. https://gurobi.com (cited on pages 63, 65)
13. Lumbreras S, Ramos A (2013) Optimal design of the electrical layout of an offshore wind farm applying decomposition strategies. IEEE Trans Power Syst 28:1134–1441. https://doi.org/10.1109/tpwrs.2012.2204906 (cited on page 51)
14. Pruckner M, Thurner C, Martin A, German R (2014) A coupled optimization and simulation model for the energy transition in Bavaria. In: Fischbach K, Großmann M, Krieger UR, Staake T (eds) Proceedings of the international workshops SOCNET 2014 and FGENET 2014, pp 97–104 (cited on page 51)
15. Schlechte T, Borndörfer R, Erol B, Graffagnino T, Swarat E (2011) Micro-macro transformation of railway networks. J Rail Transp Plann Manage 1(1):38–48. https://doi.org/10.1016/j.jrtpm.2011.09.001 (cited on page 50)
16. Shapiro JF (1984) A note on node aggregation and Benders' decomposition. Math Program 29:113–119. https://doi.org/10.1007/bf02591733 (cited on page 51)
17. Statistisches Bundesamt S (2017) Website of the Federal Statistical Office, June (2017). https://www.destatis.de (cited on page 48)
18. Trukhanov S, Ntaimo L, Schaefer A (2010) Adaptive multicut aggregation for two-stage stochastic linear programs with recourse. Eur J Oper Res 206:395–406. https://doi.org/10.1016/j.ejor.2010.02.025 (cited on page 51)
19. Zipkin PH (1977) Aggregation in linear programming. PhD thesis, Yale University (cited on page 51)

Chapter 4
Freight Train Routing

Torsten Klug

Abstract This chapter is about strategic routing of freight trains in railway transportation networks with mixed traffic. A good utilization of a railway transportation network is important since in contrast to road and air traffic the routing through railway networks is more challenging and the extensions of capacity are expensive and long-term projects. Therefore, an optimized routing of freight trains have a great potential to exploit remaining capacity since the routing has fewer restrictions compared to passenger trains. In this chapter we describe the freight train routing problem in full detail and present a mixed-integer formulation. Wo focus on a strategic level that take into account the actual immutable passenger traffic. We conclude the chapter with a case study for the German railway network.

T. Klug (✉)
Department of Optimization, Zuse Institute Berlin, Berlin, Germany
e-mail: klug@zib.de

R. Borndörfer et al. (eds.), *Handbook of Optimization in the Railway Industry*, International Series in Operations Research & Management Science 268,
https://doi.org/10.1007/978-3-319-72153-8_4

4.1 Introduction

Rail transport volume in Germany has been increasing for years, while a corresponding expansion of the infrastructure is rather reluctant, since construction projects are always capital-intensive and long-term projects. Being a transition country in central Europe, Germany faces great traffic challenges in the next years. In particular, this applies for the rail freight transportation sector. Recent estimates assume an increase up to 80% by 2025 IFMO-Studie [16]. Therefore, it is necessary to analyze the existing network to estimate and make the best use of the available capacity. In this context, the major German railroad company, Deutsche Bahn AG (DB), uses a simplified (macroscopic) transport network for the rail *freight train routing* at a strategic planning level. The major aim is to determine routes for freight trains by taking into account the available railway infrastructure and the already planned and invariant passenger traffic.

The routing of freight trains is quite different from passenger trains since departure and arrival time windows are less strict and routes are not limited by several intended intermediate stops. Nevertheless, passenger and freight trains in Germany share the same infrastructure to a large extent. Therefore, the railway system has to be considered as a whole including passenger transport and infrastructure.

The problem we discuss in this chapter lies in between well studied problems in the planning process of railway transportation systems. On the one end are problems concerning the infrastructure, i.e., network design, capacity assessment, and long term traffic forecasts. These are topics on a strategic level with a macroscopic scale of the transportation network and the traffic flows. On the other end are problems like line planning, track allocation and timetabling that provide information on the passenger transportation with an accurate schedule of trains. In these cases the networks are very detailed, for instance, a station is described by several tracks and platforms instead of a single node. However, the solvable problem size are very different. In the case of network design, it is possible to consider the entire German railway network, while state-of-the-art timetabling approaches are able to solve instance for corridors or notable smaller subnetworks.

The *freight train routing problem (*FTRP*)* is investigated from a strategic perspective, calculating the routes in a macroscopic transportation network. In this context "macroscopic" means that complex structures are aggregated into fewer elements and the departure and arrival times of freight trains are approximated. The problem has a strategic character, since it asks only for a rough routing through the network without precise timings, i.e., in particular most of the input data consists only of coarse estimates. The day is partitioned into a small number of time slices, which reflects different traffic situations over the day, for instance, a higher utilization by passenger trains or constructions sites that affects the capacity of a track. The freight trains are given by origin destination pairs together with a departure time and train type for each train. The train type defines the driving characteristics of a train. A standard model day is considered with the assumption that the demand is equal to the day before and after. The actual timetable for passenger trains is mapped to the macroscopic network and given by the number of trains per track and time slice. Since the actual schedule of the freight trains on a track is not part of the problem the congestion is measure by a nonlinear function. The main part of this measure is the CR-function, which will be described in Sect. 4.2.2. The task is to find a route for each train in the transportation network. The determined routes should minimize the sum of all expected delays and the subordinate criteria running time and length. Capacity limitations of the arcs are implicitly handled by the congestion function, i.e., potential conflicts of trains using the same infrastructure element result in larger congestion values. Hence, by minimizing the capacity congestion function we directly aim to produce timetables where the probabilities of delays are smaller.

It follows a short review of the related literature. A formal description of the problem will be given in Sect. 4.2. In Sect. 4.3 a solution approach by mixed-integer programming is presented. A summary of computational results will be discussed in Sect. 4.4. Finally, Sect. 4.5 conclude the chapter and discuss promising further developments.

A similar and closely related problem is considered in Cacchiani et al. [8], where passenger trains are given as fixed and freight trains have to be scheduled as well. The main difference is that in contrast to our problem formulation the level of detail is higher, i.e., the time windows for departure and arrival are discretized with a much higher granularity, and, as a consequence, more specific capacity restrictions are given. On the other hand, the network size that can be handled by this approach is much smaller compared to the results in this chapter; the authors present computational results for several corridors and some major stations of the Italian railway network.

Many recent contributions from research concentrate on the subsequent step, the timetabling or track allocation which assumes inter alia freight train routes as an input, see Chaps. 6 and 7. Caimi [9] presents a top-down approach and uses it to handle the complete Swiss network by an a-priori decomposition of the network into different zones. In contrast to that, Schlechte et al. [25] presents a bottom-up approach to define a macroscopic railway model based on microscopic simulations. A similar approach can be found in Kettner et al. [20]. There an automated generation of macroscopic data from a microscopic basis is described for the Austrian Federal Railways (*ÖBB*). Extensive literature surveys on train timetabling problems and railway track allocation can be found in Cordeau et al. [10] and in the more recent ones Lusby et al. [24] and Cacchiani and Toth [7]. The authors of Bussieck et al. [6], Huisman et al. [15], and Gorman [14] give an overview on optimization in public railway transportation.

Besides the special application context, our problem has similarities to the broader class of *network design* problems, see Balakrishnan et al. [3] for a general survey. A network design problem tested on the German railway network is presented in Chap. 3. A framework for a general class of network design problems is presented by Kim and Barnhart [21] and applied to the blocking problem in railroad traffic in the US; see Barnhart et al. [4]. Integrated service network design for rail freight transportation in the US is considered in Zhu et al. [27], Ahuja et al. [2], and Jha et al. [18].

Recent developments for the freight train routing of single cars is presented in Chap. 8. Small customers order only 1–5 cars instead of complete train with more than 20 cars. In such cases it is too expensive to pull these groups of cars each by a single locomotive through the network. Instead, the cars are only pulled to the next classification yard. There they are grouped with the cars from other customers, and then as new trains pulled to the next classification yard. There the trains are disassembled, and the cars are again re-grouped with others until each car has reached its final destination. This problem gives rise to a natural network design question, i.e., where are the classification yards located and how to route between them. Fügenschuh et al. [11, 13] discuss the whole system of *single wagon freight transportation*, show the positive effect of bundling cars, and compare the problem to other freight transportation concepts mentioned in the literature, e.g., the *railroad blocking problem* in the US or Canada.

The railroad blocking problem can be formulated as a very large-scale, multi-commodity, flow-network-design and routing problem with billions of decision variables; see Jha et al. [18] and Barnhart et al. [4]. Ahuja et al. [2] presents an algorithm using an emerging technique known as very large-scale neighborhood search to support major US railway companies that transfers millions of cars over its network annually. The authors report that their heuristic approach is able to solve the problem to near optimality using 1–2 h of computer time on a standard workstation computer.

Modeling railway capacity is technically very complex and hence the prediction of congestion and waiting times is a major challenge. Nevertheless, the crucial relation is that there is almost no waiting time, as long the mixture of allocated trains can be handled by the infrastructure capacity. Once the capacity limit is reached, congestion starts and smooth operation is not possible anymore. The closer it gets to the capacity limit, the more delay occurs for each train. As soon as the number of trains goes further beyond the capacity limit, the average delay for each train grows even faster. There are many different ways to model and compute capacity values for tracks. For a survey on this complex issue we refer to Abril et al. [1] and the references therein. For a recent account of the theory we refer the reader to Chap. 2.

The approach we chose to model the functional relationship between the number of trains passing a certain infrastructure is to introduce a *capacity restraint (CR) function*. These functions are designed to give a reasonable measure of the expected average delay. One of the earliest appearances of CR-functions in the literature is due to Irwing and Cube [17]. Wohl [26] uses CR-functions to describe the travel performance or travel time and delay as a function of the flow using properties of the infrastructure and its capacity during the trip distribution and assignment phases of a travel forecasting process. Most applications of CR-functions are tailored to road traffic. Only recently, Lieberherr and Pritscher [23] and Borndörfer et al. [5] use CR-functions in railway transport.

In the case of road traffic, Köhler et al. [22] presents a theoretical analysis of the mathematical aspects of flow dependent cost functions. A major difference is that in road traffic the routing is decentralized, arbitrarily partitionable, and assumed to be selfish. In contrast to that railway systems are centralized and we are aiming for a system optimum. In addition, the train flow cannot be partitioned arbitrarily and thus the routing and timetable is a more rigid system in comparison to the flow of cars.

4.2 The Freight Train Routing Problem

In this section we formally describe the problem, hereafter referred to as the Freight Train Routing Problem (FTRP) and model the problem as a time slice expanded graph.

4.2.1 Transportation Network

The transportation network is a directed graph $\hat{G} = (\hat{\mathscr{V}}, \hat{\mathscr{A}})$ in which $\hat{\mathscr{V}}$ and $\hat{\mathscr{A}}$ are the node set and the arc set, respectively. Each node $v \in \hat{\mathscr{V}}$ represents a station, a junction or some other infrastructure element where a train can start, branch or end. Each arc $a \in \hat{\mathscr{A}}$ represents a connecting railway track. For convenience, for each node $v \in \hat{\mathscr{V}}$, we will let $\delta^+(v) \subseteq \hat{\mathscr{A}}$ denote the set of arcs leaving v, and $\delta^-(v) \subseteq \hat{\mathscr{A}}$ denote the set of arcs entering v.

The day is partitioned into a small number of time slices. Let $\mathscr{S}$ be the set of time slices, where τ_s denotes the time span of the time slice $s \in \mathscr{S}$. The starting time and end time of a time slice s is denoted as $\underline{s}$ and $\overline{s}$, respectively. They are arranged in a cyclic order, such that a train at the end of the day (last time slice) can continue at the beginning of the day (first time slice). The end time of a time slice is always equal to the start time of the next time slice, the only exception is the last time slice which ends at 86,400 s and the next time slice starts at 0 s. For simplicity in notation, we identify the start time of the first slice with the end time of the last time slice.

The trains are classified into a set of standard *train types* $\mathscr{T}$. With these definitions we can define the arc parameters as follows. For each arc $a \in \mathscr{A}$, we have:

l_a: length of the arc in meters;
$\tau_{t,a}$: running time in seconds depending on the train type $t \in \mathscr{T}$;
κ_a^s: approximated number of trains that could use arc a within time slice $s \in \mathscr{S}$;
ρ_a^s: number of passenger trains traversing arc a within time slice $s \in \mathscr{S}$.

A *route* in the graph is denoted as p, where the corresponding set of nodes and arcs are denoted as $\mathscr{V}_p$ and $\mathscr{A}_p$, respectively. Furthermore, each arc $a \in \mathscr{A}_p$ is assigned to a time slice. Let $s_p(a)$ be the assigned times slice and $\mathscr{S}_p = (s_1, s_2, \ldots, s_k)$ be the sequence of time slices used by route p, where s_1 is the time slice the route starts and s_k is the time slice the route ends. For a route p we allow only time slice sequences $\mathscr{S}_p = (s_1, s_2, \ldots, s_k)$ such that for all $i, j \in \{1, \ldots, k\}$, $s_i \neq s_j$ and $\overline{s}_i = \underline{s}_{i+1}$ hold.

Then the *running time* of route p for train type t and departure time τ^{dep} is defined by

$$\tau_p(t, \tau^{\text{dep}}) = \underline{s}_k - \tau^{\text{dep}} + \sum_{\substack{a \in \mathscr{A}_p: \\ s_p(a) = s_k}} \tau_{t,a},$$

The running time of a route is not only the sum of the arc running times. It is possible that a route uses only one arc or even no arc at all within a time slice. A train using this route will simply wait the remaining time of the time slice duration. But this waiting time is part of route running time. The difference of the time span of the time slice and the running times of used arcs within the time slice is called *transit time*, since it is the time spend to transit to the next time slice. The definition of the running time and the restriction on the time slice sequence of a route implies that a route is no longer than a day.

The *length* of a route is independent from the time slices and is defined by

$$l(p) := \sum_{a \in \mathscr{A}_p} l_a.$$

4.2.2 Freight Train Demand and Objective

The *freight train demand* is given by a set of trains $\mathscr{R}$. For each *train* $r \in \mathscr{R}$, we have: origin node; destination node; τ_r^{dep} departure time; and t_r train type.

The constraints of the problem are determined by the definition of a feasible routing. It is required that the routes are not far apart from the shortest and fastest paths. Therefore, the running time and length of a possible routing is restricted to a multiple of the fastest or shortest path. The so-calculated maximum length and maximum running time of a routing for request r is denoted as Δ_{dist}^r and Δ_{time}^r, respectively. A route p for $r \in \mathscr{R}$ is called *feasible* if

1. it starts at the origin node;
2. ends at the destination node;
3. has a length less than or equal to Δ_{dist}^r;
4. has a running time less than or equal to Δ_{time}^r;
5. the sum of arc running times for each time slice is less than or equal to the time slice duration.

Constraint (5) implies that each arc must traversed completely within one time slice. It is forbidden to traverse a fraction of an arc in one time slice and the remaining part in the next slice.

The *objective function* has two components; the routing cost which enclose the routes lengths and running times and the total congestion cost.

For each arc $a \in \mathscr{A}$ we define the congestion cost as follows: Let n be the number of trains on an arc, then the congestion cost is defined as:

$$\tau\left(1+\alpha\left(\frac{n}{\kappa\gamma}\right)^{\beta}\right), \quad \alpha,\beta \in [0,\infty[,\ \gamma \in]0,\infty[,$$

where the running time τ and the capacity κ depends on the arc. This function is an undamped variant of the CR-function presented in Lieberherr and Pritscher [23]. In this work, a justification for the exponential growth of the CR-function is also given. The shape of the CR-function can be controlled by the parameters α,β,γ. The multiple of the running time that a train is penalized, if the capacity is reached, is defined by α. We choose $\alpha = 1$, which means we must pay the running time of a train if we reach the capacity. The capacity is scaled with γ and could be used to keep an amount of reserve capacity. Since we do not want to keep any capacity, we choose $\gamma = 1$. This simplifies the CR-function to

$$\tau\left(1+\left(\frac{n}{\kappa}\right)^{\beta}\right).$$

The rapidness of penalization is controlled by β: a large value for β means a big slope near the capacity; a small value leads to a moderate slope. Since the running time is already part of the routing cost, we take only the surplus as congestion cost

$$f(n) = \tau\left(\frac{n}{\kappa}\right)^{\beta}. \tag{4.1}$$

We use function (4.1) to estimate the congestion on each arc. A small example for one arc and a set of curves for different values of β is shown in Fig. 4.1. We scale

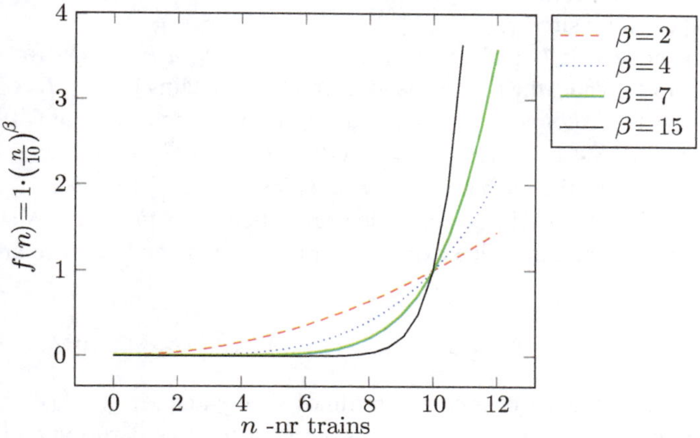

Fig. 4.1: CR-functions with arc capacity $k = 10$, average running time $\tau = 1$ and $\beta \in \{2,4,7,15\}$

the different objective function components with $\lambda_{\text{time}}, \lambda_{\text{dist}}, \lambda_{\text{wait}}$ for the running time, the length and the congestion cost, respectively. Thus, the objective function is defined as follows:

$$f(\mathscr{P}) := \lambda_{\text{time}} \left[\sum_{r \in \mathscr{R}} \tau_{p_r}(t_r, \tau_r^{\text{dep}}) \right] + \lambda_{\text{dist}} \left[\sum_{r \in \mathscr{R}} l(p_r) \right] + \lambda_{\text{wait}} \left[\sum_{s \in \mathscr{S}} \sum_{a \in \mathscr{A}} \tau_a \left(\frac{n_a^s}{\kappa_a^s} \right)^{\beta} \right],$$

where for each arc a: τ_a is the average running time over all train types; ρ_a^s is the number of passenger trains ρ_a^s; and n_a^s is the number of freight trains from the actual routing $\mathscr{P}$. The routing $\mathscr{P}$ contains a feasible route $p_r \in \mathscr{P}$ for each train $p \in \mathscr{P}$.

To summarize we give a compact definition of the FTRP as follows:

Definition 4.1. Given an transportation network $\hat{G} = (\hat{\mathscr{V}}, \hat{\mathscr{A}})$, a set of time slices $\mathscr{S}$, a set of train types $\mathscr{T}$, arc length l_a, arc running times $\tau_{t,a}$, arc passenger traffic ρ_a^s, arc capabilities κ_a^s, and a set of trains $\mathscr{R}$ with an origin and destination node and the tuple $(\tau_r^{\text{dep}}, \Delta_{\text{dist}}^r, \Delta_{\text{time}}^r)$ for each request $r \in \mathscr{R}$. The task is to find a feasible routing for each train $\mathscr{P}$ such that $f(\mathscr{P})$ is minimized.

It is important to notice that the capacities of the arcs are not constraints. The capacities are only considered by the cost function. If the cost function is linear then the problem decompose into routing problems for single freight trains.

4.2.3 Time Slice Expanded Graph

To model the problem we construct a *time slice expanded graph* G. Many real-world problems have been formulated as time-space network models, see Kennington and Nicholson [19] for a survey.

For each node $v \in \hat{\mathscr{V}}$ and for each arc $a \in \hat{\mathscr{A}}$ we have a copy for each time slice $s \in \mathscr{S}$ in G. Thus, the time expanded graph G contains $|\mathscr{S}| =: L$ copies of the original graph $\hat{G}$. To connect the subgraphs of different time slices in G further arcs are added as follows: Let $v_{s_1}, v_{s_2}, \dots v_{s_L}$ be the nodes in G for node $v \in \hat{\mathscr{V}}$ and the time slices $s_1, \dots, s_L$. For each node $v \in \hat{\mathscr{V}}$ the arcs $(v_{s_i}, v_{s_{i+1}})$ for $i = 1, \dots, L-1$ and (v_{s_L}, v_{s_1}) are added to G. The so added arcs are called *transition arcs*. We denote the nodes and arcs of the time slice expanded graph G by $\mathscr{V}$ and $\mathscr{A}$. The arc set of G can be partitioned into

$$\mathscr{A} = \mathscr{A}_{s_1} \dot\cup \mathscr{A}_{s_2} \dot\cup \dots \dot\cup \mathscr{A}_{s_L} \dot\cup \mathscr{A}_{s_1,s_2} \dot\cup \mathscr{A}_{s_2,s_3} \dot\cup \dots \dot\cup \mathscr{A}_{s_{L-1},s_L} \dot\cup \mathscr{A}_{s_L,s_1},$$

where $\mathscr{A}_s \subseteq \mathscr{A}$ is the set of arcs within times slice s and $\mathscr{A}_{s_1,s_2} \subset \mathscr{A}$ is the set of transition arcs from time slice s_1 to time slice s_2. The same partition by time slices is used for the nodes of G:

$$\mathscr{V} = \mathscr{V}_{s_1} \dot\cup \mathscr{V}_{s_1} \dot\cup \dots \dot\cup \mathscr{V}_{s_L}.$$

The lengths, running times, passenger traffic and capacities of non-transition arcs are taken from $\hat{G}$. Whereas, a transition are has an unlimited capacity and the length, running time, and passenger traffic are zero. Since the time slice of an arc is described by the time slice expanded graph structure we omit the time slice subscript for all arc parameters.

For each request $o_r \in \mathscr{V}$ is the origin node in the time slice expanded graph. Since the arrival time slice of a train is not restricted, we have a destination node for each time slice. The set of destination nodes of request r is denoted by $D_r \subset \mathscr{V}$.

The running time of a route p with train type t and departure time τ^{dep} in terms of the time slice expanded graph is defined as follows:

$$\tau_p(t, \tau^{\text{dep}}) := \overline{s}_k - \tau_r^{\text{dep}} + \sum_{a \in \mathscr{A}_p \cap \mathscr{A}_{s_k}} \tau_{t,a}$$

The following Example 4.1 demonstrates the construction of the graph and the definition of a route in G.

Example 4.1. The graph $\hat{G} = (\hat{\mathscr{V}}, \hat{\mathscr{A}})$ on the left contains the running times. In this example there are equal for all trains. All time slices have a time span of 10 min. On the right hand side is the time slice expanded graph $G = (\mathscr{V}, \mathscr{A})$ of $\hat{G}$. Each slice $(\mathscr{V}_s, \mathscr{A}_s)$ is a copy of $\hat{G}$. And the copies are cyclic connected by transition arcs.

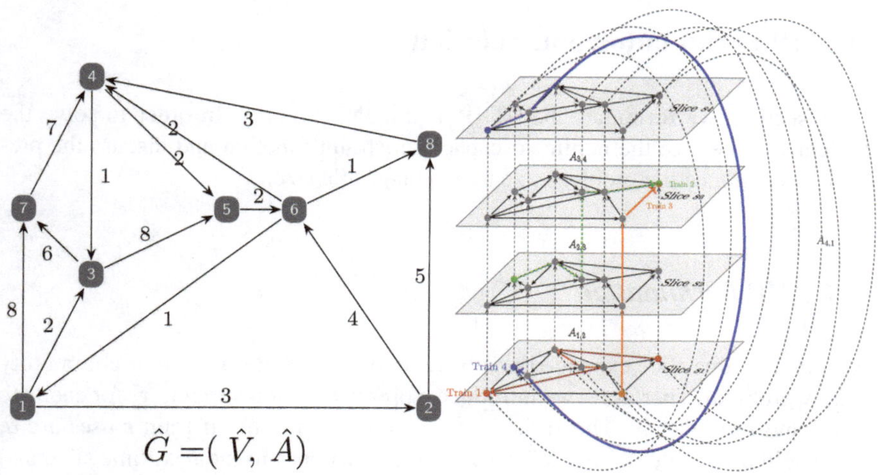

In the following the route for each train is listed. The numbers above the arcs are the arc running times and the numbers below are the transition times.

$$\text{Train 1:} \quad 8_{s_1} \xrightarrow{3} 4_{s_1} \xrightarrow{2} 5_{s_1} \xrightarrow{2} 6_{s_1} \xrightarrow{1} 1_{s_1}$$

$$\text{Train 2:} \quad 7_{s_2} \xrightarrow{7} 4_{s_2} \xrightarrow{2} 5_{s_2} \xrightarrow[1]{} 5_{s_3} \xrightarrow{2} 6_{s_3} \xrightarrow{1} 8_{s_3}$$

$$\text{Train 3:} \quad 2_{s_1} \xrightarrow[10]{} 2_{s_2} \xrightarrow[10]{} 2_{s_3} \xrightarrow{5} 8_{s_3}$$

$$\text{Train 4:} \quad 1_{s_4} \xrightarrow{2} 3_{s_4} \xrightarrow[8]{} 3_{s_1} \xrightarrow{6} 7_{s_1}$$

Thus we get the values in the table where; the second column is the sum of the arc running times; the second column is the sum of the transition times; and the last column is the sum of running times and transition times which defines the path running time.

	Arc running time	Transit time	Route running time
Train 1	8	0	8
Train 2	12	1	13
Train 3	5	20	25
Train 4	8	8	16

4.3 MIP Formulation and Solution

In this section, we formulate the FTRP as a MINLP model. In order to solve the problem we linearize the nonlinear capacity restraint function and discuss the pre-solving that reduce the problem size to a manageable size.

4.3.1 MIP Formulation

Based on the time slice expanded graph we model the FTRP as a multi-commodity arc flow problem. Therefore, we introduce a binary decision variable x^r_a for each arc $a \in \mathscr{A}$ and each $r \in \mathscr{R}$. The variable equals one if and only if train r uses arc a, otherwise the variable is zero. Whenever a train switches to the next time slice and the running time is less than the time span of the time slice, the train has to wait until the time slice ends. Therefore, we introduce continuous variables z^r_s for each train $r \in \mathscr{R}$ and time slice $s \in \mathscr{S}$ which represent the difference between the running time of the train and the time slice time span.

The objective function contains the total nonlinear congestion cost for each arc and the sum of all running times and lengths. $\lambda_{\text{time}}, \lambda_{\text{running}}, \lambda_{\text{length}}$ are the normalized cost factors of each part. τ_a is the average running time of this arc considering all train types, i.e., $\tau_a = \frac{\sum_{t \in \mathscr{T}} \tau_{t,a}}{|\mathscr{T}|}$.

The nonlinear objective function is defined as follows:

$$\min \quad \underbrace{\lambda_{\text{wait}} \sum_{\forall a \in \mathscr{A}} \tau_a \left(\frac{\sum_{r \in \mathscr{R}} x_a^r + \rho_a}{\kappa_a} \right)^{\beta}}_{\text{congestion cost}} + \underbrace{\lambda_{\text{time}} \sum_{r \in \mathscr{R}} \sum_{a \in \mathscr{A}} x_a^r \tau_{t_r,a}}_{\text{running time}} + \underbrace{\lambda_{\text{length}} \sum_{r \in \mathscr{R}} \sum_{a \in \mathscr{A}} x_a^r l_a}_{\text{length}}$$

We define a flow balance function for train $r \in \mathscr{R}$ as follows:

$$b(v) = \begin{cases} 1 & \text{if } v \in o_r \\ -1 & \text{if } v \in D_r \\ 0 & \text{otherwise} \end{cases} .$$

The constraints are

$$\sum_{a \in \delta^+(v)} x_a^r - \sum_{a \in \delta^-(v)} x_a^r = b(v) \quad \forall r \in \mathscr{R}\ \forall v \in \mathscr{V} \tag{4.2}$$

$$\sum_{a \in \mathscr{A}_s} \tau_{t_r,a} x_a^r \le \tau_s \quad \forall s \in \mathscr{S}\ \forall r \in \mathscr{R} \tag{4.3}$$

$$\sum_{a \in \mathscr{A}} l_a x_a^r \le \Delta_{\text{dist}}^r \quad \forall r \in \mathscr{R} \tag{4.4}$$

$$\sum_{a \in \mathscr{A}} \tau_{t_r,a} x_a^r + \sum_{s \in \mathscr{S}} z_s^r \le \Delta_{\text{time}}^r \quad \forall r \in \mathscr{R} \tag{4.5}$$

$$\tau_s \sum_{a \in \mathscr{A}_{s,s+1}} x_a^r - \sum_{a \in \mathscr{A}_s} \tau_{t_r,a} x_a^r \le z_s^r \quad \forall s \in \mathscr{S}\ \forall r \in \mathscr{R} \tag{4.6}$$

$$x_a^r \in \{0,1\} \quad \forall a \in \mathscr{A}\ \forall r \in \mathscr{R}$$

$$z_s^r \ge 0 \quad \forall s \in \mathscr{S}\ \forall r \in \mathscr{R}$$

We have the common flow constraints (4.2) for each train: the outflow must be one at the origin; the inflow must be one at exactly one of the destination nodes in the time slice expanded graph; and at the remaining nodes, flow conservation is required. A train must change to the succeeding time slice if the running time is larger than the time span of the time slice (4.3). Constraints (4.4) and (4.5) handle the length and running time restrictions. The waiting time to switch a slice z_s^r is part of the running time. The constraints (4.6) ensure that the variables z_s^r are set to the difference of time span and running time of the corresponding time slice only if the train switch to the succeeding time slice.

4.3.2 *Solving the* FTRP

The mixed-integer nonlinear programming (MINLP) model contains a binary variable for each arc and train. In terms of the considered instances of DB this amounts to up to 25 million binary variables. Setting up such a model consumes an enormous amount of computer memory and solving such a model takes a daunting amount of time. In the following we describe our efforts to reduce the resource demand. We focus on presolving; that is, we only generate those parts of the model that are really necessary because they contain an optimal solution, and remove all the others. Furthermore we describe the linearization of the nonlinear objective function that transforms the MINLP to a MILP (mixed-integer linear program).

4.3.3 Presolving

For the preprocessing we try to identify, for each train, arcs and nodes that cannot be part of a feasible solution.

Obviously, all ingoing arcs of the origin node and all outgoing arcs of destination nodes can be ignored. The running time restriction is less than 24 h for all trains. Thus, we can assume that a train cannot enter its starting time slice again by running through all four time slices. This means all transition arcs into the starting time slice of a train will also be ignored.

The major part of the preprocessing is to reduce the network for a train to the subset of arcs and nodes that are elements of a path from the origin to one of the destination nodes and observe the length and running time restrictions. To find the relevant arc and node subsets we construct a shortest path tree from the origin node and one tree from the destination node in $\hat{G}$. We can stop in the leafs of the tree if the distance from the root to the leaf is larger than the length restrictions. Hence, for each node, we check if the distance from the origin node plus the distance to the destination node is less than the length restriction. If the sum is less than the length restriction, then the node could be in a feasible solution. Otherwise the node cannot be contained in a feasible path without violating the length restriction. The feasible arcs and nodes for $\hat{G}$ can then be transferred to the time slice expanded graph G, since a shortest path in $\hat{G}$ remains a shortest path in the copies of each slice. A transition arc is feasible if its head and tail node are feasible.

In case of the running time we have to consider one time slice of the time slice expanded graph after another. If the feasible nodes are determined then the arcs could be checked easily. We restrict our discussion to the nodes. From the fastest path trees of $\hat{G}$ we get for each node the minimum running time from the origin and to the nearest destination, respectively. A node in the time slice expanded graph can only be feasible if the corresponding node is feasible for $\hat{G}$. Thus, we could restrict the feasible nodes set in the time slice expanded graph to the copies of such nodes.

For each slice the following procedure is executed: We have a set of possible starting nodes and an available amount of time to reach the destination. For the departure time slice the set of starting nodes consists only of the origin node. Furthermore, the available amount of time is the value of the running time restriction. A node v in the current time slice is feasible, if and only if the available amount of time is larger than or equal to the fastest running time, i.e., the sum of the fastest running time from one of the starting nodes to the node v and the fastest running time from the node v to the destination. Both running times, from a starting node and to the destination node, are provided by fastest path trees of $\hat{G}$.

If one slice is finished, the set of starting nodes for the next slice is given by the set of nodes, that are adjacent to feasible nodes of the preceding time slice, i.e., connected by a transition arc. The available amount of time for the next time slice is the available amount of time of the current time slice minus its time span. The procedure terminates if the available amount of time is less than the time slice and therefore no time would be left for the next slice. Thus, we get a significant reduction of flow variables by identifying for each train its relevant subset of arcs.

Since the length and running time bounds are rather strict with 150% of the shortest and fastest path, respectively, the reductions of variables is significant. With this procedure the instance set in Borndörfer et al. [5] could be reduced to about 5% of the original problem size.

4.3.4 Linearization

In order to solve the FTRP with standard MIP solvers we linearize the nonlinear terms of the objective function. We apply the linearization technique used in Fügenschuh et al. [12]. Since we are not allowed to split trains, the total number of trains traversing an arc a is always integer. Hence, we need only the function values for feasible integer input values. We introduce for each arc a an artificial continuous variable y_a. Without loss of generality we assume that the total number of trains per arc is bounded by some value N. Then the constraints

$$\Gamma_1^a(m)\left(\sum_{r\in\mathscr{R}} x_a^r + \rho_a\right) + \Gamma_2^a(m) \leq y_a \quad \forall a \in \mathscr{A}\ \forall m \in \{1,\dots,N\}$$

describe the convex hull of all feasible integer points. $\Gamma_1^a(m)$ is the slope and $\Gamma_2^a(m)$ the y-intersection of the linear function through the points $(m, f(m))$ and $(m-1, f(m-1))$. The slope is defined by

$$\Gamma_1^a(m) = f_a(m) - f_a(m-1) = \frac{\alpha\tau_a}{\kappa_a^\beta}\left((m+\rho_a)^\beta - (m-1+\rho_a)^\beta\right)$$

and the y-intersection by

$$\Gamma_2^a(m) = f_a(m) - \Gamma_1^a(m)m$$

An example of this linearization is depicted in Fig. 4.2. The transformed cost function is:

$$\min \quad \underbrace{\lambda_{\text{wait}} \sum_{\forall a \in \mathscr{A}} y_a}_{\text{congestion cost}} + \underbrace{\lambda_{\text{time}} \sum_{r \in \mathscr{R}} \sum_{a \in \mathscr{A}} x_a^r \tau_{t_r,a}}_{\text{running time}} + \underbrace{\lambda_{\text{length}} \sum_{r \in \mathscr{R}} \sum_{a \in \mathscr{A}} x_a^r l_a}_{\text{length}}$$

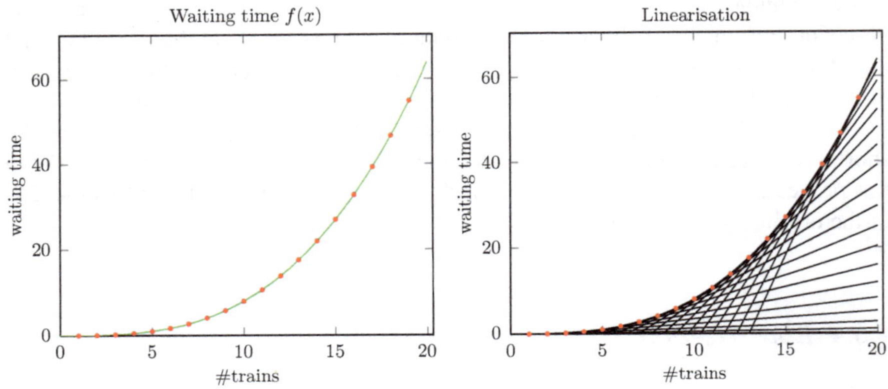

Fig. 4.2: Linearization of a CR-function where only integer points are interesting

4.4 Computational Results

In this section we give a short summary of the computational results presented in Borndörfer et al. [5].

The industrial partner provided data for the entire macroscopic railway network of Germany and a corresponding freight train demand forecast. The transportation network $\hat{G}$ is constructed from a given base graph with additional turning restrictions. These restrictions ensure that a route across a node in the macroscopic network can be mapped to a feasible route in the microscopic network. Turning restrictions are integrated into the directed graph $\hat{G}$ such that they are implicitly encoded by the graph structure, see Borndörfer et al. [5] for more details. In general the arcs are directed, but two opposite arcs share the same attributes and their capacity as well.

The train set $\mathscr{R}$ is generated from real-world data collected in 2010. We consider four types of freight trains with varying speeds. One can distinguish between four basic loads which are given by the typical distribution of passenger trains during the day: morning rush hour, midday, evening rush hour, and night. Therefore four time slices will be enough to determine the capacity on a strategic level for the freight train routes.

We considered 33 instances (1 east, 18 south, 13 west) from networks of three important geographical areas; see Fig. 4.3. The complete macroscopic graph of Ger-

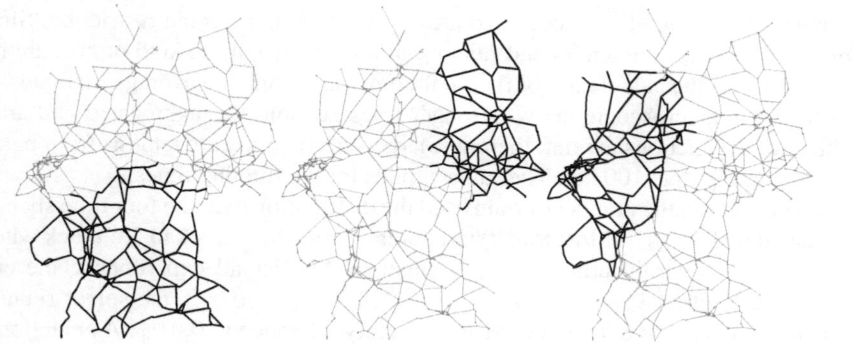

(a) (b) (c)

Fig. 4.3: Macroscopic infrastructure graphs for three geographical areas of Germany. (**a**) South. (**b**) East. (**c**) West

Table 4.1: Instance size of the considered transportation network

Instance	# Nodes	# Arcs	Length (m)			Running time (s)		
			Min	Avg	Max	Min	Avg	Max
East	1935	3599	62	10,067	121,527	3	403	4899
South	3350	6583	72	8955	121,811	2	349	5466
West	4479	8874	24	6366	106,285	1	250	5466

Table 4.2: Variety of the instances

Instance	Train types	# Trains	Train departure slice			
			5–9	9–16	16–20	20–5
East	3	437	69	89	46	233
South	4	300–700	36–84	54–132	41–104	162–394
West	4	300–600	39–99	74–154	37–76	136–315

The second column is the number of train types followed by the number of trains. The last four columns depict the number of trains that depart in each time slice

many has 1620 nodes and 5162 arcs. The given demand forecast contains 3389 trains. For each area we retain only the trains where the origin and the destination node within the corresponding area. All other trains are ignored for this area. For each area we create several instances with an increasing number of trains that are randomly selected from the pool of possible trains. Instances for the same area and with the same number of trains differ in the subsets of randomly selected trains. Table 4.1 contains the magnitude and characteristics of the networks. The number of nodes and arcs in this table are the sizes of the transportation network $\hat{G}$. The arc lengths are given in meters, arc running times in seconds. By construction $\hat{G}$

contains several auxiliary nodes and arcs to integrate the turning restriction. Since for almost all arcs the length and running times are zero, these kind or arcs are not considered by the calculation of the minimum, maximum and average in Table 4.1. Since in the macroscopic network longer track sections are aggregated and areas with a lot of junctions are not, there is a large variety in the arc lengths from below 100 m to more than 100 km. The same applies for the running times.

The variety in the number of trains and the distribution over the four time slices is presented in Table 4.2. Most trains start at night from 20 o'clock to 5 o'clock where the network is hardly utilized by passenger traffic. For all experiments, the cost values were set to $\lambda_{\text{wait}} = 1000, \lambda_{\text{time}} = 0.028, \lambda_{\text{length}} = 0.002$, the solver running time limit was set to 6 h and the MIP optimality tolerance to 0.01%. As mentioned in the presolving section the problem size could be reduced to 2%, 5% or 10% of the original problem size depending on the allowed detour factor for the routes lengths and running times. Thus, the instances become manageable for a state of the art MILP solver.

The carried out experiments comprise a variation of the length and running time restrictions and different values for $\beta = [2,4,7,15]$. The results are compared with a shortest path routing of the freight trains. For all but two of the instances a gap below 0.1% could be reached within the 6 h time limit.

The results show that already for small length and running time restrictions the distribution over the network can be optimized and balanced using the presented MIP model. As you could expect an increase of length and running time limits leads to a better utilization of the network and decrease the number trains that are routed above the capacity limit of an arc.

Table 4.3 give an impression of the network utilization and demonstrates the potential of the model to provide more balanced solutions. The result are averaged over the instances of the corresponding area. The running times and length are rounded to the nearest integer, whereas the capacity values are rounded to two decimal places.

Table 4.3: Comparison of the solution in which each train takes the shortest path and the solution with CR-function defined by $\beta = 4$ and a detour factor of 1.5

Instance	Shortest path				1.5			
	Length	Time	Over capacity		Length	Time	Over capacity	
	(km)	(h)	# Arcs	# Trains	(km)	(h)	# Arcs	# Trains
East	41,473	434	17	40	42,277	442	2	3
South	49,249	522	48.05	80.77	53,937	574	11.11	13.33
West	37,086	381	70.77	112.77	40,209	421	11.00	12.69

4.5 Conclusion

In this chapter we present the FTRP and propose a MINLP model based on a time slice expanded graph formulation of the problem. An algebraic approximation of the delays of the trains by capacity restraint functions leads to a nonlinear model. We reduced the MINLP to a mixed-integer linear model by piecewise linear approximation and we were able to tackle large-scale instances with state-of-the-art MILP solvers by utilizing several graph preprocessing techniques. The experiments conclude that CR-functions are a reasonable tool to penalize waiting times and to provide a balanced solution for the requested distribution of the traffic through a macroscopic railway network.

References

1. Abril M, Barber F, Ingolotti L, Salido MA, Tormos MP, Lova A (2008) An assessment of railway capacity. Transp Res E: Log Transp Rev 44(5):774–806. https://doi.org/10.1016/j.tre.2007.04.001 (cited on page 76)
2. Ahuja RK, Jha KC, Liu J (2007) Solving real-life railroad blocking problems. Interfaces 37(5):404–419. ISSN: 0092-2102. https://doi.org/10.1287/inte.1070.0295 (cited on pages 75, 76)
3. Balakrishnan A, Magnanti TL, Mirchandani P (1997) Network design. In: Dell'Amico M, Maffioli F, Martello S (eds) Annotated bibliographies in combinatorial optimization. Wiley, London, pp 311–334. ISBN: 978-0-471-96574-9 (cited on page 75)
4. Barnhart C, Jin H, Vance PH (2000) Railroad blocking: a network design application. Oper Res 48(4):603–614. ISSN: 0030-364X. https://doi.org/10.1287/opre.48.4.603.12416 (cited on pages 75, 76)
5. Borndörfer R, Klug T, Schlechte T, Fügenschuh A, Schang T, Schülldorf H (2016) The freight train routing problem for congested railway networks with mixed traffic. Transp Sci 50(2):408–423. https://doi.org/10.1287/trsc.2015.0656 (cited on pages 76, 85, 86)
6. Bussieck MR, Winter T, Zimmermann UT (1997) Discrete optimization in public rail transport. Math Program 79B(1–3):415–444 (cited on page 75)
7. Cacchiani V, Toth P (2012) Nominal and robust train timetabling problems. Eur J Oper Res 219(3):727–737. Feature clusters. ISSN: 0377-2217. https://doi.org/10.1016/j.ejor.2011.11.003 (cited on page 75)
8. Cacchiani V, Caprara A, Toth P (2010) Scheduling extra freight trains on railway networks. Transp Res B Methodol 44(2):215–231. https://doi.org/10.1016/j.trb.2009.07.007 (cited on page 75)
9. Caimi GC (2009) Algorithmic decision support for train scheduling in a large and highly utilised railway network. PhD thesis, Eidgenössische Technische Hochschule Zürich. https://doi.org/10.3929/ethz-a-005947637 (cited on page 75)

10. Cordeau J-F, Toth P, Vigo D (1998) A survey of optimization models for train routing and scheduling. Transp Sci 32(4):380–404. https://doi.org/10.1287/trsc.32.4.380. transci.journal.informs.org/cgi/reprint/32/4/380.pdf (cited on page 75)
11. Fügenschuh A, Homfeld H, Huck A, Martin A, Yuan Z (2008) Scheduling locomotives and car transfers in freight transport. Transp Sci 42(4):1–14 (cited on page 76)
12. Fügenschuh A, Homfeld H, Schülldorf H, Vigerske S (2010) Mixed-integer nonlinear problems in transportation applications. In: Rodrigues H et al (eds) Proceedings of the 2nd international conference on engineering optimization (+CD-rom), pp 319–320. https://www1.dem.ist.utl.pt/engopt2010/ (cited on page 85)
13. Fügenschuh A, Homfeld H, Schülldorf H (2015) Single-car routing in rail freight transport. Transp Sci 49(1):130–148. https://doi.org/10.1287/trsc.2013.0486 (cited on page 76)
14. Gorman MF (2010) Rail research since deregulation: past trends and evolution. The newsletter of the railway applications section of INFORMS 2010, pp 6–7 (cited on page 75)
15. Huisman D, Kroon LG, Lentink RM, Vromans MJCM (2005) Operations research in passenger railway transportation. Statistica Neerlandica 59:467–497 (cited on page 75)
16. IFMO-Studie (2005) Zukunft der Mobilität - Szenarien für das Jahr 2025. Tech. rep. Institut für Mobilitätsforschung (cited on page 74)
17. Irwing N, Cube HV (1962) Capacity restraint in multi-travel mode assignment programs. Highw Res Board Bull 347:258–287 (cited on page 76)
18. Jha KC, Ahuja RK, Şahin G (2008) New approaches for solving the block-to-train assignment problem. Networks 51(1):48–62. ISSN: 0028-3045. https://doi.org/10.1002/net.v51:1 (cited on pages 75, 76)
19. Kennington JL, Nicholson CD (2010) The uncapacitated time-space fixed-charge network flow problem: an empirical investigation of procedures for arc capacity assignment. INFORMS J Comput 22(2):326–337 (cited on page 80)
20. Kettner M, Sewcyk B, Eickmann C (2003) Integrating microscopic and macroscopic models for railway network evaluation. In: Proceedings to the European transport conference (cited on page 75)
21. Kim D, Barnhart C (1997) Transportation service network design: models and algorithms. In: Wilson NHM (ed) Proceedings of the seventh international workshop on computer-aided scheduling of public transport (CASPT), Boston, 1997. Lecture notes in economics and mathematical systems, vol 471. Springer, Berlin, Heidelberg, pp 259–283 (cited on page 75)
22. Köhler E, Möhring RH, Skutella M (2009) Traffic networks and flows over time. In: Lerner J, Wagner D, Zweig KA (eds) Algorithmics of large and complex networks. Design, analysis, and simulation. Lecture notes in computer science, pp 166–196. ISBN: 978-3-642-02093-3. https://doi.org/10.1007/978-3-642-02094-0_9 (cited on page 76)

23. Lieberherr J, Pritscher E (2012) Capacity-restraint railway transport assignment at SBB-Passenger. In: Proceedings of the 12th Swiss transport research conference (cited on pages 76, 79)
24. Lusby RM, Larsen J, Ehrgott M, Ryan D (2011) Railway track allocation: models and methods. OR Spectr 33(4):843–883. https://doi.org/10.1007/s00291-009-0189-0 (cited on page 75)
25. Schlechte T, Borndörfer R, Erol B, Graffagnino T, Swarat E (2011) Micro-macro transformation of railway networks. J Rail Transp Plann Manage 1(1):38–48. https://doi.org/10.1016/j.jrtpm.2011.09.001 (cited on page 75)
26. Wohl M (1968) Notes on transient queing behavior, capacity restraint functions, and their relationship to travel forecasting. Pap Reg Sci 21(1):191–202. https://doi.org/10.1007/bf01952729 (cited on page 76)
27. Zhu E, Crainic TG, Gendreau M (2009) Integrated service network design in rail freight transportation. Research report CIRRELT-2009-45. CIRRELT, Montréal, Canada (cited on page 75)

Chapter 5
Robust Train Timetabling

Valentina Cacchiani and Paolo Toth

Abstract Nowadays railway systems are highly affected by disturbances, occurring in daily operations, and causing train delays and passenger inconvenience. Not only they negatively affect the passengers satisfaction, but they also cause additional operational costs, since the planned schedule needs to be modified in real-time. Train timetabling is a particularly critical phase in railway system management, since, in real-time operations, all the changes applied to the planned timetable impact on platform assignment, rolling stock circulation and crew scheduling. Therefore, in the strategic planning, it is an important issue to determine *robust timetables*, i.e., timetables that "perform well" under disturbances, avoiding delay propagation as much as possible. In this chapter, we present state-of-the-art methods that achieve robust timetables, and discuss their advantages and drawbacks.

V. Cacchiani (✉) · P. Toth
DEI, University of Bologna, Viale Risorgimento 2, 40136 Bologna, Italy
e-mail: valentina.cacchiani@unibo.it; paolo.toth@unibo.it

R. Borndörfer et al. (eds.), *Handbook of Optimization in the Railway Industry*, International Series in Operations Research & Management Science 268,
https://doi.org/10.1007/978-3-319-72153-8_5

5.1 Introduction

Train timetables are determined in a *planning phase*. Traditionally, the goal of this phase (called *nominal problem*) was to derive *efficient* train schedules, i.e., to have a service with high frequency and short waiting and travel times, while no importance was given to the effects that a "tight" schedule could have on real-time operations (i.e., long travel times due to delays, missed train connections). Recently, robustness has received increasing attention in the literature and in real-world applications for the optimization of a railway system. The aim of robustness is to determine timetables that "perform well" under disturbances, avoiding delay propagation as much as possible. It is crucial to provide a good quality service to the passengers, by running trains with high punctuality and short travel times, in order to attract the passengers to use this transportation mode. Railway networks are more and more utilized and traffic congestion inevitably causes delays, that not only affect the passengers, but also the freight trains. When a delay occurs, the planned train schedule can become infeasible and has to be changed in real-time, causing operational costs as well as inconvenience to the passengers. Furthermore, changes in the train schedule most likely imply changes to the platform assignment, rolling stock circulation and crew schedules.

For these reasons, train timetabling turns out to be one of the most critical phases in the optimization of a railway system. In addition, at an operational level, train delays often cause delay propagation, thus making the entire train schedule ineffective. *Robust train timetabling* calls for deriving, in the planning phase, train schedules that show a good trade-off between *efficiency* and *robustness*, so as to simplify and make smooth the actions to be executed in real-time, while maintaining a good quality service to the passengers. More precisely, robust train timetabling aims at determining timetables that, on one hand, optimize the nominal objective (e.g., minimize the total passenger travel time, or, in a competitive market, minimize the changes to the timetables requested by different train operators) and, on the other hand, minimize delay propagation that can occur at the operational level, while respecting all the nominal constraints (e.g., infrastructure capacity, minimum travel and dwell times, safety, connections, etc.).

In this chapter, we present state-of-the-art methods to derive robust train timetables. For a complete survey on the nominal and robust train timetabling problems, we refer the reader to Cacchiani and Toth [9]. In Sect. 5.2, we describe the Nominal Train Timetabling Problem (TTP) and the Robust Train Timetabling Problem (RTTP). Section 5.3 is devoted to present recent effective methods to solve RTTP. We conclude this overview with some comments on computational results in Sect. 5.4 and open perspectives in Sect. 5.5.

5.2 Problem Description

We first briefly introduce the nominal TTP, as its description is helpful for a better understanding of the robust TTP.

5.2.1 Nominal TTP

Traditionally, two versions of TTP have been studied, namely the periodic and the non-periodic problems, also known as cyclic and non-cyclic TTP, respectively. In the former, each train is operated in a cyclic way, based on a period of usually 1 h: this is very useful for the passengers, as they can easily memorize the train schedule. However, it is not always doable to build a periodic schedule, especially in a competitive market where several train operators access the same railway infrastructure. In this case, non-periodic TTP is the most appropriate version. Even though the latter is known as non-periodic, train schedules are repeated in the same way every day. For this reason, mathematical models for periodic TTP can also be applied to non-periodic TTP and vice versa. In both problems, given a railway network, characterized by a set of stations and a set of tracks connecting them, one needs to schedule arrivals and departures of trains at the stations, while respecting minimum travel and stopping times, headway times and track capacity constraints (related to overtaking and crossing of trains that must be avoided due to the structure of the railway network).

We report in Sects. 5.2.1.1 and 5.2.1.2, respectively, two formulations of the nominal TTP, that are later modified to take into account robustness issues. We refer the reader to Lusby et al. [29] and Cacchiani and Toth [9] for recent surveys on TTP, to Cacchiani et al. [12] for a tutorial on the non-periodic version of TTP and to Liebchen [26] for its periodic version.

5.2.1.1 Periodic TTP

The most well-known model for the periodic TTP is the Periodic Event Scheduling Problem (PESP) model, introduced by Serafini and Ukovich [35]. Let P be the cycle time (e.g., 1 h). Let $G = (N, A)$ be the so-called *constraint graph*, i.e., G is a directed graph with a node set N, corresponding to the *event* set (i.e., the arrivals and departures at the stations of the trains that have to be scheduled), and an arc set A, where arc (i, j) corresponds to a cyclic constraint imposed on events i and j and represents a *process*. For each event $i \in N$, we introduce an integer variable $v_i \in \{0, \ldots, P-1\}$, representing the time instant at which event i takes place. Each cyclic constraint deals with a pair of events $i, j \in N$, and with a time window, which imposes, respectively, lower and upper bounds l_{ij} and u_{ij} on the time interval between the two events. By introducing a binary variable p_{ij}, with $p_{ij} = 1$ if $v_j < v_i$ and $p_{ij} = 0$ otherwise, we can write the PESP model as follows:

$$\min F(v)$$

$$l_{ij} \leq (v_j - v_i) + Pp_{ij} \leq u_{ij}, \quad (i,j) \in A \tag{5.1}$$

$$v_i \in \{0,\ldots,P-1\}, \quad i \in N$$

$$p_{ij} \in \{0,1\}, \quad (i,j) \in A,$$

where $F(v)$ represents, for example, the passenger travel time (e.g. $F(v) = \sum_{(i,j)\in A}(v_j - v_i + Pp_{ij} - l_{ij})$. Constraints (5.1) are used to model the feasibility of the solution, i.e., they impose to satisfy minimum travel and stopping times, headway times and track capacity constraints, but also passengers transfers between connecting trains.

5.2.1.2 Non-periodic TTP

The formulation we describe for the non-periodic TTP was proposed in Caprara et al. [13], and is then enhanced to derive robust solutions in Cacchiani et al. [11]. This model is used in the context of a competitive market, in which several train operators ask for scheduling trains according to their preferred timetables, and the infrastructure manager has to change "as little as possible" these timetables to obtain an overall feasible schedule. If no feasible schedule can be derived, train cancellations are allowed.

Let T be the set of all trains to be scheduled. Let $G = (V,A)$ be a time-space graph: nodes in set V represent time instants at which some train can arrive at and depart from a station. In addition, V includes an artificial source node σ and an artificial terminal node τ. The arc set A is partitioned into sets $A^1,\ldots,A^{|T|}$, one for each train $t \in T$. These arcs represent either the travel or the stop of a train, or correspond to artificial arcs connected to σ and τ. Each arc $a \in A$ is assigned a profit p_a which takes into account the priority given to the train by the train operator, and the penalties due to potential timetable changes that need to be performed to obtain a feasible timetable. We introduce, for each train $t \in T$ and each arc $a \in A^t$ a binary variable x_a equal to 1 if, and only if, arc a is selected in an optimal solution. We denote by $\mathscr{C}$ the (exponentially large) family of maximal subsets C of pairwise arcs which are incompatible due to headway time or track capacity violation. The corresponding Integer Linear Programming (ILP) model reads:

$$\max \sum_{t\in T} \sum_{a\in A^t} p_a x_a \tag{5.2}$$

$$\sum_{a\in\delta_t^+(\sigma)} x_a \leq 1, \quad t \in T, \tag{5.3}$$

$$\sum_{a\in\delta_t^-(v)} x_a = \sum_{a\in\delta_t^+(v)} x_a, \quad t \in T, v \in V \setminus \{\sigma,\tau\}, \tag{5.4}$$

$$\sum_{a \in \mathscr{C}} x_a \leq 1, \quad C \in \mathscr{C}, \tag{5.5}$$

$$x_a \in \{0,1\}, \quad a \in A. \tag{5.6}$$

The goal is to maximize the sum of the arc profits, thereby minimizing the changes to the preferred timetables. Constraints (5.3) impose to choose at most one path, which corresponds to a timetable, for each train (note that train t is cancelled if the corresponding left-hand side of (5.3) is equal to zero). Constraints (5.4) require that the arcs are selected to form a path in graph G (i.e., a timetable) for each scheduled train. Finally, constraints (5.5) are used to ensure that the obtained train schedules respect the headway time and track capacity constraints.

As mentioned above, a mathematical model for the periodic TTP can also be adapted to the non-periodic TTP. This is what is done, for example, in Fischetti et al. [20], where the PESP model is adapted to take into account penalties for modifying the given preferred timetables provided by the train operators.

5.2.2 Robust TTP

Finding robust yet efficient solutions to optimization problems is a major practical issue. Roughly speaking, a TTP solution is considered to be robust if it avoids delay propagation as much as possible. A common and practical way to obtain robust timetables is to introduce, in the planning phase, *buffer times* that can absorb possible delays occurring at the operational level. Buffer times correspond to empty time slots, inserted in the schedule of the trains, to mitigate delay propagation. RTTP calls for determining *where* the buffer times should be inserted and *how long* they should be to guarantee a good trade-off between the nominal *efficiency* and the *delay resistance*.

The elements that come into play in the RTTP are the same as in the TTP. We are given a railway network, with a set of stations and tracks connecting them, and a set of trains to be scheduled on the network, while satisfying constraints on the travel and stopping times of the trains, as well as headway times between consecutive trains and track capacity constraints. As opposed to the TTP, the RTTP has two objectives that are in contrast with each other, namely efficiency and robustness of the solution.

Several works from the literature consider to have on input a feasible timetable for the trains and propose methods that allow to improve the timetable robustness (see, e.g., Kroon et al. [25], Fischetti et al. [20]). Other works, on the contrary, propose methods to build a robust timetable from scratch (see, e.g., Cacchiani et al. [11]). Robustness can be achieved by applying various techniques. In Sect. 5.3, we give an overview of the state-of-the-art methods.

5.3 Robustness in Train Timetabling

The seminal work presented in Soyster [36], which can be referred to as *strict robustness* and was later extended in Ben-Tal and Nemirovski [2], introduces a first approach to deal with the uncertain data present in a mathematical model. This method calls for determining a solution that is feasible for all the considered scenarios, with the goal of minimizing the worst-case performance of a solution. These methods tend to be overconservative, as they require to find a solution that is feasible for *all* the considered scenarios. In Ben-Tal and Nemirovski [3, 4], ellipsoidal uncertainties are considered in order to try to limit overconservative solutions. In Bertsimas and Sim [5, 6], a new concept of robustness is introduced. The uncertainties of the data are represented by letting each coefficient assume a value in an interval centered in its nominal value. The number of coefficients that can simultaneously take their worst-case value is limited: in particular, their method requires to define a robust model such that its optimal solution is feasible for every change of at most Γ_i coefficients in each row i of the constraint matrix. The parameters Γ_i are used to determine the quality of robustness of the solution: basically, if $\Gamma_i = 0$ then constraint i corresponds to the nominal constraint, while $\Gamma_i = n$ (where n is the number of variables) corresponds to the strict robustness described in Soyster [36]. Even though this method overcomes most of the drawbacks of the previous approaches, it is not suitable for the TTP, since most of the coefficients of the corresponding model are used to define the "structure" of the problem, and the number of uncertain coefficients in each row is very small, as discussed in Fischetti and Monaci [19].

Several models and algorithms, which are very effective for deriving robust solutions to the TTP, have been recently defined. In the following sections, we describe the most relevant ones. We start by describing, in Sect. 5.3.1, an alternative method to robust optimization, i.e., *stochastic programming*, applied to the TTP. New robustness concepts, namely *recoverable robustness*, *recovery to optimality*, and *light robustness*, presented for general Mixed Integer Linear Programming (MILP) problems, and effectively used for the RTTP, are described in Sects. 5.3.2, 5.3.3, 5.3.4, respectively. Finally, in Sect. 5.3.5, we describe a Lagrangian-based approach to derive good quality heuristic solutions to the RTTP, which also has the advantage of being a simple modification of an effective method for the nominal problem.

5.3.1 Stochastic Programming

One of the first methods used to improve the robustness of a timetable is based on the stochastic programming method [8]. In Kroon et al. [25], a two-stage stochastic model is applied to a given cyclic timetable of NS Reizigers, the main operator of passenger trains in the Netherlands, with the aim of minimizing the weighted delays of the trains. In the following, we assume that the period (cycle) is of 1 h. The first stage of the model consists of the timetabling part, while the second stage consists of a simulation part for evaluating the robustness of the timetable under construction. The key idea is to handle, in the second stage, a set of delay scenarios, and to determine, in the first stage, a timetable such that the average weighted sum of the delays in all these scenarios is minimized. Therefore, the first stage corresponds to the deterministic part of the model, while the second stage manages the stochasticity that depends on the specific scenario. Time supplements are variables, introduced in the first stage, that can be allocated to specific processes with the aim of absorbing disturbances occurring in the scenarios of the second stage.

The timetabling part of the model is very similar to the PESP model described in Sect. 5.2.1.1, but the cyclic orders of the trains are taken as fixed in the given timetable, i.e., variables p_{ij} are given parameters. In addition, the event time variables are not restricted to be in the interval $\{0,\ldots,P-1\}$, but can take any integer value: this is allowed to avoid restrictions for the event times (e.g., an event planned at time 0 could only move forward if the event times were restricted in $\{0,\ldots,P\}$). The constraints to impose the time duration of the processes can be expressed as follows:

$$l_{ij} + s_{ij} = v_j - v_i + p_{ij}P, \quad (i,j) \in A,$$

where s_{ij} represents the amount of time supplement that is allocated to process $(i,j) \in A$. As in the PESP model, an upper bound can be imposed on the process duration:

$$v_j - v_i + p_{ij}P \leq u_{ij}, \quad (i,j) \in A,$$

which is equivalent to limit the time supplement of this process. Since the event time variables are not restricted in the interval $\{0,\ldots,P-1\}$, but the timetable must be cyclic, the following constraints are added to the model:

$$0 \leq v_{e_2} - v_{e_1} \leq P-1,$$

where e_1 and e_2 are, respectively, the first and the last events in an hour on the same part of infrastructure. Finally, in order to appropriately allocate the time supplements, Q disjoint subsets $A_1,\ldots,A_Q$ of processes are selected, and each subset A_q is connected with a global amount of time supplement S_q to be allocated to the processes in A_q:

$$\sum_{(i,j)\in A_q} s_{ij} \leq S_q, \quad q = 1,\ldots,Q.$$

The simulation part of the model is based on R independent realizations, each corresponding to a single day, of the timetable operated under selected stochastic disturbances. Each realization covers H consecutive hours. It is assumed that the orders of the events are the same in the realizations as planned in the timetable, i.e., the simulation part of the model does not include traffic control decisions. A process

(i,j) with $p_{ij} = 0$ has $v_i < v_j$, thus it is planned within a single hour. On the contrary, a process (i,j) with $p_{ij} = 1$ has $v_j < v_i$: since it is clearly impossible that the process ends before it started, it is assumed that, if process (i,j) starts in hour h of realization r, then it ends in hour $h+1$ of the same realization. Let $\tilde{v}_{i,r,h}$ be the realized event time of event i of realization r in hour h, and $\delta_{ij,r,h}$ be the disturbance of process (i,j) of realization r in hour h. The following constraints are imposed to respect the minimum process times when disturbances occur:

$$l_{ij} + \delta_{ij,r,h} \leq \tilde{v}_{j,r,h+p_{ij}} - \tilde{v}_{i,r,h}, \quad (i,j) \in A, r = 1,\ldots,R, h = 1,\ldots,H. \tag{5.7}$$

Let E_d and E_a be, respectively, the departure and the arrival event sets. Furthermore, constraints are imposed that link the departure event times with the corresponding realized event times:

$$v_i + hP \leq \tilde{v}_{i,r,h}, \quad i \in E_d, r = 1,\ldots,R, h = 1,\ldots,H, \tag{5.8}$$

i.e., a realized event cannot take place before its planned time. In addition, constraints are used to compute the delays of arrival events in all the realizations:

$$\tilde{v}_{i,r,h} - (v_i + hP) \leq D_{i,r,h}, \quad i \in E_a, r = 1,\ldots,R, h = 1,\ldots,H,$$

where $D_{i,r,h}$ is a non-negative variable representing the delay of arrival event i of realization r in hour h. The average weighted sum (with weights w_i, $i \in E_a$) of the delays is minimized in the objective function of the model:

$$\min \sum_{i \in E_a} \sum_{r=1}^{R} \sum_{h=1}^{H} w_i D_{i,r,h} / (|E_a| RH). \tag{5.9}$$

Weight w_i can be chosen, for example, based on the number of passengers that reach their destination with the arrival event i. Note that, in the objective function (5.9), an approximation is used: instead of considering the expected weighted delays of the trains, a random sample of R independent vectors of disturbances is considered. We refer the reader to Fischetti et al. [20] and Meng and Zhou [30] for other applications of stochastic programming in the TTP. In particular, Fischetti et al. [20] proposes two alternative stochastic programming models, called "fat" and "slim", respectively. The fat model has one variable for each event in each delay scenario, thus it becomes very time consuming to solve. The slim model has just one copy of the original variables, and additional variables which take into account the unabsorbed delay and whose sum is minimized in the objective function. This model has much less variables than the fat one. In Meng and Zhou [30], a rolling horizon approach, based on stochastic programming, is developed.

One drawback of the Stochastic Programming methods is that they require to know probabilities on the delay scenarios, that are not easily available. In addition, the size of these models increases rapidly with the number of the considered scenarios. However, they have the advantage of embedding the evaluation of the realization timetables, and become suitable for the periodic TTP, in which the time horizon for the planned timetable is usually short.

5.3.2 *Recoverable Robustness*

Recoverable robustness was introduced in Liebchen et al. [27] and applied in the TTP context. It integrates the notion of robustness and recoverability (i.e., *delay management*) into a common framework. The key idea is that *recovery* actions can be used to recover (i.e., to make feasible) a plan through limited changes in every *likely* scenario. Therefore, besides the nominal problem and the set of likely scenarios, one needs to define a set of recovery algorithms. A solution is recovery-robust if, in all the situations that may occur in the set of considered scenarios, one can recover the solution, by means of one of the given recovery algorithms. This is a two-stage approach that overcomes the drawbacks of the stochastic programming and strict robustness methods: the former becomes quickly intractable for large-scale instances, while the latter is too conservative for the TTP.

In the TTP, typical recovery actions consist of delaying events or cancelling connections: the former causes delay propagation, while the latter limits it but affects the passengers. Other recovery actions correspond to cancelling train services or rerouting trains. We describe the model, presented in Liebchen et al. [27], which takes into account the recovery actions of delaying events and cancelling connections. The model considers the nominal problem, which is formulated by using the PESP model, a set of delay scenarios taking into account small disturbances, and limited recovery possibilities (delaying events, i.e., propagating delays through the network, and cancelling passenger connections to avoid delay propagation) that allow to make the scenarios feasible through a limited effort. The latter feature is what mainly distinguishes the recoverable robustness approach from the stochastic programming model proposed by Kroon et al. [25]. In particular, additional variables and constraints are introduced to model and limit the accepted recovery actions.

The authors consider the periodic version of the TTP and the PESP model (see Sect. 5.2.1.1). A set S of likely scenarios, defined by train delays, is considered. Similar to the stochastic programming model, described in Sect. 5.3.1, variables v_i $(i \in E)$, representing the event times of the planned timetable, and variables $\tilde{v}_{is}$ $(i \in E, s \in S)$, representing the event times of the realized timetable of a specific scenario s, are introduced. Besides the classical PESP constraints (5.1) that are used to model the feasibility of the planned timetable, constraints that ensure the timetable feasibility for all the scenarios in S are imposed (see (5.7)), as well as constraints that require the departure realized events to be not earlier than the planned ones (see (5.8)). In addition, constraints are imposed to model the recovery action of event delaying. In particular, let λ_1 and λ_2 be two variables to be minimized in the objective function. The constraints read as follows:

$$\sum_{i \in E_a} w_i(\tilde{v}_{is} - v_i) \leq \lambda_1 \qquad (5.10)$$

$$\tilde{v}_{is} - v_i \leq \lambda_2, \quad i \in E_a. \tag{5.11}$$

Constraint (5.10) limits the weighted sum of the delays of all the arrival events (with weights based on the number w_i of passengers that reach their final destination with the arrival event $i \in E_a$). Constraints (5.11) limit the delay for each arrival event separately, i.e. they ensure that no passengers will experience a delay larger than λ_2.

Let A_{TF} be the subset of transfer arcs, i.e., of the arcs modelling the transfer of passengers from one train to another one at the same station. To model the recovery action of cancelling connections, additional binary variables x^s_{ij} are introduced for each transfer arc $(i,j) \in A_{TF}$ and each scenario $s \in S$: variable x^s_{ij} assumes value 1 if transfer (i,j) is cancelled, and 0 otherwise. The following additional constraints are enforced:

$$\tilde{v}_{js} - \tilde{v}_{is} + Mx^s_{ij} \geq l_{ij}, \quad s \in S, (i,j) \in A_{TF},$$

with $x^s_{ij} \in \{0,1\}$, and with M a large positive constant. To limit connection cancellations, one can impose a maximum number of such cancellations, which can be weighted based on the number g_{ij} of passengers travelling along the arc. Let λ_3 be a variable to be minimized in the objective function. The constraints limiting connection cancellations can be written as:

$$\sum_{(i,j) \in A_{TF}} g_{ij} x^s_{ij} \leq g_3 \lambda_3.$$

The considered objective function

$$min \sum_{(i,j) \in A} g_{ij}(v_j - v_i) + g_1\lambda_1 + g_2\lambda_2 + g_3\lambda_3$$

calls for minimizing the weighted sum of the passenger travel times (i.e., it maximizes the efficiency of the solution), and *recovery costs*, represented by the recovery actions consisting of delaying events and canceling connections.

Recoverable robust approaches to the TTP are proposed in Caprara et al. [14], Cicerone et al. [15, 16], D'Angelo et al. [17]. In Caprara et al. [14], a recoverable robust method, in which the only recovery action consists of propagating the delay over the given network, is proposed. It is applied to the Train Platforming Problem (TPP) that considers a railway station and a set of trains whose entry and exit points at the station are specified, and calls for assigning each of these trains a stopping platform, an arrival path from its entry point to the platform and a departure path from the platform to its exit point. In Cicerone et al. [15] complexity results are derived for special classes of recoverable robust TTPs. In Cicerone et al. [16], recoverable robust models are presented for shunting and timetabling problems: in particular, the recoverable robust TTP on corridors is studied, and the price of robustness is analyzed. In D'Angelo et al. [17], the recoverable robust TTP on tree networks is studied.

The main advantage of the recoverable robustness is that it combines robustness and delay management, thus overcoming the drawbacks of strict robustness and stochastic programming. However, the recovery algorithms have to be included in the optimization model, and, therefore, only limited recovery actions can be taken into account.

5.3.3 Recovery-to-Optimality

The concept of *recovery-to-optimality* is proposed in Goerigk and Schöbel [22] (see also Goerigk and Schöbel [21]), and applied to the non-periodic TTP. The goal is to determine a solution that can be recovered to an *optimal* solution with small recovery costs. In particular, the goal is to minimize either the sum or the maximum of the recovery costs. To this aim a *distance* function, which represents the recovery costs, is defined. In the TTP, a distance function can be, for example, the increase in the passenger travel times from one timetable to another one.

Recovery-to-optimality is a two-stage approach that combines recoverable robustness and stochastic programming. In the first stage, a solution has to be found, and it can be recovered when the scenario realizes (second stage). With respect to recoverable robustness, in which a set of recovery actions are chosen, recovery-to-optimality replaces these actions with the notion of the distance function measuring the recovery costs. Furthermore, in recovery-to-optimality, an optimal solution, with respect to a feasible solution in recoverable robustness, has to be determined. Recovery-to-optimality has similarities with stochastic programming too: however, on one hand, no probability distribution is needed, and, on the other hand, the objective function is not the sum of the nominal objective and the recovery costs. Indeed, optimality of the second stage variables is imposed as a hard constraint.

In Goerigk and Schöbel [22], a setting similar to the one in Liebchen et al. [27] is considered: the nominal problem is formulated as an adaptation of the PESP model to the non-periodic case, and a set of scenarios corresponding to train delays are considered. However, the goal in Goerigk and Schöbel [22] is to minimize the maximum recovery costs to determine an optimal solution for all the considered scenarios instead of to just determine a feasible solution. This is achieved by minimizing, for all the scenarios, the maximum distance (recovery costs) of the optimal solution from the solution of the nominal problem.

Let $\mathscr{X}$ be the decision space, d the distance function, S the scenario set and $Opt(s)$ the set of optimal solutions to the problem corresponding to scenario $s \in S$. We use $x(s)$ to specify one optimal solution of scenario $s \in S$. The recovery-to-optimality model, when the goal is to minimize the worst case costs over all scenarios, reads:

$$\min \sup_{s \in S} d(x, x(s))$$

$$x(s) \in Opt(s), \; s \in S,$$

$$x \in \mathscr{X}. \tag{5.12}$$

Let $\bar{s}$ correspond to the nominal scenario and $\mathscr{F}$ to the set of feasible solutions. Constraint (5.12) can be replaced by $x \in \mathscr{F}(\bar{s})$ to guarantee nominal feasibility.

A sampling algorithm is proposed in Goerigk and Schöbel [22] to derive recovery-to-optimality robust solutions. A discrete set of scenarios $\{s^1,\ldots,s^N\} \subseteq S$ is considered and the optimal solution x^i for each scenario s^i $(i = 1,\ldots,N)$ is computed. Then, the solution $x^* \in \mathscr{X}$ minimizing the distance to the obtained optimal solutions $\{x^1,\ldots,x^N\}$ is determined by solving a location problem. In particular, if the goal is to minimize the worst case costs, one needs to determine x^* such that

$$\max_{i=1,\ldots,N} d(x^*,x^i)$$

is minimal.

In Goerigk and Schöbel [22], the non-periodic TTP is modelled by an event-activity network, i.e., a directed graph $G = (V,A)$. The node set V consists of departure and arrival events, while the set of arcs is used to represent either the travelling of a train from one station to another, or the stopping of a train at a station, or the transfer of passengers from one train to another at the same station, or the headway constraints between trains. This graph is very similar to the constraint graph presented in Sect. 5.2.1.1. The model used for the TTP is an adaptation of the PESP model to the aperiodic case.

A relevant difference that distinguishes recovery-to-optimality from other robustness methods for the TTP is that recovering a solution to optimality may also mean to let events take place *earlier* than in the nominal timetable. Clearly, rescheduling events in the timetable to occur earlier by changing, for example, the train departures cannot be done within a short-time horizon, because this would cause confusion for the passengers. Therefore, this approach is useful for long-term disruption scenarios. An example of practical application is provided in Goerigk and Schöbel [22], and reported here: the railway company plans a major construction area for modernizing some tracks, but it is not yet decided which part of the tracks should be under work. In particular, three different scenarios are possible. Therefore, the timetable to be determined should be as close as possible to each of the optimal ones in the three scenarios, in order to keep the recovery costs as small as possible.

5.3.4 Light Robustness

In Fischetti and Monaci [19], a general heuristic scheme for robustness, called *light robustness*, is proposed. Instead of requiring feasibility of the solution in all the considered scenarios (as in strict robustness), slack variables are introduced that allow to relax the feasibility constraints. The goal is to minimize the sum of the slack variables, while imposing a limit on the worsening of the nominal objective function value. Light robustness is applied to the TTP in Fischetti et al. [20]. The idea is to improve the robustness of the timetable while imposing a maximum increase of the solution cost with respect to the nominal one. This is achieved by imposing a constraint on the objective function value to be not larger than a given percentage of the nominal one, and by requiring a certain protection level, i.e., a buffer time, for the events (departures and arrivals of trains). This protection level is allowed to be violated and the goal is to minimize the total violation.

The authors adapt the PESP model to the non-periodic TTP. Similar to what is done in Kroon et al. [25], all the trains have to be scheduled and all the event precedences are fixed according to a given nominal timetable. The constraints used to improve the robustness of the timetable read as follows:

$$v_j - v_i + \gamma_{ij} \geq l_{ij} + \Delta_{ij},\ \gamma_{ij} \geq 0\ \forall (i,j) \in A$$

where variable v_i (with $i \in N$) represents the time instant of event i, l_{ij} (with $(i,j) \in A$) is a given parameter specifying the minimum time difference between the two consecutive events i and j, γ_{ij} and Δ_{ij} are, respectively, the slack variables and the required protection level parameters with respect to uncertainty in the process durations. Slack variables are used to relax the constraints on the solution feasibility under uncertainties. In particular, γ_{ij} takes a strictly positive value if the corresponding robust constraint is violated. The objective function of the light robust model then becomes:

$$\min \sum_{(i,j) \in A} \gamma_{ij}.$$

To find a good compromise between the robustness of the solution with respect to uncertainties and the quality of the solution with respect to the objective function (i.e., the efficiency of the solution), a constraint on the maximum worsening δ of the objective function value with respect to the optimal nominal solution value F^* is imposed:

$$F(v) \leq (1+\delta)F^*.$$

In this way, the light robust model calls for finding the most robust solution among those which are "not too far" from the optimal solution of the nominal problem.

Light robustness for the TTP is also applied in Goerigk et al. [23]: in particular, the authors consider robust *timetable information*, i.e., how to identify a path that will bring the passengers to the planned destination even in the case of delays. They analyze both the concepts of strict and light robustness. The latter calls for finding a path with "reasonable length", i.e. its length should not exceed the length of a nominal optimal path by "too much". The performance of strict and light robustness are evaluated according to the "price of robustness", showing that light robustness is more promising.

A slight extension of the light robustness approach is proposed in Liebchen et al. [28]: delay resistant timetables are computed by considering an objective function which takes into account the nominal objective and a simplified *delay management* objective, that counts the expected number of missed connections, while considering a strict no-wait policy and fixed driving times.

In Schöbel [33], the so-called generalized light robustness is introduced for continuous problems: it extends the concept of light robustness to arbitrary optimization problems and arbitrary uncertainty sets. In the generalized light robust counterpart formulation, the objective is to minimize a given norm of the grade of infeasibility of a solution. The original concept of light robustness turns out to be a special case of the generalized light robustness concept, in which the norm is chosen as the 1-norm of the slack variables. It is shown that the obtained robust counterpart is computationally tractable for several types of uncertainty sets including polyhedral and ellipsoidal sets. In addition, the author discusses how the concept of generalized light robustness can be used to control the trade-off between the nominal quality and the robustness of a solution, and shows how all Pareto solutions with respect to nominal quality and robustness can be determined.

The concept of light robustness is very effective in the TTP context. In Fischetti et al. [20], the light robustness method is compared with two stochastic programming models (fat and slim), described in Sect. 5.3.1 on the studied application of the non-periodic TTP. The light robustness approach turns out to be, on one hand, the fastest one and, on the other hand, accurate in terms of quality of the robust solutions obtained (it usually produces good results that are only slightly worse than those obtained with the slim model).

5.3.5 Lagrangian Robustness

The methods described in the previous sections are all based on the PESP model, directly applied to the periodic TTP or adapted to the non-periodic case. In this section, we present a method that is based on the model described in Sect. 5.2.1.2 for the non-periodic case. This method for deriving robust solutions to TTP is proposed in Cacchiani et al. [11]. Many of the approaches developed to introduce robustness lead to a significant change in the problem formulation with respect to the nominal case. This causes often major changes of the associated software and a much larger computational effort. On the contrary, the method proposed in Cacchiani et al. [11] is derived as a *simple modification* of an existing method for the nominal TTP (see Caprara et al. [13], Cacchiani et al. [10]). Despite its simplicity, it turns out to be very effective compared to other robust methods for the TTP.

This approach is related to bi-objective optimization Ehrgott [18]. Indeed, the objective function consists of two terms that appear to be in contrast: on one hand, the goal is to maximize the efficiency of a solution (i.e., the nominal objective) and, on the other hand, to maximize its robustness (i.e., the buffer times between pairs of departures or arrivals). The model proposed in Cacchiani et al. [11] consists of the model for the nominal TTP with a different objective function, namely a weighted sum of efficiency and robustness. The idea is to solve, in a heuristic way, the modified model by an iterative approach that improves efficiency in the first iterations and robustness in the following ones. A key feature of the method is that it derives many heuristic solutions, that can then be evaluated so as to choose the subset showing the best trade-off between efficiency and robustness. Therefore an approximated Pareto front is obtained for the bi-objective problem.

The method consists of a heuristic algorithm based on Lagrangian optimization: these algorithms approximately solve a Lagrangian relaxation of the studied problem through an iterative Lagrangian optimization scheme (e.g., a subgradient optimization procedure), and iteratively apply a heuristic algorithm driven by the Lagrangian costs in order to obtain a good quality feasible solution to the problem.

To deal with robustness a simple modification of this Lagrangian optimization scheme is applied. It consists of two features:

- the problem formulation is modified through the introduction of artificial parameters intended to "control" the solution robustness;
- the weight of the *control parameters* is changed during the Lagrangian optimization process to progressively increase the importance of robustness. This leads to the determination of a set of heuristic solutions that have a different trade-off between robustness and efficiency. This is another major advantage of this method, as it allows a decision maker to select among a large set of solutions.

The Lagrangian robustness method consists of inserting buffer times corresponding to longer stops of trains at the stations (train travel times are taken as fixed), so as to reduce delay propagation: short delays can be absorbed by these idle time slots and feasibility of the planned timetable can be retained. To this aim, an artificial *prize* is assigned to the arcs of the time-space graph that represent a longer stop of a train at a station with a stopping time not greater than a maximum given value M (15 min in Cacchiani et al. [11]). In this way, stops longer than the minimum ones will be inserted in the planned timetable, but efficiency will be also preserved. Efficiency corresponds to the maximization of the timetable profits, i.e., to the construction of timetables that are as close as possible to those preferred by the train operators. Other efficiency measures could also be considered. With respect to the light robustness approach, it is not necessary to fix the protection level a priori: indeed, this is determined on the fly by the approach itself. In addition, it is not imposed to have a maximum worsening of the solution value with respect to the nominal one: solutions with different levels of worsening are obtained by the Lagrangian robustness approach.

To deal with robustness, model (5.2)–(5.6) is modified by considering the new objective function

$$\max \sum_{t\in T}\sum_{a\in A^t} p_a x_a + F_k \sum_{t\in T}\sum_{a\in A^t} b^t_a x_a$$

and constraints (5.3), (5.4), (5.5), and (5.6), where F_k is a weighting factor dynamically-updated according to the iteration counter k, and $b^t_a \geq 0$ is a parameter giving an additional profit depending on the amount of buffer time associated with arc a for train t, corresponding to the stop of t at a station. Both F_k and b^t_a are the control parameters used to implement robustness. In particular, F_k depends on the subgradient iteration counter k: it starts with a small value (0.1) and then it is increased, in a non-linear way (up to 2.0), so as to concentrate on efficiency in the first iterations and on robustness later on (see Cacchiani et al. [11] for further details). Parameter b^t_a is used to determine where buffer times should be placed, i.e. which are the most critical bottlenecks (in space and time) of the timetable. In Cacchiani et al. [11], the optimal distribution of buffers proposed in Kroon et al. [25] is used to assign proper prizes to the arcs:

$$b^t_a := \min\{q^t_a, M\} \cdot (1 - e^{-\lambda p})(len(t) - p). \tag{5.13}$$

In (5.13), q^t_a is the number of minutes of buffer used by train t for arc a (i.e., the number of minutes for which t stops at the station corresponding to arc a above the minimum stopping time), M is the maximum value for considering the buffer a useful one, λ is a parameter (set to 3 in Cacchiani et al. [11]), $len(t)$ is the number of stations visited by train t, and p is the position of arc a for train t along its path. Roughly speaking, buffer times are favored especially in the "middle" of the train path: indeed, it turns out that buffer times may be unused in the very beginning of the train path because the probability to face a delay in the early sections is low, nor in its very end since it may be too late for the buffer times to be useful.

The Lagrangian robustness method is obtained as a simple modification of the Lagrangian heuristic algorithm for the nominal TTP. The Lagrangian-based heuristic algorithm for the nominal TTP proposed in Caprara et al. [13], which is taken as basis for the robust algorithm, relaxes constraints (5.5) in a Lagrangian way: the resulting Lagrangian problem calls for determining a set of maximum Lagrangian profit paths (i.e., timetables) for the trains, where the Lagrangian profit takes into account both the original profit and the penalties for the relaxed constraints (i.e., for the conflicting trains). Violated constraints of type (5.5) are dynamically added to the pool of active constraints. A subgradient optimization procedure is applied to derive near-optimal Lagrangian multipliers. A heuristic algorithm, combined with a local search procedure, is iteratively executed during the subgradient optimization procedure to obtain a good feasible solution to the nominal TTP.

As in the nominal algorithm, constraints (5.5) are relaxed in a Lagrangian way, and a subgradient optimization procedure is applied. Also in this case, the Lagrangian relaxation calls for determining a set of maximum Lagrangian profit paths for the trains, which now additionally takes into account, for each train and each arc corresponding to a train stop, the prize $F_k b^t_a$. At each iteration of the subgradient procedure, the same heuristic algorithm as in the nominal case, based on the new Lagrangian profits, is executed.

The Lagrangian robustness method derives many heuristic solutions, each characterized by an efficiency quality and a robustness level. Among the obtained solutions, it is useful to select "Pareto optimal" ones that have a different trade-off between robustness and efficiency. Thus, different efficiency thresholds (e.g., 99%, 95% and 90%) are considered with respect to the best nominal objective value.

Among the solutions that respect these efficiency values, the "most robust" ones are chosen, based on the following sum, defined in Fischetti et al. [20]:

$$\sum_{(i,j)\in E} \max\{0, w_{ij}(\Delta - (v_j - v_i - l_{ij}))\}, \tag{5.14}$$

which evaluates that a solution is robust if, for each pair of trains, there is "enough time" (at least Δ time units) between the departure/arrival time instants (v_j, $v_i \in E$) at each station that they both visit. In (5.14), E is the set of all the pairs of events occurring at the same station, $w_{ij} := w_i + w_j$ is the weight of the pair of events v_i and $v_j \in E$, where $w_i := (1 - e^{-\lambda p})(len(t) - p)$, already used in (5.13), and Δ is the protection level. For each efficiency threshold, the solution having the smallest value of (5.14) is selected. If no solution found is above the efficiency threshold, the one with the highest efficiency is selected. The selected solutions are then proposed to the decision maker. Other selection criteria are also possible, and the decision maker is allowed to choose among a set of different solutions.

Other bi-criteria approaches are proposed in Schlechte and Borndörfer [32], Schöbel and Kratz [34] and Bešinović et al. [7]. In Schlechte and Borndörfer [32], a model with exponentially many variables is proposed, including the bi-objective of efficiency (i.e., number of feasible scheduled trains) and robustness maximization. The model is solved by applying column generation and a hybrid method which combines a weighted sum and an ε-constraint approaches. In Schöbel and Kratz [34], Pareto optimal timetables are derived for the non-periodic robust TTP. In Bešinović et al. [7], a micro-macro hierarchical framework is developed to derive robust solutions for the periodic TTP. Microscopic and macroscopic models interact iteratively to generate a set of timetables that are feasible at the microscopic level. Then, a number of different delay scenarios are generated, each one characterized by a random delay for each train: on each of these scenarios, a local search procedure is applied, that tries to eliminate conflicts by retiming trains. The quality of each timetable is evaluated based on the nominal cost and on the robust cost, which takes into account the unresolved conflicts and the time to absorb the delays.

A main advantage of the Lagrangian robustness is that it has been derived as a simple modification of an existing method. Despite of that, as shown in Cacchiani et al. [11], the obtained solutions often dominate those computed by light robustness and the computing times are generally shorter. However, light robustness can impose an efficiency threshold to be achieved, while Lagrangian robustness allows the method itself to derive appropriate solutions, without guaranteeing to provide adequately efficient solutions. Another main advantage of the Lagrangian Robustness is that, in relatively short computing times, it obtains several solutions, among which a decision maker can choose the best one, also according to additional objectives.

5.4 Comments on the Computational Results

Although it is very hard to make a computational comparison of the described methods, since each one has been applied to different real-world instances, we have tried to analyze the advantages and drawbacks of each method. It is clear that there is no "best robust method", but each one can be appropriate for specific TTP applications, and all of them can be classified as state-of-the-art methods for the RTTP.

5.4.1 Validation Tool

A major issue of the RTTP is how to assess the quality of the determined robust solutions. In Stochastic Programming, validation is carried out inside the model itself, since the different timetable realizations are considered within the model. On the contrary, the authors of Fischetti et al. [20] developed an external *validation tool*, which is a simulation-based validation module that is independent of the optimization model itself. Its main advantage is that it can be applied to evaluate solutions determined with different methods and it can be used to compare them. The validation tool estimates the *total cumulative delay* of the trains, while assuming limited recovery actions in response to delays. In particular, the tool assumes that train cancellations are not allowed, and all the train precedences are fixed. It takes a TTP solution and considers one delay scenario at a time. It updates the given timetable to make it feasible under the occurred perturbation, and computes the total delay incurred. The average total delay across all the scenarios is finally computed, and used as a measure of the robustness of the solution. The validation tool consists of a Linear Programming model and is very similar to the simulation part of the Stochastic Programming model described in Sect. 5.3.1. It has the advantage of analyzing many different scenarios in short computing times. The validation tool is also used in Cacchiani et al. [11] to evaluate the robustness of the selected solutions, and to compare the results with those reported in Fischetti et al. [20].

Validation can also be achieved through optimization methods, such as delay management integer programming models (see, e.g., Schachtebeck and Schöbel [31], Bauer and Schöbel [1]), as in Liebchen et al. [28]. However, as observed in Fischetti et al. [20], the complexity of such models grows rapidly as soon as complex decisions can be made.

5.4.2 Real-World Instances

In this section we want to provide an overview of the real-world instances considered by the robust methods described in the previous sections. This is not meant to compare the methods each other. For each method, we specify the studied railway network (network or corridor), the size of the instances solved and the time horizon.

In Kroon et al. [25] (stochastic programming) instances, provided by NS Reizigers, representing the northern part of North-Holland in the Netherlands are studied. The timetable is cyclic with a cycle time of 60 min. Almost all train lines are operated twice per hour with a cycle time of 30 min. The authors consider 420 h of timetable operations split into $R = 20$ independent days of $H = 21$ h each.

In Liebchen et al. [27] (recoverable robustness), although models for the TTP are presented, instances of train platforming are tested. Recoverable robustness applied to the TTP is tested in Cicerone et al. [16]: the authors consider corridors of the Italian railway network with up to 54 stations and 291 trains in a time horizon of 1 day.

In Goerigk and Schöbel [22] (recovery-to-optimality) an instance based on the intercity train network of the German IC/ICE railway system is considered. The corresponding event-activity network contains 379 activities and 377 events. The time horizon is 8 h.

In Fischetti et al. [20] (light robustness) corridors of the Italian railway system are considered, with up to 48 stations and 127 trains, in a time horizon of 1 day.

In Cacchiani et al. [11] (Lagrangian robustness) corridors of the Italian railway system and realistic instances obtained by combining two or more instances are considered, with up to 102 stations and 1000 trains, in a time horizon of 1 day.

5.4.3 Practical Considerations

The application of optimization techniques to derive robust timetables is still very limited in railway companies. Traditionally, in practice, time supplements are added to the minimum travel times in the construction of the train timetables: this can be viewed as a sort of basic robust measure against delay propagation. However, time supplements are assigned to the trains independently of the timetables of the other trains and of the bottlenecks of the network. Optimization methods allow one to better allocate buffer times and to produce timetables that can better absorb delays. We report the main reasons why, in our opinion, these methods are still not applied in practice.

First of all, robustness is in contrast to efficiency in terms of capacity utilization of the railway network and of minimum travel times. Railway operators and infrastructure managers need to focus mainly on the income of the company and therefore to have a high utilization of the network. Consequently, robustness becomes a secondary goal. However, as a consequence of the high utilization of the network, delays are more frequent and punctuality plays a crucial role to convince passengers to use railways as a mode of transport. Moreover, the costs, such as crew salary or passenger reimbursement, due to the real-time rescheduling of the trains when delays occur, are very relevant. This means that we can expect robustness to become more and more relevant also from a practical point of view. This is also testified by the recent European Projects in which robustness in railways has been investigated (see, e.g., ARRIVAL [37] and ON-TIME [38]). In addition, robustness can be introduced when the timetables are constructed in the planning phase: therefore, the optimization methods can be easily applied, since the corresponding computing times do not need to be very short. Furthermore, embedding robustness in the planning process can also be achieved by simple modifications of existing methods Cacchiani et al. [11].

Robustness is a relatively recent concept and more results might be needed before effective research methods (even demonstrated on real-world instances) can become "real-life practice". In the past few years, several railway companies have started using more advanced tools for timetable planning, a topic that has been deeply studied in the literature since many years. Therefore, we will hopefully see a similar trend for what concerns robust timetabling.

Finally, it is not straightforward to assess the quality of the robust timetables. Indeed, the results are very much related to the considered delay scenarios. However, it is not always possible to have reliable historical data for generating appropriate delay scenarios and it is not easy to completely reproduce the real-life situations. Nevertheless, the methods described in this chapter show that significant improvements can be achieved on real-world instances and we believe that an effort should be made to make them applied in railway companies.

5.5 Conclusions and Open Perspectives

The methods described in the previous sections are very effective in producing robust timetables. They have all solved real-world instances of significant size and have reduced delay propagation. However, the application of optimization techniques to derive robust timetables is still very limited in railway companies. As mentioned in Sect. 5.4.3, this could be mainly due to the difficulty of showing the practical effectiveness of the methods. Beside making optimization approaches for robust train timetabling applied in real-life practice, further improvements can still be achieved in several directions. *Sustainability* of railway systems is one of the main factors that play a role nowadays in the achievement of an advanced railway system: it includes several aspects such as reducing environmental impact, improving energy efficiency (see, e.g., Goverde et al. [24] and ON-TIME [38]), becoming more customer-driven, maximizing the system capacity, etc. All these elements characterize the performance of a railway system and it would be important to take them into account when the timetabling problem is solved. All the described methods tackle the so-called *disturbances*, i.e., relatively small delays, as it is often over-conservative to consider large-scale disruptions, such as track unavailability or rolling stock breakdowns, when building robust timetables. However, in congested railway systems, these situations occur very commonly and it would improve the system reliability to take them into account already in the timetabling planning phase. Finally, the integration of more planning phases, such as train platforming, rolling stock circulation, crew scheduling, when determining the train timetables, would definitely improve the quality of the efficiency of the entire system.

References

1. Bauer R, Schöbel A (2014) Rules of thumb: practical online-strategies for delay management. Public Transp 6(1–2):85–105. ISSN: 1866-749X. https://doi.org/10.1007/s12469-013-0082-8 (cited on page 110)
2. Ben-Tal A, Nemirovski A (1998) Robust convex optimization. Math Oper Res 23(4):769–805. https://doi.org/10.1287/moor.23.4.769 (cited on page 98)
3. Ben-Tal A, Nemirovski A (2000) Robust solutions of linear programming problems contaminated with uncertain data. Math Program 88(3):411–424. https://doi.org/10.1007/PL00011380 (cited on page 98)
4. Ben-Tal A, Nemirovski A (2002) Robust optimization–methodology and applications. Math Program 92(3):453–480. https://doi.org/10.1007/s101070100286 (cited on page 98)
5. Bertsimas D, Sim M (2003) Robust discrete optimization and network flows. Math Program 98(1):49–71. https://doi.org/10.1007/s10107-003-0396-4 (cited on page 98)
6. Bertsimas D, Sim M (2004) The price of robustness. Oper Res 52(1):35–53. https://doi.org/10.1287/opre.1030.0065 (cited on page 98)
7. Bešinović N, Goverde RMP, Quaglietta E, Roberti R (2016) An integrated micro-macro approach to robust railway timetabling. Transp Res B Methodol 87:14–32. https://doi.org/10.1016/j.trb.2016.02.004 (cited on page 109)
8. Birge JR, Louveaux F (2011) Introduction to stochastic programming. Springer Science & Business Media, New York. https://doi.org/10.1007/978-1-461-40237-4 (cited on page 98)
9. Cacchiani V, Toth P (2012) Nominal and robust train timetabling problems. Eur J Oper Res 219(3):727–737. Feature clusters. ISSN: 0377-2217. https://doi.org/10.1016/j.ejor.2011.11.003 (cited on pages 94, 95)
10. Cacchiani V, Caprara A, Toth P (2010) Scheduling extra freight trains on railway networks. Transp Res B Methodol 44(2):215–231. https://doi.org/10.1016/j.trb.2009.07.007 (cited on page 106)
11. Cacchiani V, Caprara A, Fischetti M (2012) A Lagrangian heuristic for robustness, with an application to train timetabling. Transp Sci 46(1):124–133. https://doi.org/10.1287/trsc.1110.0378 (cited on pages 96, 97, 106, 107, 108, 109, 110, 111, 112)
12. Cacchiani V, Galli L, Toth P (2015) A tutorial on non-periodic train timetabling and platforming problems. EURO J Transp Logist 4(3):285–320. https://doi.org/10.1007/s13676-014-0046-4 (cited on page 95)
13. Caprara A, Fischetti M, Toth P (2002) Modeling and solving the train timetabling problem. Oper Res 50(5):851–861. https://doi.org/10.1287/opre50.5.851.362 (cited on pages 96, 106, 108)
14. Caprara A, Galli L, Stiller S, Toth P (2014) Delayrobust event scheduling. Oper Res 62(2):274–283. https://doi.org/10.1287/opre.2014.1259 (cited on page 102)
15. Cicerone S, D'Angelo G, Di Stefano G, Frigioni D, Navarra A (2009) Recoverable robust timetabling for single delay: complexity and polynomial algo-

rithms for special cases. J Comb Optim 18(3):229–257. https://doi.org/10.1007/s10878-009-9247-4 (cited on page 102)
16. Cicerone S, D'Angelo G, Di Stefano G, Frigioni D, Navarra A, Schachtebeck M, Schöbel A (2009) Recoverable robustness in shunting and timetabling. In: Ahuja RK, Möhring RH, Zaroliagis CD (eds) Robust and online largescale optimization. Springer, Berlin, pp 28–60. https://doi.org/10.1007/978-3-642-05465-5_2 (cited on pages 102, 111)
17. D'Angelo G, Di Stefano G, Navarra A (2009) Evaluation of recoverable-robust timetables on tree networks. In: Fiala J, Kratochvíl J, Miller M (eds) Combinatorial algorithms. Springer, Berlin, pp 24–35. https://doi.org/10.1007/978-3-642-10217-2_6 (cited on page 102)
18. Ehrgott M (2006) Multicriteria optimization. Springer Science & Business Media, Berlin. https://doi.org/10.1007/3-540-27659-9 (cited on page 106)
19. Fischetti M, Monaci M (2009) Light robustness. In: Ahuja RK, Möhring RH, Zaroliagis CD (eds) Robust and online large-scale optimization. Springer, Berlin, pp 61–84. https://doi.org/10.1007/978-3-642-05465-5_3 (cited on pages 98, 104)
20. Fischetti M, Salvagnin D, Zanette A (2009) Fast approaches to improve the robustness of a railway timetable. Transp Sci 43(3):321–335. https://doi.org/10.1287/trsc1090.0264 (cited on pages 97, 100, 104, 106, 109, 110, 111)
21. Goerigk M, Schöbel A (2010) An empirical analysis of robustness concepts for timetabling. In: Proceedings of the 10th workshop on algorithmic approaches for transportation modelling optimization, and systems (ATMOS 2010). OpenAccess series in informatics (OASIcs), vol 14. Schloss Dagstuhl, Wadern, pp 100–113. https://doi.org/10.4230/OASIcs.ATMOS.2010.100 (cited on page 103)
22. Goerigk M, Schöbel A (2014) Recovery-to-optimality: a new two-stage approach to robustness with an application to aperiodic timetabling. Comput Oper Res 52:1–15. https://doi.org/101016/j.cor.2014.06.025 (cited on pages 103, 104, 111)
23. Goerigk M, Schmidt M, Schöbel A, Knoth M, Müller-Hannemann M (2013) The price of strict and light robustness in timetable information. Transp Sci 48(2):225–242. https://doi.org/10.1287/trsc.2013.0470 (cited on page 105)
24. Goverde RMP, Bešinović N, Binder A, Cacchiani V, Quaglietta E, Roberti R, Toth P (2016) A three-level framework for performance-based railway timetabling. Transp Res C: Emerg Technol 67:62–83. https://doi.org/10.1016/j.trc.2016.02.004 (cited on page 112)
25. Kroon L, Maróti G, Helmrich MR, Vromans M, Dekker R (2008) Stochastic improvement of cyclic railway timetables. Transp Res B Methodol 42(6):553–570. https://doi.org/10.1016/j.trb.2007.11.002 (cited on pages 97, 98, 101, 105, 108, 111)
26. Liebchen C (2008) The first optimized railway timetable in practice. Transp Sci 42(4):420–435. https://doi.org/10.1287/trsc1080.0240 (cited on page 95)
27. Liebchen C, Lübbecke M, Möhring R, Stiller S (2009) The concept of recoverable robustness, linear programming recovery and railway applications. In: Ahuja RK, Möhring RH, Zaroliagis CD (eds) Robust and online

large-scale optimization. Springer, Berlin, pp 1–27. https://doi.org/10.1007/978-3-642-05465-5_1 (cited on pages 101, 103, 111)
28. Liebchen C, Schachtebeck M, Schöbel A, Stiller S, Prigge A (2010) Computing delay resistant railway timetables. Comput Oper Res 37(5):857–868. https://doi.org/10.1016/j.cor.2009.03.022 (cited on pages 105, 110)
29. Lusby RM, Larsen J, Ehrgott M, Ryan D (2011) Railway track allocation: models and methods. OR Spectr 33(4):843–883. https://doi.org/10.1007/s00291-009-0189-0 (cited on page 95)
30. Meng L, Zhou X (2011) Robust single-track train dispatching model under a dynamic and stochastic environment: a scenario-based rolling horizon solution approach. Transp Res B Methodol 45(7):1080–1102. https://doi.org/10.1016/j.trb.2011.05.001 (cited on page 100)
31. Schachtebeck M, Schöbel A (2010) To wait or not to wait-and who goes first? Delay management with priority decisions. Transp Sci 44(3):307–321. https://doi.org/10.1287/trsc.1100.0318 (cited on page 110)
32. Schlechte T, Borndörfer R (2010) Balancing efficiency and robustness–a bi-criteria optimization approach to railway track allocation. In: Ehrgott M, Naujoks B, Stewart TJ, Wallenius J (eds) Multiple criteria decision making for sustainable energy and transportation systems. Springer, Berlin, pp 105–116. https://doi.org/10.1007/978-3-642-04045-0_9 (cited on page 109)
33. Schöbel A (2014) Generalized light robustness and the trade-off between robustness and nominal quality. Math Meth Oper Res 80(2):161–191. https://doi.org/10.1007/s0018601404749 (cited on page 106)
34. Schöbel A, Kratz A (2009) A bicriteria approach for robust timetabling. In: Ahuja RK, Möhring RH, Zaroliagis CD (eds) Robust and online large-scale optimization. Springer, Berlin, pp 119–144. https://doi.org/10.1007/978-3-642-05465-5_5 (cited on page 109)
35. Serafini P, Ukovich W (1989) A mathematical model for periodic scheduling problems. SIAM J Discret Math 2(4):550–581. ISSN: 0895-4801. https://doi.org/10.1137/0402049 (cited on page 95)
36. Soyster AL (1973) Convex programming with set-inclusive constraints and applications to inexact linear programming. Oper Res 21(5):1154–1157. https://doi.org/10.1287/opre.21.5.1154 (cited on page 98)
37. Website of the ARRIVAL EU Project (June 2017). https://arrival.cti.gr (cited on page 111)
38. Website of the OnTime EU Project (June 2017). https://ontime-project.eu (cited on pages 111, 112)

Chapter 6
Modern Challenges in Timetabling

Laura Galli and Sebastian Stiller

Abstract Timetabling is a central step in the planning of public transport and important for the quality of service. Thereby, it also faces requirements like punctuality, cost efficiency, flexibility and minimization of travel time. We show the state-of-the-art techniques and their extensions to new challenges, in particular, multi-period timetables and robustness. We conclude with a case study from the Italian Railways that shows the effectiveness of our robustness methods.

L. Galli (✉)
Dipartimento di Informatica, Università di Pisa, Largo B. Pontecorvo 3, 56127 Pisa, Italy
e-mail: laura.galli@unipi.it

S. Stiller
Institut für Mathematische Optimierung, Technische Universität Braunschweig, Pockelsstr. 14, 38106 Braunschweig, Germany
e-mail: sebastian.stiller@tu-braunschweig.de

R. Borndörfer et al. (eds.), *Handbook of Optimization in the Railway Industry*, International Series in Operations Research & Management Science 268,
https://doi.org/10.1007/978-3-319-72153-8_6

6.1 Introduction

Timetabling is a planning phase of public transport usually between line planning and vehicle scheduling. Among its results one may find the arguably most prominent outcome of planning and optimization in public transport, namely, the timetable published to the users. In addition timetabling also determines the timing for events that are important for the technical functioning of the system. The assignment of trains to tracks in the stations can also be part of the timetabling problem.

Timetabling problems in public transport are mostly considered on a level of some abstraction. For example driving events are considered with fixed driving times. Acceleration and deceleration dynamics are not explicit in the optimization model. Further, for timetabling models that omit some details of the underlying infrastructure are used. These models are mesoscopic in the sense that they are still more detailed than macroscopic models that only have cities or stations as nodes of the networks.

Many real-world problems require events to be scheduled periodically. For these problems the events are given in groups together with a period. After the timing for the first event in a group is planned, it has to be repeated periodically for the other events in that group.

The underlying mathematical problem is of course not confined to public transport but occurs, e.g., in production planning or CPU scheduling.

While traditionally timetables have been constructed by human experts, today it is clear that timetabling is a task to be carried out by means of mathematical optimization. The main challenges for optimization in timetabling can be grouped into three areas:

- Versatility and comprehensiveness of the optimization model.
- Optimization with respect to disturbances in operations.
- Integration of timetabling with other planning steps.

In this chapter we give topical examples for the first two areas. In Sect. 6.2 we show how periodic timetabling can be extended to the multi-period case, and in Sect. 6.3 we present a quite general mathematical framework for robust timetabling. In Sect. 6.3 we also present a case study to test the effectiveness of our robust mathematical framework using real-world instances from the main Italian Railway Infrastructure Manager.

6.2 Periodic Timetabling with Multiple Periods

6.2.1 *Aperiodic Timetabling*

In timetabling we are given a set of n events for which we have to choose planned times $t_i, i \in [n]$. Typically in public transport, these are departure and arrival events at stations or specific points on the network, e.g., a switch or a section of a track.

The times must be chosen to respect a set of scheduling constraints and optimize an objective function. Typical constraints are headway constraints, minimum or maximum dwell times, and minimal driving times (minimal driving times are often specified as a percentage of the technical driving time, e.g., 107%. Confer [17] for details on technical and planned driving times.) Constraints can also refer to the maximal or minimal transfer time of passengers who make a connection in their travel. This bounds the time between arrival of one train and departure of another at the same station.

Usually, the objective is to minimize the sum of travel times of all passengers. In case, driving times are fixed, this amounts to minimizing the sum of the transfer and waiting times over all connections weighted by the number of passengers on a connection.

A second important objective is to minimize the rolling stock. In an indirect way this can be attained by including waiting times at stations and turn-around times at endpoints of a line in the objective to be minimized (cf. [8] for details).

In such a basic, aperiodic model all constraints and the objective can be expressed by bounds on differences of planned event times: $t_j - t_i \geq \ell_{ij}$. In such a simple timetabling model all constraints and the objective are linear in the variables t, and the problem can be solved easily, e.g., by a linear program.

Still, some constraints are naturally of an alternating form: Consider a headway constraint of 5 min between two trains A and B that pass the same switch. Then either train A has to clear the switch 5 min before train B enters it or train B has to clear the switch 5 min before train A enters it. Thus, there are two constraints that can be expressed in the form $t_j - t_i \geq 5$ of which exactly one must be fulfilled. As a consequence, even simple aperiodic timetabling models involve combinatorial decisions, by means of which they often pose NP-hard optimization problems.

A similar effect arises for transfer times. Connecting passengers will use the first train on a line they can connect to. But, before the timetable is fixed it is not clear which train on a line is first after passengers have arrived from the other line. Thus, the cost function also involves a combinatorial decision, namely, which train of the line to connect to.

6.2.2 Periodic Timetabling

Many railway systems and other public transport providers around the world operate periodic or almost periodic timetables. That means for a certain stretch of time during a day all events are planned to recur periodically or with a slight, negligible jitter.

Probably the most important reason for periodic timetabling is the dependability for the customer. Even in times where online information about transport is ubiquitous, it is a relieve to know that there is a bus every 4 min, so one can go ahead with one's errants without worrying about when to catch the bus.

Prima facie, periodic timetabling can be modeled by a small add-on to aperiodic timetabling. In such a model we roll out all the recurring events of the planned timespan as if they are independent aperiodic events, and construct the suitable aperiodic timetabling instance. We write t_{ai} as the time of the ith recurrence of the periodic event a and P_a for its period. Adding simple linear equalities enforces periodicity for the timing of related events t_{ai} for all i: $t_{a0} + i \cdot P_a = t_{ai}$.

This approach preserves the polynomial solvability of a linear program but contains a variable for each recurrence of a periodic event, which in some instances makes the model rather large.

Still, as most timetabling models require combinatorial decisions anyway, a more compact, combinatorial model stands to reason: the Periodic Event Scheduling Problem. The beauty of this model is that it covers decisions on which trains goes first on a switch, or to which train of a line one connects, and the periodicity of the events with the same, natural tool.

6.2.3 The PESP

The standard mathematical optimization model for periodic timetabling is called Periodic Event Scheduling Problem (PESP). It was introduced by Serafini and Ukovich in [15].

In the PESP we are given a global period P, a finite set of events $i \in V$, each of which periodically recurs after P time units, and a set of periodic, pairwise precedence constraints. The goal is to periodically schedule the events, i.e., to determine for each event $i \in V$ an offset π_i, meaning that for every integer k a realization of the periodic event i occurs at time $\pi_i + kP_i$. A pair of periodic precedence constraints for periodic events a and b has the form

$$\ell_{(a,b)} \leq \pi_b - \pi_a + k_{(a,b)} P \leq u_{(a,b)}.$$

A typical example of such a pair of constraints is the dwell time of a train at a station. Consider π_a as the periodic arrival time of a periodically recurring train and π_b as its periodic departure time.

For example, an ICE train from Berlin to Basel arrives at Braunschweig π_a minutes past the hour and departs towards Hanover π_b minutes past the hour. Or the subway line U2 in Berlin arrives at station *Stadtmitte* every 10 min, always π_a minutes past the full 10 min.

By the pair of constraints the input parameter $\ell_{(a,b)}$ models the minimum dwell time of the train at the station and $u_{(a,b)}$ a hard upper bound on this dwell time.

For a given solution (π, k) to a PESP instance the values π are called potentials and the values of the terms $x_{(a,b)} = \pi_b - \pi_a + k_{(a,b)} P$ are called periodic tensions. A PESP instance naturally gives rise to a directed graph $G(V,A)$ where potentials live on the nodes, and constraints and tensions live on the arcs. For example, the tension $x_{(a,b)}$ is on the directed arc from a to b. We assume w.l.o.g. that G is connected.

The objective function of a PESP is the minimization of a weighted sum of these tensions. For example, weighing the tensions on travel and transfer arcs with the number passengers projected to use it, yields a minimization of travel times.

Note, given the origins and destinations of all passengers, the number of passengers on a line and on a transfer still depends on the routes the passengers choose. This choice can change due to the timetable. Thus, strictly speaking the weights cannot be known a priori to the timetabling. This effect is often neglected, but poses a further challenge in integrated planning. In particular, for dense networks or systems with small periods, i.e., very frequent trains allowing passengers to choose different paths through the network, this is likely to be of importance.

In total a PESP instances reads as follows:

Definition 6.1. Given a digraph $G(V,A)$ together with two rational functions on the arc set, $\ell : A \to \mathbb{Q}$ and $u : A \to \mathbb{Q}$, a natural number $P \in \mathbb{N}$, and a cost vector $c \in \mathbb{Q}^{|A|}$, the following mixed integer program is called a *periodic event scheduling problem* (PESP):

$$\begin{aligned} \min \sum_{(i,j)=e\in A} & c_e(\pi_j - \pi_i + k_e P) \\ \ell_e \leq \pi_j - \pi_i + k_e P_e \leq u_e & \quad \forall e(=(i,j)) \in A \\ \pi \in \mathbb{Q}^{|V|}, k \in \mathbb{Z}^{|A|}. & \end{aligned}$$

The PESP has been shown to be NP-complete in [11, 13], even MAXSNP-hard in [8], and to have an unbounded Chvatal rank in [9].

The PESP model is very versatile. One can include certain rolling stock, crew scheduling, and line planning decisions into this timetabling model without compromising the characteristic structure of its IP-formulation, see [8]. Thus, it also represents a method for the third type of challenges managed in the introduction, namely, the partial integration of planning steps.

6.2.3.1 Solving the PESP

To solve the problems formulated as PESP three approaches are predominant. A classical approach for periodic timetabling is the CADANS algorithm based on the work of Schrijver and Steenbeck [14]. A second line of methods to solve large, real-world PESP instances are mathematical programming methods mostly based on special cuts and carefully chosen IP-formulations, see [8]. We will follow this line in some detail here. Recently, in [6] it has been proposed to formulate PESP for SAT-solvers to significantly increase the size of solvable instances.

To solve a PESP by integer programming methods such as cuts and strong formulations the following structural properties are pivotal. We will prove them later in a more general setting.

The first property is that there are optimal solutions with a tree structure, i.e., where k_a can be chosen equal to 0 on the arcs a of a spanning tree in the network.

In case G is a tree, solving the PESP instance is apparently trivial. The following generalizes this observation.

- Let (π,k) be a feasible solution with tension x. For any tree $T \subseteq G$ there is a solution (π',k') with the same tension x—thus maintaining feasibility and optimality—but having $k_a = 0$ for all $a \in T$.
- Let (π,k) be a feasible solution and the resulting tension x. For any spanning tree $T \subseteq G$ we can construct a feasible solution from x only in the following way: For an arbitrary node i set $\pi_i = 0$. Propagate the potentials π from node i through the tree T using the tension without period, i.e., setting $\pi_b - \pi_a = x_{(a,b)}$ iteratively along the tree arcs (a,b).

The second important concept are cycle bases. A cycle basis is a basis for the linear subspace spanned by the incidence vectors of cycles in the vector space $\mathbb{Q}^{|A|}$ spanned by the arc incidence vectors. Note that a cycle may have forward and backward arcs. For the latter the incidence vector of the cycle has entry (-1). We give some basic facts on cycle bases without proof.

A cycle basis is called integral, if all cycles are integer linear combinations of the elements of the basis. Given a spanning tree T in a graph G, for each non-tree arc $e = (a,b) \in G \setminus T$ there is a unique path from a to b in T. This path together with e form a so-called fundamental cycle $C(e,T)$. All $|A(G)| - |A(T)| + 1$ fundamental cycles of T together form a cycle basis, called the *strictly fundamental cycle basis* of T. Every strictly fundamental cycle basis is integral. For more details on cycle basis confer [7, 8].

Next, we arrive at a reformulation of the PESP, the *cycle basis formulation*. To this end we substitute the variables for the potentials π by those for the tensions, i.e., $\pi_j - \pi_i + k_a P = x_{ij}$. Summing these equalities along a directed cycle C yields a cycle equality:

$$\sum_{e \in C} x_e = k_C P$$

where $k_C = \sum_{e \in C} k_e$. Note, we can replace any arc $e = (a,b)$ by its inverted arc $e' = (b,a)$ and multiplying the corresponding constraints by (-1), without changing the PESP instance, effectively exchanging the lower bound ℓ_e and the upper bounds u_e for the inverted arc. Therefore, in case C is not directed, we can substitute it with a cycle where all arcs e in one direction of C are replaced by their inverted arc e'. Note this re-orientation of arcs might require to consider an arc with different orientations for different cycles.

A tension vector x fulfills these cycle constraints for all cycles of an integral cycle basis, if and only if it is the tension x of a solution (π,k) of the original PESP formulation. If in addition $x_e \in [\ell_e, u_e]$, then it is the tension of a feasible solution (π,k). In other words, for uniform modules any integral cycle basis gives rise to an equivalent IP-formulation. This cycle basis formulation has computationally proven significantly stronger than the original arc formulation, see [8].

Finally, one can use cycles in a PESP to arrive at a special kind of rounding cuts. Consider again a directed cycle C possibly using backward arcs e'. The following two inequalities are valid.

$$\left\lceil \frac{\sum_{e\in C} \ell_a - \sum_{e'\in C} u_a}{P} \right\rceil \le k_C \quad \text{and} \quad \left\lfloor \frac{\sum_{e\in C} u_a - \sum_{e'\in C} \ell_a}{P} \right\rfloor \ge k_C$$

These so-called Odijk inequalities [13] further improve the performance of the integer program. Such cycle inequalities have recently been shown to be separable in pseudo-polynomial time in [1]. That paper [1] also demonstrates the high practical value of the approach.

When solving PESP instances it is helpful to use Odijk inequalities together with a cycle basis formulation using a short integral cycle basis. The length of a cycle basis is the sum of the lengths of its cycles. Thereby, length of cycle C is the sum of the spread of the constraints over all arcs, i.e., $\sum_{e\in C} |u_e - \ell_e|$. Choosing a short cycle basis with respect to this length is a heuristic approach to strengthen the IP-formulation. The rational can be seen from the Odijk inequalities: the smaller the numerators, the larger the rounding effect after dividing by the period. Short cycle base formulations appear to be strong IP-formulations.

6.2.4 PESP with Multiple Periods

While in many cases periodic timetables operate by a single period P, there are relevant, real-world examples of systems with different lines operating at different periods. A public transport system, for example, is often comprised of lines with different periods, e.g., subways running every 10 min, some buses running every 30 min, others every 5 min, and regional trains every hour. Also traffic lights in the same urban area may well have different periods. Therefore, one has to plan events with different periods in the same network. The model for this we discuss here has first been presented in [12] under the name of Extended Periodic Event Scheduling Problem, short EPESP.

For two events a and b with periods P_a and P_b the closest follow-up of an event of b after an event of a is given by the minimum over $k_{(a,b)}$ of

$$\pi_b - \pi_a + k_{(a,b)} P_{(a,b)} (= x_{(a,b)}),$$

where $P_{(a,b)} = \gcd(P_a, P_b)$.

As an example consider a transfer from Line A to Line B at a station where Line A always arrives π_a minutes past the full period P_a and Line B always departs π_b minutes past the full period P_b. There is a train of Line A such that one can connect to Line B with only $x_{(a,b)}$ minutes.

Therefore, in multi-period event scheduling we focus on the following constraints:

$$\ell_{(a,b)} \le \pi_b - \pi_a + k_{(a,b)} P_{(a,b)} \le u_{(a,b)},$$

and the following objective

$$\min \sum_{(a,b)} c_{(a,b)}(\pi_b - \pi_a + k_{(a,b)}P_{(a,b)}).$$

Together we arrive at the following multi-period PESP model:

Definition 6.2. Given a digraph $G(V,A)$ together with two rational functions on the arc set, $\ell : A \to \mathbb{Q}$ and $u : A \to \mathbb{Q}$, a third natural valued function on the arcs, $P : A \to \mathbb{N}$, and a cost vector $c \in \mathbb{Q}^{|A|}$, the following mixed integer program is called a *multi-periodic event scheduling problem* (mPESP):

$$\min \sum_{(i,j)=e\in A} c_e(\pi_j - \pi_i + k_e P_e)$$

$$\ell_e \leq \pi_j - \pi_i + k_e P_e \leq u_e \quad \forall e(=(i,j)) \in A$$

$$\pi \in \mathbb{Q}^{|V|}, k \in \mathbb{Z}^{|A|}.$$

Note for a natural instance the arc-function $P : A \to \mathbb{N}$ is derived from a node-function $P' : V \to \mathbb{N}$ where P on an arc a from v to w is the greatest common divisor of the P' values of the nodes v and w, i.e., $P(a = (v,w)) = \gcd(P'(v), P'(w))$. But technically this will be irrelevant in what follows.

The minimum difference between two consecutive events of a and b is given by $x_{(a,b)}$. The maximum difference between two consecutive events of a and b is bounded from above trivially by the lesser of P_a and P_b. Still, within these bounds every multiple of $\gcd(P_a, P_b)$ plus $x_{(a,b)}$ will occur for some combination of trains from Line A and B. Thus, we do not optimize the transfer times for all passengers, but only for those on the best combination of trains.

One can argue that this kind of optimization is best possible for mixed periods. Put differently, the longer transfer times for other combinations of trains from lines A and B is a result of the incompatible periods rather than the timetabling for these periods. Nevertheless, two remarks of caution are in place.

First, in case passengers transfer twice on their journey, the mPESP may calculate transfer times not attainable with any combination of trains: Using a train of first line that has a fast connection to the second line may always lead to using a train on the second line, that has a bad connection to the third line.

Second, also upper bounds u only bound those pairs of events with smallest difference. This must be kept in mind, in particular, when technical constraints are modeled.

6.2.5 Solving PESP with Multiple, Nested Periods

The algebraic behavior of multi-period event scheduling is considerably less convenient. For PESP with a single period we stated that one can choose any spanning tree and set all k on its arcs to zero without changing the tensions of a feasible solution. In particular, each fundamental cycle has at most one arc with non-zero k.

With multiple periods it is easy to construct instances where more than one arc on a circle must have non-zero k to achieve a feasible solution. Some peculiarities of multiple periods have been collected in [5].

The key notion to recover the machinery of PESP solving for mPESP is that of a *sharp tree*, see [5].

Definition 6.3. For an mPESP instance and its graph G, a spanning tree T is called *sharp*, if for each of its fundamental cycles $C(e,T)$ the set of periods on the arcs of $C(e,T)$ has greatest common divisor equal to P_e, the period the non-tree arc e in $C(e,T)$.

In an mPESP with only one period, i.e., a standard PESP, all spanning trees are sharp trees.

The central lemma for mPESP and sharp trees is the following:

Lemma 6.1. *If T is a sharp tree spanning G and (π,k) a feasible solution with tension x, then there is a feasible solution with the same tension and $k=0$ on all arcs of T.*

Proof. Consider the linear (diophantine) equation system defining the tension vector x by equalities of the form $\pi_j - \pi_i + k_{(i,j)}P_{(i,j)} = x_{(i,j)}$. W.l.o.g. we can assume its matrix M to be ordered such that the following holds: The matrix M starts with the $n-1$ rows corresponding to the arcs in T. Restricted to the columns affecting the π variables, these rows form a lower triangular matrix (the first column omitted). The columns affecting k form a diagonal matrix.

Index the nodes according to their column and arcs according to their rows in M. For two nodes v and w denote by $\mathscr{P}(v,w,T)$ the unique path from v to w in T.

Let $(\pi,k)^0$ be some solution. We construct a solution $(\pi,k)^{n-1}$ over $n-1$ stages denoted $(\pi,k)^i, i \in \{1\dots n-1\}$. For each $i \in \{1\dots n-1\}$ successively with increasing row index we take four steps:

1. Set $k_i^i = 0$.
2. Re-establish the correctness of the ith equation by changing the node value π_i corresponding to $M_{i,i}$, the right most non-zero entry in the first $n-1$ columns, i.e., $\pi_i^i := \pi_i^{i-1} + M_{i,i}k_i^0 P_i$. Let $\ell(i)$ be minimal with $M_{i,\ell(i)} \neq 0$, i.e., the other node of arc i.
3. Propagate the new node value downwards along the tree. Formally: For all remaining three rows $t \in \{i+1\dots n-1\}$ successively with increasing row index set $\pi_t^i := \pi_t^{i-1} + M_{i,i}k_i^0 P_i$ in case $\ell(i) \notin \mathscr{P}(i,t,T)$.
4. Re-establish correctness (in arbitrary order) for the equations of non-tree arcs by adjusting their arc variables. Formally: For all $j \in \{n,\dots,m\}$ let $1 \le r < s \le n-1$ be the nodes of arc j, i.e., $M_{j,r}$ and $M_{j,s} \neq 0$. Set $k_j^i := k_j^{i-1} - \frac{M_{j,r}(\pi_r^i - \pi_r^{i-1}) + M_{j,s}(\pi_s^i - \pi_s^{i-1})}{P_j}$.

(We were a bit sloppy dropping exceptional handling of first row and column, and omitting when $\pi_j^i := \pi_j^{i-1}$ and likewise for k.)

Obviously, each $(\pi,k)^i$ fulfills all equalities. Observe, we touch the arc variable k_t of any tree row only in stage t. Therefore, $(\pi,k)^i$ the solution of any stage $i \in \{1\dots n-1\}$ has $k^i_t = 0$ for all $t \in \{1\dots i\}$. It remains to show for the non-tree arcs $j \in \{n\dots m\}$ that every k^i_j is an integer, in particular, that

$$P_j \mid \left[M_{j,r}(\pi^i_r - \pi^{i-1}_r) + M_{j,s}(\pi^i_s - \pi^{i-1}_s)\right].$$

For all nodes s we have $\pi^i_s - \pi^{i-1}_s = \mid k^0_i P_i \mid$. Now, distinguish whether $\ell(i) \in \mathscr{P}(r,s,T)$ or not. If $\ell(i)$ is in, so is i and the ith arc is on the fundamental cycle of j in T. Thus, $P_j \mid P_i$ by condition of the lemma and we are done.

In case, $\ell(i) \notin \mathscr{P}(r,s,T)$ both nodes are changed by the same value. Therefore, $M_{j,r}(\pi^i_r - \pi^{i-1}_r) + M_{j,s}(\pi^i_s - \pi^{i-1}_s) = 0$, which completes the proof.

From this lemma we get the following theorem of Galli and Stiller [5] stating that for sharp trees the cycle basis formulation is equivalent to the original arc formulation.

Theorem 6.1. *Let G be the graph of an mPESP instance. Assume there exists a sharp tree T spanning G and let $\mathscr{B}$ be the strictly fundamental cycle basis resulting from T. For an arc vector x the following three statements are equivalent:*

1. *The vector x is the arc tension of a node potential π, i.e., there is π and k such that $\pi_b - \pi_a + k_{(a,b)}P_{(a,b)} = x_{(a,b)}$.*
2. *The vector x fulfills the cycle equality for every cycle C in G, i.e., there is $k_C \in \mathbb{Z}$ such that $\sum_{a\in C} x_a = k_C \cdot \gcd(C)$.*
3. *The vector x fulfills the cycle equality for every cycle $C \in \mathscr{B}$.*

Proof. The inclusion $2 \Rightarrow 3$ being trivial, we show $1 \Rightarrow 2$ and $3 \Rightarrow 1$.

$1\Rightarrow 2$: According to (1) we have $x_{(i,j)} = \pi_j - \pi_i + k_{(i,j)}P_{(i,j)}$ for all arcs $(i,j) \in A$. Summing along a cycle C (multiplying the equation of a with (-1) for arcs a that lie in C contrary to its orientation) we get $\sum_{(i,j)\in C} x_{(i,j)} = k_{(i,j)}P_{(i,j)}$.

$3\Rightarrow 1$: Since T is sharp and x fulfills the cycle inequalities, we get from Lemma 6.1 that there is a node potential π corresponding to x with $k_a = 0$ for all arcs in T. Setting w.l.o.g. $\pi_s = 0$ for some node s we can thus propagate the x value along the sharp tree T and construct such a node potential π. It remains to show that for every non-tree arc $(i,j) \in A \setminus T$ there is $k_{(i,j)} \in \mathbb{Z}$ such that

$$\pi_j - \pi_i + x_{(i,j)} = k_{(i,j)}P_{(i,j)}. \tag{6.1}$$

Since $C := (\mathscr{P}(i,j,T),(i,j))$ is in $\mathscr{B}$, T is sharp, and $P_{(i,j)} = \gcd(C)$ Eq. (6.1) holds.

6.2.6 Finding Sharp Trees

In general, mPESP instances do not contain sharp trees (cf. [5] for examples). We say an mPESP instance has nested periods, if for each pair of periods $P_a < P_b$ we have $P_a|P_b$. For nested periods the graph G always contains a sharp spanning tree. This can be seen constructively by the following algorithm.

For a given mPESP instance let $r(P_e)$ be the number of *different* values for periods that are larger than P_e. Assign to the arcs in G an artificial cost function $h(e) = r(P_e) \cdot X + |u_e - \ell_e|$, where X is a sufficiently large number. Now construct a minimum spanning tree for G according to h. In other words, we construct a minimum spanning tree with respect to a lexicographic order, where the arcs with large periods are preferred as tree arcs.

It is well-known for a minimum spanning T that each non-tree arc e is among the most expensive in its fundamental cycle $C(e,T)$. By, construction and because the periods are nested, this implies that the periods on all tree arcs must be equal to P_e or a multiple of P_e. Therefore, the resulting spanning tree is sharp. (The reason for the secondary ordering by the difference of lower and upper bound will be discussed below.)

6.2.7 Accelerating mPESP Instances

With the above consideration one can use the same techniques to strengthen IP-formulations for mPESP as for simple PESP, provided the instances contains a sharp tree, e.g., if all periods are nested.

In the multi-period case Odijk inequalities have the following form:

$$\left\lceil \frac{\sum_{e \in C} \ell_a - \sum_{e' \in C} u_a}{\gcd(\{P_e : e \in C\})} \right\rceil \leq k_C \quad \text{and} \quad \left\lfloor \frac{\sum_{e \in C} u_a - \sum_{e' \in C} \ell_a}{\gcd(\{P_e : e \in C\})} \right\rfloor \geq k_C$$

Thus, to get a strong rounding, one would prefer cycle basis that have large $\gcd(\{P_e : e \in C\})$ and small length. To this end, in [5] the authors use strictly fundamental bases found by the above minimum spanning tree algorithm. The lexicographically posterior ordering leads the algorithm to find a sharp tree for which the strictly fundamental cycle basis heuristically is as short as possible.

The results reported in [5] show that these techniques for most instances yield a significant reduction of the running time. But, more important with the chosen cycle bases formulation several test instances could be detected as infeasible, for which a standard arc formulation of the mPESP exceed the assigned running time without result.

6.3 Delay-Robust Event Scheduling: A Mathematical Framework

As applied mathematicians we like to solve real world problems as well as develop new theories. Optimization problems that arise in transport are diverse, ranging from railways, airlines, maritime, urban transport and many others. These are complex problems, where resources need to be allocated to human activities that span a given planning horizon.

Modeling real-world problems is a challenging task, and the process usually goes through different, iterated, phases. Yet, one can think of two main "levels": (i) problem *abstraction*, and (ii) *mathematical* modeling. The choices made at the first level strongly influence the other, and vice-versa. The first step we take when abstracting is identifying the "relevant" *entities*, assessing their importance for the optimization objective, and determining the "relevant" *relations* between them, establishing which are closely connected to the optimization process. There is no given rule for such a process, and as much trouble must be taken in reasoning from facts to recognize the relevant ones, as it is in establishing their relation. The second step of abstraction, which naturally leads to mathematical modeling, is trying to remove any dependence on real world objects with which it was originally connected, and "generalizing" it so that it has wider applications or matching among other abstract descriptions of equivalent problems.

Nowadays, a rich literature provides us with a wide array of models, but systematic classification in the literature can make the process of abstraction harder. Indeed, over classification, and the forcing of facts into the compartments provided for them, whether they fit or not, may prevent us from reaching the right level of "generality". For this reason, the availability of *mathematical frameworks* can be very helpful. By mathematical framework, aka "class" of models, we mean a very general mathematical model. In other words, a mathematical framework originates from removing as much dependence as possible on the application objects in order to allow a broader matching with other applications. Defining a mathematical framework requires a certain amount of abstraction, or ought to be. Clearly, the more general the framework, the wider its applicability. Yet, the downside of too much generality, is the lack of characterization.

In this chapter, we present a mathematical framework called *delay-robust event scheduling*, that can be used to model a rather broad class of *scheduling* problems when subject to *time uncertainty*. Most railway optimization problems belong to this class, thus the interest for the book.

Our goal is twofold: on one hand we tell our experience, recollecting how the framework was devised and successfully applied to a real-world railway optimization problem; on the other we believe it could be a profitable and useful study to apply it to other optimization problem, for instance other railway applications, to avoid the *post hoc ergo propter hoc* argument. The section is organized as follows. In Sect. 6.3.1 we present a class of problems called *event scheduling* and a type of network model called *delay propagation* network. In Sect. 6.3.2 we recall the concept of *recoverable robustness* and derive a specific class called *delay-robustness*. Finally, in Sect. 6.3.3 we apply our framework to a real-world application taken from the optimization of a railway system, namely Train Platforming Problem.

6.3.1 Event Scheduling and Delay Propagation Networks

When modeling problems where *resources* are allocated *over time*, it is often worth paying attention to the way we are thinking about *events* in our application. The system is subject to *activities* that change its "status" and implicitly define the set of events of the optimization problem. Typically, the decision to be made is on how to perform such activities, i.e., which resources to use and the corresponding time instant. Clearly, resources can be human, machinery, financial etc., and more generally represent something that the system needs in order to achieve some goal. On one hand, a system tends to saturate or exceed the resources available. On the other, resources are costly, therefore the corresponding cost is minimized in the objective function. For example, in train timetabling problems, the main entities are trains and the corresponding activities are arrival, departure or stop. Resources can be platforms, tracks and other elements of the railway network. Therefore, events correspond to arrival and departure at/from the stations or specific points of the stations such as paths, stopping platforms, switches, crossings, etc.

Identifying events is key to verifying whether our analysis is sound or unsound, whether we have any faults of thinking, whether we are unconsciously making assumptions, which are either too specific or illogical, and seeing if we can improve the quality and effectiveness of our model. Yet, the quantity and variety of events in a real world application are of quite a different order from those that we actually need to optimize. Therefore, one needs to recognize the *activities* from which the events are generated, and whether those activities are subject to the optimization process. Clearly, this process may not be straightforward when activities are interrelated, making it hard to pin down the correct order of relation. In what follows, we are therefore considering problems under the headings of "activities" and "events", in their most general meaning, but without any other formal classification.

In other words, these are problems featuring time-related activities, called *events*, having some "properties" that we need to decide upon, namely, roughly speaking, "when" and "where" they take place, i.e., *scheduling*. This is what we call an *event scheduling* problem.

The main difference with respect to *standard scheduling problems* is twofold. First, in classical scheduling theory we often aim at minimizing the total completion time, so the main demand is on time. In event scheduling, this is not true in general. In fact, the objective may well be a weighted sum of the resources that are used and perhaps "how" they are used too (e.g., "when", penalties for resource "conflicts" etc.), so the stress is either on the resources or resources *and* "something more". Second, a distinguishing feature in event scheduling is that the constraints on resources can be different from the classical processing constraints of a standard scheduling problem.

There exist many classical frameworks to represent the scheduling of events. These frameworks rely on a graph representation (aka. *network*), and differ on the semantic of nodes and edges of such graphs. Indeed, *network flows* can be used to represent a surprisingly large number of applications in logistics and scheduling, including event scheduling problems. One of the archetype models are *time-space* (aka *time-expanded*) networks. This model is based roughly on formal logic, which is used unconsciously in much of our reasoning and it is meant to avoid the drawback of explicitly considering all possible connections between actions. The idea is to exploit the transitivity property of partial ordered sets which says that for activities i, j, k the following conclusion applies: $(i \prec j) \wedge (j \prec k) \Longrightarrow i \prec k$. A time-space network is a directed graph where each node n is associated to a specific time period $t(n)$. An arc from node n_1 to node n_2 exists if and only if $t(n_2) \geq t(n_1)$. In other words, the *flow* of events is only between nodes in the same time period or to nodes in a future time period. Clearly, a *flow* in the network represents a *sequence of events*. The structure of a *time-space* network depends on the application considered, yet, we can identify the following main features:

- time is divided into *discrete periods*
- each *vertex* in the network is characterized by a *spatial* index and a *time* index
- there are three main classes of arcs:
  1. *arc in space*, i.e., from one location to another with no change of timetables
  2. *arc in time*, i.e., from one time period to another, with no change of location
  3. *arc in time-space*, i.e., from one location in one time period to another location in another location in another time period

The main idea behind time-space network is replicating resources *over time and space*, each node of the network represents *what-where-when* and flow variables represent the choice of a particular scheduling of events, i.e., the corresponding problem can be modeled as some kind of network flow problem. Depending on the application, different side-constraints and objectives will be needed. Nevertheless, these problems are usually variants or generalizations of static single/multi-commodity network flow problems.

Another type of model used to represent the scheduling of events is the so called *time-activity* network. This is a different type of network, in that the resources are not replicated over time, because *time* is a "property" of the node. Each node represents *what-where* and time is a decision variable associated to the node (aka as *potential*). From a mathematical point of view, this is a different type of problem, whose structure strongly depends on whether the events are *periodic* or *non-periodic*. On one hand, non-periodic timetable problems are generally easier to solve because they can be modeled as some kind of shortest path problem. On the other, for fixed interval timetables, where all departure and arrival events must be repeated periodically, models are more complex, yet more realistic. Interestingly, *time-activity* networks were first defined for the *Periodic Timetabling* problem. Indeed, most results on periodic timetabling are based on the *Periodic Event Scheduling* model, because the associated periodic event activity network allows a flexible modeling of timetables in public transport. Another class of problems that can be easily modeled using time-activity networks is *Delay Management*. Note that periodic models are not well-suited for delay management problems, because delays are in general non-periodic.

Finally, we present our network model, called *delay propagation* network, because, as we will see, it is used to model delay propagation in event scheduling problems. We start observing that, in event-activity networks, a *schedule* assigns event *times* $\pi_e \in \mathbb{R}$ to all events $e \in E$. Hence, an activity $a : i \rightarrow j$ is a time consuming process between events i and j, which consumes the amount $\pi_j - \pi_i$ of time. We take this idea further, observing that "where" can represent a physical *resource* allocated to the corresponding event for a time period defined by the schedule. Thus, an event can be thought of as some time-space resource allocation for one of the entities involved in the system. In other words, a delay propagation network is a more "compact" representation with respect to the two previous ones, in that a node only represents *what* (i.e., the event), while *where-when* are both properties of the node itself. The optimization problem consists of deciding upon the property values of each event according to some objective function and constraints. A *schedule* assigns event *times* $\pi_e \in \mathbb{R}$ and resources $r_e \in R$ to all events $e \in E$. As a consequence, events are no more replicated over time nor space. The other observation is that an activity or arc $a : i \rightarrow j$ in the network is a resource and time consuming process between events i and j that can be subject to delay propagation. Delay propagation means that a change on time of node i may affect time of node j. Clearly, an arc of the network represents a *precedence relationship* between two events and therefore also represents how delays can propagate. Note that the arcs are *buffered* in the sense that they may absorb delays up to a given amount. The corresponding buffer values are uniquely determined by the property values of the events. The network structure is not given "a priori", but it is implicitly defined by the choices on the event properties (i.e., scheduling). In other words, the ability of the network to absorb delays depends on the property values (i.e., *where*, *when*) assigned to the events, i.e., the buffers are functions of the scheduling plan.

This kind of network is relevant in event scheduling problems for three reasons:

- It provides a *compact* representation. If resources and time periods are many, replicating nodes may lead to extremely large networks.
- It provides a *flexible* representation. If the scheduling constraints are complex, the network may be also complex.
- It provides a *general* representation. The network structure is independent from the time and space requirements (i.e., scheduling model), as it only represents precedence relationships.

6.3.2 Delay-Recovery Robustness

The delay-propagation network model naturally lends itself to a *recovery robust* model. The goal is to compute a plan (i.e., resources and time) for a given system. The system is subject to external disturbances that can propagate from one event to another generating delays. Yet, the plan may involve some buffer times, which can absorb delays and ought to be selected carefully as function of the plan. In robust modelling, the evaluation of the plan does not only involve the nominal cost (e.g., platform usage, nominal waiting times of passengers and/or trains), but also a *worst-case estimate* on its performance for some *scenarios*) (e.g., disturbances that are applied to it). Disturbances are of course propagated according to well-defined rules represented by the delay-propagation network. In this context, the importance of delay propagation is twofold. First, the amount of delay that is propagated provides a way to measure the performance of the system when subject to disturbances. Second, in a recovery-robust model the plan may "react" by some specific recovery actions and delay propagation is the simplest recovery action one can think of. Indeed, delay propagation is a recovery action as it involves a change in the original plan, more specifically a change in the *time* property of the scheduling plan. Thus, neither the scenarios nor the propagation themselves do appear in the plan explicitly, but rather are just used for its advanced evaluation. The plan only comes with the buffer times as well as with the most appropriate recovery actions. Delay propagation is the simplest recovery action. Of course one can think of more complex recovery actions, such as additional resources, or other rules like the possibility of breaking a transfer, inverting train order, changing track or platform. All this can be expressed using a delay propagation network. Clearly, the more complex the rules, the more complex the network. In what follows, we will restrict ourselves to the simplest recovery action (i.e., delay propagation) that corresponds to the simplest recovery-robust model.

Delay-Recovery Robustness is based on the *recoverable robust* paradigm and moves from the so called *network buffering* approach, both presented in [10, 16]. Recoverable robustness joins *classical robust optimization* and *2-stage stochastic programming*. In particular, we focus on a class of problems aimed at scheduling events with time-distance requirements on the allocated resources, thus forming a precedence relationship network, that we call *delay propagation* network. From a recoverable robustness point of view, the *recovery action* is represented by *delay propagation*, and our goal is to minimize the total delay propagation among all the scheduled events. Indeed, a recovery action is a way to modify the nominal solution in order to restore feasibility after a scenario realization. Minimizing delay propagation means optimally distributing buffers on the network arcs. Indeed buffer times can absorb delays, which otherwise would spread through. A good schedule concentrates the buffers on strategic points allowing limited worst-case propagation. The problem of distributing buffers in a given network to minimize the maximal total delay that can be caused by initial disturbances was first presented in [3, 4, 10, 16] as the robust *network buffering* problem, which is a special case of recoverable robustness. The framework we present here is more complex, as the definition of the buffers is not direct but *implicit* in the definition of the nominal solution. Indeed, buffer values are not variables themselves, but depend on the properties of the nodes, i.e., on the scheduling. Formally, the framework can be defined as follows:

- A set $\mathscr{E}$ of *events* to be scheduled, i.e., for which we have to define *when* and/or *where* they will take place with respect to an *ideal timetable* $\mathscr{I}$.

- Denoting by F the set of *feasible* schedules of events $\mathscr{E}$, i.e., the feasible region, a *nominal problem* of the form $\min\{c(x) : x \in F\}$ aims at scheduling the set of events $\mathscr{E}$ so as to minimize a suitable nominal objective function $c : F \to R$.
- A set $\mathscr{S}$ of possible *scenarios*, where each scenario $s \in \mathscr{S}$ is defined by the external *disturbance* $\delta_e^s \geq 0$ on the ideal timetable $\mathscr{I}$, assigned to each event $e \in \mathscr{E}$ and is represented by vector $\delta^s \in R_+^{|\mathscr{E}|}$.
- Clearly, the events are not independent from each other, because different events may be scheduled on the same or *incompatible* resources (in different time instants), causing a disturbance on one event to propagate to others. We represent this using a *delay-propagation network* $N = (\mathscr{E}, A)$, with $(e', e) \in A$ if a delay $d_{e'}$ on event e' may propagate to a delay d_e on event e.
- Given $N = (\mathscr{E}, A)$ and arc $(e', e) \in A$, if the two events e' and e are scheduled on the same or incompatible resources, delay propagation may happen. Delay propagation not only depends on *where*, but also *when* events are scheduled, since a (some) delay may be *absorbed* if the events are sufficiently *detached* in time. Mathematically, a delay $d_{e'}$ on event e' may propagate to a delay d_e on event e according to the relation $d_e \geq d_{e'} - b((e', e), x)$, where $b((e', e), x)$ is a *buffer time* function $b : A \times F \to R_+$, depending on the scheduling of the two events e' and e in the solution $x \in F$.

Note that buffer values are "implicit", in the sense that they are function of the scheduling choice. Buffers also represent the link between scheduling (nominal model) and delay propagation (delay-recovery model), when the system is subject to disturbance. Since buffer times model the absorption of delays, we assume $b((e', e), x)$ to be infinite in case the events e' and e do not interact in the solution x.

The network N is not necessarily acyclic (i.e., the order of the events may not be fixed a priori), because it represents delay propagation *before* the "real" timetable (scenario) is realized. The "real" timetable corresponds to the ideal timetable $\mathscr{I}$ *after* the disturbances δ^s take place according to some scenario s. Finally, $\mathscr{S}$ is not necessarily finite nor countable.

We denote by d_e^s the delay of event e for scenario s, which depends on the buffer times (and hence on the solution $x \in F$). From a robust point of view, our objective is to minimize the *cumulative delay* over all events associated with scenario s given by $\sum_{e \in \mathscr{E}} d_e^s$.

A rather straightforward goal of the *delay-recovery* counterpart of an event scheduling problem minimizes the nominal objective function $c(x)$ as well as the maximum cumulative delay over all scenarios in $\mathscr{S}$. The maximum cumulative delay is a standard objective that can be easily expressed as a linear program, as shown below. Yet, the framework allows one to use more complex objectives, such as the expected delay or other delay functions (e.g., Kleinrock) which come at the cost of more complex models.

The delay-robust problem can then be formulated as:

$$\min c(x) + D, \tag{6.2}$$

subject to

$$x \in F, \tag{6.3}$$

$$D \geq \sum_{e \in \mathscr{E}} d_e^s, \qquad s \in \mathscr{S}, \tag{6.4}$$

$$d_e^s \geq \delta_e^s, \qquad e \in \mathscr{E},\ s \in \mathscr{S}, \tag{6.5}$$

$$d_e^s \geq d_{e'}^s - b((e',e),x), \qquad (e',e) \in A,\ s \in \mathscr{S}, \tag{6.6}$$

Here, vector x represents the variables of the nominal problem. D is the maximum cumulative delay over all scenarios, and d_e^s is the delay of event e for scenario s, which depends on solution x according to $b((e',e),x)$, i.e., the buffer time of arc (e',e) for solution x. Note that the buffer times do not depend on the scenario, but only on solution x. Constraints (6.4) define the value of variable D, while constraints (6.5) and (6.6) express the fact that the delay of event e is the maximum between the external disturbance δ_e^s and the delay propagated from each event e' such that $(e',e) \in A$, which is partly absorbed by buffer $b((e',e),x)$. We stress again that since our recovery action consists in delay propagation, "delay" variables are in fact "recovery" variables. As already pointed out, delay variables are used to measure the performance of the system as well as to express the system reaction when subject to disturbance. Note that the *disturbance* δ_e^s is a parameter, hence an input that only depends on the chosen scenario. The *delay* d_e^s, instead, is a variable whose value depends on how the external disturbances are propagated, which in turn depends on the chosen scheduling plan (i.e., buffers $b((e',e),x)$).

6.3.2.1 The Scenario Set

Our scenario set is based on the following observation. For any scenario $s \in S$, the total amount of external disturbances is bounded. This is formalized by $\sum_{e \in \mathscr{E}} \delta_e^s \leq \Delta$, for a given parameter Δ. In other words, the set of scenarios is defined by the vectors:

$$\{\delta^s \in R_+^{|\mathscr{E}|} : \|\delta^s\|_1 \leq \Delta\} \tag{6.7}$$

This clearly leads to *uncountably* many scenarios, which can however be handled easily using the proposition below.

Proposition 6.1. *For the delay-robust problem* (6.2), (6.3), (6.4), (6.5), *and* (6.6), *the* compact *scenario set defined by* (6.7) *is equivalent to the* finite *scenario set S defined by the vectors*

$$\{\delta^s \in \{0,\Delta\}^{|\mathscr{E}|} : \|\delta^s\|_1 = \Delta\}, \tag{6.8}$$

which contains only $|\mathscr{E}|$ scenarios.

The reader can refer to [3] for a proof. The importance of this proposition lays in the fact that despite considering implicitly uncountably many scenarios, we can construct a compact model that considers only the $|\mathscr{E}|$ scenarios in which the maximum disturbance Δ is applied to *one single* event. This is clearly due to the *worst-case* nature of a robust model. Our method applies to any set of scenarios that defines a polytope, by restricting attention to one scenario for each vertex of the polytope. If the polytope is a simplex, as in the current version, we simply get as many scenarios as events. Otherwise, the number of vertices may grow exponentially, as it happens, e.g., if we put different upper bounds on the delays of the events.

6.3.3 A Real-World Study: Delay Robust Platforming

The Train Platforming Problem (TPP) belongs to the class of event scheduling problems since it involves both a decision on resources (railway station elements such as platforms and paths), as well as time. TPP is a special case of Train Timetabling. One is given a major railway station and a set $\mathscr{T}$ of trains whose arrival and departure times and entry and exit points at the station are specified. TPP calls for assigning each of these trains a *stopping platform*, an *arrival path* from its entry point to the platform and a *departure path* from the platform to its exit point, as well as the corresponding times. Indeed, one can modify, a.k.a. *shift*, the arrival and departure times within a given range. Note that a platform consists of a single track inside the train station where the train stops. The platform does not include sequences of tracks, (aka as *paths*) that the train uses to enter or leave the station. To guarantee the existence of feasible solutions, we allow the trains to be assigned to *dummy* platforms (i.e., platforms which do not exist in the considered station). Any train assigned to a dummy platform is considered as a *non-scheduled* train leading to a practically infeasible solution.

In the application we consider, an explicit list of arrival and departure paths is given, along with the associated *occupation time*, i.e., the time taken to travel along them (possibly dependent on the train), and a list of incompatible path pairs, representing the fact that the two paths physically intersect each other. Note that path incompatibility is static and not time-expanded. In other words, paths are physically incompatible because they share common elements (e.g., tracks, switches, etc). Two trains cannot occupy incompatible paths for a time window of duration larger than a so-called *conflict threshold*. Also, no two trains can be simultaneously present at the same platform. Finally, after a train leaves the platform, a minimum so-called *headway time* (possibly dependent on the platform and the two trains considered) must elapse before another train arrives at the platform.

6.3.3.1 Events and Delay-Propagation Network

A scheduling plan for TPP looks as follows. For each train $t \in \mathscr{T}$ the nominal solution defines (i) the instant in which t occupies (i.e., starts to occupy) its arrival path, (ii) the instant in which t frees (i.e., ends to occupy) its arrival path, which coincides with the instant in which t occupies its platform, (iii) the instant in which t frees its platform, which coincides with the instant in which t occupies its departure path, and (iv) the instant in which t frees its departure path. All these "events" in principle may be subject to external disturbances, which however in practice occur essentially only on two of them, namely (1), in case the train arrives late at the station (i.e., disturbance on event of type (i)), and (2), in case the platform operations take longer than required (i.e., disturbance on event of type (iii)). In other words, it is widely realistic to assume that the travel times along the arrival and departure paths are not affected by external disturbances, but of course may be affected due to propagation. Accordingly, to model TPP within our framework, the following two events are associated with each train $t \in \mathscr{T}$: *arrival* a_t, meaning that the train occupies a given arrival path at a given instant, and *departure* p_t, meaning that the train frees its platform at a given instant. This defines $\mathscr{E}$.

The associated *delay-propagation network* $N = (\mathscr{E}, A)$ consists of the following arcs in A joining events of the same train $t \in \mathscr{T}$:

- (a_t, p_t), meaning that a delay in a_t may cause a delay in p_t, depending on the travel time along the arrival path and the stopping time at the platform in the nominal solution and the minimum possible values of these two times.

Moreover, there are the following arcs in A joining events of two trains $t', t \in T$:

- $(a_{t'}, a_t)$, meaning that a delay for t' in occupying its arrival path may cause a delay for t in occupying its arrival path, in case the arrival times of t' and t are sufficiently close and t' and t are scheduled on incompatible arrival paths or on the same path;
- $(a_{t'}, p_t)$, meaning that a delay for t' in occupying its arrival path may cause a delay for t in freeing its platform, in case the arrival time of t' and the departure time of t are sufficiently close and t' and t are scheduled on incompatible arrival and departure paths, respectively;
- $(p_{t'}, a_t)$, meaning that a delay for t' in freeing its platform may cause a delay for t in occupying its arrival path, in case the departure time of t' and the arrival time of t are sufficiently close and t' and t are either scheduled on the same platform, or on incompatible departure and arrival paths, respectively;
- $(p_{t'}, p_t)$, meaning that a delay for t' in freeing its platform may cause a delay for t in freeing its platform, in case the departure times of t' and t are sufficiently close and t' and t are scheduled on incompatible departure paths or on the same path.

The values of the buffer time for each arc joining the events of two distinct trains are uniquely defined by the way in which these two events are scheduled in the nominal solution, as the following examples show.

Example 6.1. Assume that in the nominal solution trains t' and t stop at the same platform, event $p_{t'}$ is scheduled at 10:00, and event a_t at 10:04 with 3 min to travel along the arrival path, so that t occupies the platform at 10:07. Moreover, assume that the headway time that must elapse between t' freeing the platform and t occupying it is 2 min and that the 3-min travel time for t along its arrival path is fixed. If the departure path of t' and the arrival path of t are compatible, the buffer time $b((p_{t'}, a_t), x)$ is equal to 5 min, as a delay of up to 5 min on $p_{t'}$ does not affect a_t, whereas a larger one does.

Example 6.2. Assume that in the nominal solution trains t' and t use respectively a departure path and an arrival path that are incompatible. Event $p_{t'}$ is scheduled at 10:05, and event a_t at 10:00 with 3 min to travel along the arrival path, so that t can free the arrival path at 10:03. If the conflict threshold is 0, the two trains cannot be on the two paths at the same time, because the paths are incompatible. So the buffer time $b((a_t, p_{t'}), x)$ is equal to 2 min and a delay of up to 2 min on a_t does not affect $p_{t'}$, whereas a larger one does.

6.3.3.2 The Overall Delay-Robust Model and Its Solution

The nominal TPP objective function considered in [2] was derived after long discussion with the practitioners and had quite a few terms. So for the robust TPP we focus on the minimization of the total cumulative delay D assuming $c(\cdot)$ identically null. Our experiments show that omitting this term $c(\cdot)$ in the objective function of the robust model provides solutions whose quality in terms of duration of the path conflicts is comparable to that of the nominal ones, where $c(\cdot)$ is present.

As in [2], we represent the nominal solutions by associating *patterns* with trains. The concept of pattern is used to encapsulate all the information about the resources that are assigned to a train and the corresponding time. A pattern is a vector of properties and consists of arrival path, departure path, platform, arrival time and departure time. Specifically, for each train $t \in \mathscr{T}$ there is a collection $\mathscr{P}_t$ of patterns, each representing a feasible scheduling of both events a_t and p_t illustrated above. This has the advantage of encoding within the notion of patterns the constraints that relate these two events.

The side constraints relating the patterns are then imposed as generic incompatibilities between pattern pairs, defining an associated *pattern-incompatibility* graph having one node for each train-pattern pair (t, P), with $P \in \mathscr{P}_t$, and an edge joining each pair (t_1, P_1), (t_2, P_2) of incompatible patterns. Two patterns are deemed incompatible if this implies occupying the same platform at the same time, or also using routes that intersect at the same time or too close in time with respect to the conflict threshold. As to the definition of the buffer times, let $\varphi(e', P', e, P)$ denote the value of the buffer time $b((e', e), x)$ in case events e' and e are scheduled respectively according to patterns P' and P.

The overall delay-robust problem can be formulated by using the following variables. For each $t \in \mathscr{T}$ and $P \in \mathscr{P}_t$, variable $x_{t,P}$ is equal to 1 if train t is assigned pattern P, 0 otherwise. Moreover, for each $t \in \mathscr{T}$ and $s \in \mathscr{S}$ let $d^s_{a_t}$ and $d^s_{p_t}$ be the two continuous variables expressing the delay of events a_t and p_t in scenario s.

Letting $\mathscr{K}$ denote the collection of the maximal cliques of the pattern-incompatibility graph, the formulation reads:

$$\min D$$

subject to

$$\sum_{P \in \mathscr{P}_t} x_{t,P} = 1, \qquad t \in T, \tag{6.9}$$

$$\sum_{(t,P) \in K} x_{t,P} \leq 1, \qquad K \in \mathscr{K}, \tag{6.10}$$

$$x_{t,P} \in \{0,1\}, \qquad t \in T,\ P \in \mathscr{P}_t, \tag{6.11}$$

$$b((e',e),x) = \sum_{P' \in \mathscr{P}_{t'}} \sum_{P \in \mathscr{P}_t} \varphi(e',P',e,P)\, x_{t',P'}\, x_{t,P}, \tag{6.12}$$

and to (6.4), (6.5), and (6.6). Note that (6.3) is now given by (6.9), (6.10), and (6.11). Constraints (6.12) can be linearized as shown in [2] so that the overall model is a *Mixed-Integer Linear Program* (MILP).

We tested our framework on real-world instances from Rete Ferroviaria Italian, the main Italian Infrastructure Manager. This is the same benchmark data used in the deterministic study reported in [2], which considers the stations of Palermo Centrale, Genova Piazza Principe, Bari Centrale and Milano Centrale. The scenarios are defined by (6.8) with $\Delta = 30\,\text{min}$, i.e., a delay of 30 min on the arrival of a train at the station or on the departure of a train from its platform. The results show that the reduction of the total propagated delay with respect to the nominal solution is significant, ranging between 12% and 29% for Palermo Centrale, between 19% and 35% for Genova Piazza Principe, between 0% and 23% for Bari Centrale, and between 1% and 44% for Milano Centrale, with an average of 17%. Therefore, the new method finds solutions in which the total propagated delay is significantly smaller than the one of the previous "nominal" solutions.

6.4 Conclusion

Timetabling is a central step in railway planning. With state-of-the-art methods one can find robust solutions for large, real-world instances. Timetabling methods such as the PESP can incorporate aspects of other planning stages. For multiple periods within the same transport system planning becomes possible, provided the used periods are somewhat harmonized.

Train timetabling can also be viewed as an event scheduling problem. We proposed a framework to compute the delay-robust counterpart of event-based optimization problems. Such a framework consists of the original (nominal) optimization problem linked to a delay-propagation network. The generality and simplicity of the delay-propagation model allows one to attach it to the nominal formulation together with linking constraints to translate nominal solutions into network buffers. Our case study shows the effectiveness of the approach on real-world instances.

References

1. Borndörfer R, Hoppmann H, Karbstein M (2016) Separation of cycle inequalities for the periodic timetabling problem. In: Sankowski P, Zaroliagis CD (eds) 24th annual European symposium on algorithms, ESA. LIPIcs, vol. 57. Schloss Dagstuhl, Wadern, pp 21:1–21:13. ISBN: 978-3-95977-015-6. https://doi.org/10.4230/LIPIcs.ESA.2016.21 (cited on page 123)
2. Caprara A, Galli L, Toth P (2011) Solution of the train platforming problem. Transp Sci 45(2):246–257. https://doi.org/10.1287/trsc.1100.0366 (cited on pages 137, 138)
3. Caprara A, Galli L, Stiller S, Toth P (2014) Delayrobust event scheduling. Oper Res 62(2):274–283. https://doi.org/10.1287/opre.2014.1259 (cited on pages 132, 135)
4. Galli L (2011) Combinatorial and robust optimisation models and algorithms for railway applications. 4OR 9(2):215–218. https://doi.org/10.1007/s10288-010-0138-4 (cited on page 132)
5. Galli L, Stiller S (2010) Strong formulations for the multi-module PESP and a quadratic algorithm for graphical diophantine equation systems. In: de Berg M, Meyer U (eds) Algorithms - ESA 2010, 18th annual European symposium, Liverpool, September 6–8, 2010. Proceedings, part I. Lecture notes in computer science, vol 6346. Springer, Berlin, pp 338–349. ISBN: 978-3-642-15774-5. https://doi.org/10.1007/978-3-642-15775-2_29 (cited on pages 125, 126, 127)
6. Großmann P, Hölldobler S, Manthey N, Nachtigall K, Opitz J, Steinke P (2012) Solving periodic event scheduling problems with SAT. In: Jiang H, Ding W, Ali M, Wu X (eds) Proceedings of the 25th international conference on industrial engineering and other applications of applied intelligent systems. Lecture notes in computer science, vol 7345. Springer, Berlin, pp 166–175. ISBN: 978-3-642-31086-7. https://doi.org/10.1007/978-3-642-31087-4_18 (cited on page 121)
7. Kavitha T, Liebchen C, Mehlhorn K, Michail D, Rizzi R, Ueckerdt T, Zweig KA (2009) Cycle bases in graphs characterization, algorithms, complexity, and applications. Comput Sci Rev 3(4):199–243. https://doi.org/10.1016/j.cosrev.2009.08.001 (cited on page 122)
8. Liebchen C (2006) Periodic timetable optimization in public transport. PhD thesis, TU Berlin. dissertation.de. ISBN: 978-3-86624-150-3 (cited on pages 119, 121, 122)
9. Liebchen C, Swarat E (2008) The second Chvatal closure can yield better railway timetables. In: Fischetti M, Widmayer P (eds) Proceedings of the 8th workshop on algorithmic approaches for transportation modeling, optimization, and systems (ATMOS 2008), OASICS, vol 9. Schloss Dagstuhl, Wadern. https://doi.org/10.4230/OASIcs.ATMOS.2008.1580 (cited on page 121)

10. Liebchen C, Lübbecke M, Möhring R, Stiller S (2009) The concept of recoverable robustness, linear programming recovery, and railway applications. In: Ahuja RK, Möhring RH, Zaroliagis CD (eds) Robust and online large-scale optimization. Springer, Berlin, pp 1–27. https://doi.org/10.1007/978-3-642-05465-5_1 (cited on page 132)
11. Nachtigal K (1996) Cutting planes for a polyhedron associated with a periodic network. Technical Report, DLR (cited on page 121)
12. Nachtigall K (1996) Periodic network optimization with different arc frequencies. Discret Appl Math 69(1–2):1–17. https://doi.org/10.1016/0166-218X(95)00073-Z (cited on page 123)
13. Odijk MA (1996) A constraint generation algorithm for the construction of periodic railway timetables. Transp Res B Methodol 30(6):455–464. https://doi.org/10.1016/0191-2615(96)00005-7 (cited on pages 121, 123)
14. Schrijver A, Steenbeck A (1994) Dienstregelingontwikkeling voor Railned (Timetable construction for Railned). Technical Report, C.W.I. Center for Mathematics and Computer Science, Amsterdam (cited on page 121)
15. Serafini P, Ukovich W (1989) A mathematical model for periodic scheduling problems. SIAM J Discret Math 2(4):550–581. ISSN: 0895-4801. https://doi.org/10.1137/0402049 (cited on page 120)
16. Stiller S (2008) Extending concepts of reliability. PhD thesis, TU Berlin. https://doi.org/10.14279/depositonce-2093 (cited on page 132)
17. UIC (2000) Timetable recovery margins to guarantee timekeeping–recovery margins. ISBN: 2-7461-0223-4 (cited on page 119)

Chapter 7
Railway Track Allocation

Gabrio Caimi, Frank Fischer, and Thomas Schlechte

Abstract This chapter addresses the classical task to decide which train runs on which track in a railway network. In this context a track allocation defines the precise routing of trains through a railway network, which usually has only a limited capacity. Moreover, the departure and arrival times at the visited stations of each train must simultaneously meet several operational and safety requirements. The problem to find the "best possible" allocation for all trains is called the *track allocation problem* (TAP). Railway systems can be modeled on a very detailed scale covering the behavior of individual trains and the safety system to a large extent. However, those

---

G. Caimi
SBB AG, Bern, Switzerland
e-mail: gabrio.caimi@sbb.ch

F. Fischer
University of Kassel, Kassel, Germany
e-mail: frank.fischer@uni-kassel.de

T. Schlechte (✉)
LBW Optimization GmbH & Zuse Institute Berlin, Berlin, Germany
e-mail: schlechte@lbw-optimization.de

R. Borndörfer et al. (eds.), *Handbook of Optimization in the Railway Industry*,
International Series in Operations Research & Management Science 268,
https://doi.org/10.1007/978-3-319-72153-8_7

microscopic models are too big and not scalable to large networks, which make them inappropriate for mathematical optimization on a network wide level. Hence, most network optimization approaches consider simplified, so called macroscopic, models. In the first part we take a look at the challenge to construct a reliable and condensed macroscopic model for the associated microscopic model and to facilitate the transition between both models of different scale. In the main part we focus on the optimization problem for macroscopic models of the railway system. Based on classical graph-theoretical tools the track allocation problem is formulated to determine conflict-free paths in corresponding time-expanded graphs. We present standard integer programming model formulations for the track allocation problem that model resource or block conflicts in terms of packing constraints. In addition, we discuss the role of maximal clique inequalities and the concept of configuration networks. We will also present classical decomposition approaches like Lagrangian relaxation and bundle methods. Furthermore, we will discuss recently developed techniques, e.g., dynamic graph generation. Finally, we will discuss the status quo and show a vision of mathematical optimization to support real world track allocation, i.e. integrated train routing and scheduling, in a data-dominated and digitized railway future.

7.1 Introduction

A fundamental problem in railway planning is to assign each train a route in the railway network together with precise arrival and departure times at each station and major junction the trains visit. The railway network has only limited capacity, e.g. each physical track can only be occupied by one train at the same time and each station can only handle a limited number of trains simultaneously. The problem to find the "best possible" allocation of tracks for all trains taking the limited network capacity into account is called the *track allocation problem* (TAP).

Railway safety systems operate on the same principle all over the world. A train has to reserve infrastructure blocks (certain parts of a physical track) for some time to pass through. The situation that two trains reserve the same block of the infrastructure within the same point in time is called *block conflict.* State-of-the-art simulation systems provide accurate estimations of *running times* and *blocking times* with respect to a precise microscopic model, see Sect. 1.3.1 in Chap. 1 and Sect. 2.2.1 in Chap. 2.

Not so many years ago, timetable creation and so track allocation was a pure political and rigid manual process where only in exceptional cases the usage of software tools or mathematical models was exploited. The current practice in many European countries is a functional segregation due to the liberalization into railway passenger operators, freight operators, and railway infrastructure managing companies. The infrastructure managers are responsible for the safe operation of the timetable and have to integrate the requests of different operators simultaneously. The liberalization is one important booster of the application of mathematical models and optimization techniques in the railway industry—simply because railway operating companies are more and more confronted with competition, cost pressure, and ensuring of operational excellence.

In this chapter, we focus on the strategical or tactical planning task to produce reliable *macroscopic track allocations* years or months before operations. The available granularity and the major objective of this optimization problem depends on the taken perspective which can vary from public authorities, governments, or railway infrastructure operators to train operating companies for passengers or freight transport.

Nowadays, several commercial simulation tools exist that consider railway systems on a microscopic scale covering the behavior of trains and the safety system to a large extent, see Chap. 1. However, representing the safety layer of a railway system within a microscopic simulation tool is still a challenging goal in a currently ongoing standardization and harmonization process in the (European) railway industry. Those microscopic models of the railway system are already very large even for very small parts of the network. The reason is that all signals, incline changes, and switches around a railway station have to be modeled to allow for precise running time calculations of trains.

There are microscopic models that describe the railway system extremely detailed and thorough. A major strength of microscopic models is that almost all technical details and local peculiarities are adjustable and can be taken into account, e.g., the rolling stock characteristics. Thus, microscopic models provide the advantage that the calculation of the running times of the trains can be reasonably accurate. Nevertheless, microscopic models are inappropriate for mathematical optimization because of the size and the high level of detail.

Hence, most optimization approaches consider simplified, so called macroscopic, models. In such models major parts of the topology and layout of real stations are aggregated to single nodes with capacities. These capacities define the maximal number of trains that can operate at the same time within the station or at one gate of the stations, e.g., the number of platform tracks from the north inbound routes. Consecutive block segments where no overtaking of trains can happen are aggregated to single tracks connecting the station nodes. Moreover, the safety system in macroscopic models is handled by so called *minimum headway times* which restrict the departures of succeeding trains on these tracks.

The challenging part is to construct a reliable and condensed macroscopic model for the associated microscopic model and to facilitate the transition between both models of different scale. In Sect. 7.2 we will briefly discuss approaches to handle the interaction between microscopic and macroscopic models.

Section 7.3 will present classical optimization models in case of macroscopic railway problem formulations. Furthermore, we will highlight optimization algorithms and recent state-of-the-art techniques in Sect. 7.4. Finally, Sect. 7.5 provides a vision of mathematical optimization to support real world train routing and scheduling in the future.

7.2 On Microscopy and Macroscopy

In the following section we will discuss railway network and operation models of different scale. The level of detail of a railway infrastructure or operation model to consider depends on the quality and accuracy requirements for generating appropriate results and, of course, on the availability and reliability of the data. For long term and strategicplanning problems high accuracy data is often not manageable, might

Table 7.1: Comparison between the microscopic and the macroscopic railway model

Macroscopic element	Microscopic counterpart
General train types	Subset of (individual) train routes
Stations	Unified connected subgraph of the microscopic network
Tracks (connecting different stations)	Unified consecutive block sections, i.e., a path in the microscopic network
Running times on tracks for traintypes (in Δ)	Running times on block sections for individual routes (in δ)
Headway times on tracks for pairs of train types (in Δ)	Blocking times on sections for routes (in δ)

not exist, or cannot be provided on time without causing expenditure, e.g., [26].

Railway traffic is a high-grade complex technical system, which can be modeled in various detail. Microscopic data is for example incline, acceleration, driving power, power transmission, speed limitations, and signal positions. In particular, on a microscopic level the infrastructure consists of small elements, so called block segments. Each block segment is occupied by at most one train at the same time, see the time-space diagram on the left hand side of Fig. 7.1. The time period, when a train is physically traversing a block section, is called *running time*. Depending on the certain attributes of the train, e.g., length and speed, the segment is blocked for other trains for a certain amount of time, the *blocking time*. Hence, railway capacity not only consists of a space dimension, i.e., which are the physical infrastructure elements, but also of a time dimension composed of train movements, i.e., occupation or blocking times.

Optimization on a microscopic level, as already mentioned, is still inconceivable due to the enormous size and granularity of the data. Even more, it is not needed because the decision to run a train or to let a train wait can be done on a coarser—*a macroscopic*—level that could be based on the evaluations in a fine—*a microscopic*—level. For example, all macroscopic running times can be deduced by microscopic simulation data, assuming a standard acceleration and braking behavior of the given train and its weight. Thus, all relevant switches, inclines, curves or other velocity impacts are considered implicitly. Nevertheless, some aggregation error in particular due to the different scales of times will be introduced. Note that microscopic data is given in a finer granularity, e.g., the time discretization is often in seconds (δ) in contrast to macroscopic data which is in general assumed to be discretized in minutes (Δ). In the work [24] a definition of macroscopic elements and their original microscopic counterparts can be found. Table 7.1 lists those elements with respect to the railway safety system and the railway infrastructure resource consumption.

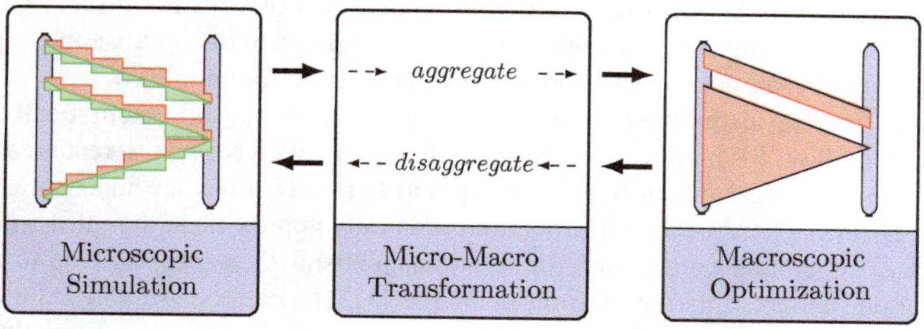

Fig. 7.1: Idealized closed loop between railway models of different scale for railway track allocation by Schlechte [24]

The literature has suggested a number of top-down approaches, e.g., the authors of [7] propose a comprehensive multi-level framework for railway scheduling starting with a high-level commercial description of intended train services and aiming at the generation of a conflict-free microscopic trains schedule as the final outcome. In order to handle large microscopic networks an a priori decomposition into condensation and compensation zones was introduced and used in [5]. Depending on the level of the intended train service, parts of the microscopic network are either classified as condensation zones where the available capacity is saturated and scarce or classified as compensation zones where capacity is available.

In [10] an optimization approach that iteratively solves a macroscopic delay management model first, and a microscopic train scheduling model afterwards, was developed. In a top-down approach to model railway systems, a coarse model of the entire system is first formulated, specifying but not detailing any "real" sub-systems. The success of such a top-down model depends on a reasonable choice of the model on the top level (macroscopic model) and a powerful and suitable feedback mechanism from the lower level (microscopic model). Otherwise, it might occur that the conflict feedback of the lower level dominates the solution process or even worse the procedure will not converge or terminate.

Hence, the question of generating "the best" macroscopic model was posed. Schlechte [24] developed a bottom-up approach for automatic construction of reliable macroscopic railway models based on very detailed microscopic ones. Figure 7.1 shows the basic idea. On the left hand side a microscopic model is shown as used in standard simulations with a time-space diagrams highlighting the block occupation of three trains. The time is running from top the bottom. Thus, the first two trains occupy six segments when running from left to right. The last train is running from right to left. The transformation produces then a macroscopic model, which aggregates the real physical block (six undirected segments) to two directed arcs. On the right hand side the capacity allocation can then be handled by minimum headway times which restrict the departures of succeeding trains on correlated arcs of the macroscopic model.

On the one hand these models are precise enough to allow for valid allocations with respect to blocking times, on the other hand they are simplified and aggregated to a coarse level, which allows for solving large scale optimization instances. As a result, a macroscopic track allocation can be re-translated to the microscopic model. Details how this automatic transformation works can be found in [25].

Based on this technique coarse macroscopic models can be automatically formulated to allow for optimization of track allocations with respect to resource consumptions, i.e., the calculation of running and headway times are incorporated in detail. However, macroscopic formulations can only approximate the infrastructure capacity. Depending on the used rounding mechanisms these models tend to overestimation, underestimation or even both, see [1]. Hence, constructing "the best" macroscopic model depends basically on the given objective, i.e., providing an upper bound of the capacity, a realistic capacity estimation or a valid and implementable track allocation. Based on those works a trend towards integrated frameworks connecting the microscopic and the macroscopic level to allow for iterative approaches has emerged, see [15]. In the remainder of the chapter we will focus on the track allocation problem for macroscopic railway models.

7.3 Macroscopic Optimization Models

The authors of [22] and [8] provide a comprehensive recent survey on the numerous problem and model formulations on railway track allocation, which all go back to the pioneer works of Brännlund et al. [4] and Caprara et al. [9] on train timetabling. A general classification according to solution methods was given by Liebchen [20] and Caimi [5]. In Fig. 7.2 the approaches on macroscopic railway timetabling are basically divided into two categories, periodic and non-periodic scheduling. The focus in this chapter is on exact solution approaches based on Mixed Integer Programming (MIP). In particular, we take the perspective of an infrastructure network provider facing the non-periodic variant with a high priority of practical operability. In contrast, the equivalent task from the perspective of a railway operator with the focus on passenger connections is called periodic event scheduling.

The *infrastructure network* is a directed graph $G^I = (V^I, A^I)$, where the nodes V^I represent stations, junctions, and crossings and the arcs A^I represent connecting railway tracks. Furthermore, we are given a set of trains R and each train $r \in R$ is associated with a subgraph $G^r = (V^r, A^r) \subseteq G^I$ of the infrastructure network. G^r represents the set of possible routes of r from its start to its destination. The aim is to determine a route as well as the arrival and departure times of each train at each of the stations on its chosen route. Between two trains $r, r' \in R$ using the same track $a \in A^r \cap A^{r'}$ there is a minimal headway time $h_a(r, r') \in \mathbb{N}$ (in minutes), which is the minimal time period that has to pass after the first train r has entered the track before the next train r' is allowed to enter the track. Note that in general the headway time may be different for any two trains and any order of trains. Furthermore, there are capacity constraints on the nodes that state that at most a certain number of trains $c_u \in \mathbb{N}$ may be at the same station $u \in V^I$ at the same time $t \in T$.

One of the most successful models in the literature for solving the track allocation problem is based on time expanded networks, see, e.g., [9]. Given a set of discrete time steps $T = \{1, \ldots, |T|\}$, we have for each train $r \in R$ a *time expanded network* $G_T^r = (V_T^r, A_T^r)$ where $V_T^r = V^r \times T$ and $A_T^r = A_{\text{run}}^r \cup A_{\text{wait}}^r$ with running and waiting arcs

$$A_{\text{run}}^r = \{((u,t),(v,t+t_{(u,v)}^r)) : (u,v) \in A^r\}$$
$$A_{\text{wait}}^r = \{((u,t),(u,t+1)) : u \in V_{\text{wait}}^r\}.$$

The running arcs A_{run}^r connect nodes $u, v \in V^r$ that succeed along the train path, i.e. $(u,v) \in A^r$, with a time difference corresponding to the running time $t_{(u,v)}^r$ of r. The waiting arcs A_{wait}^r connect nodes of succeeding time steps of a station $u \in V_{\text{wait}}^r$ at which r is allowed to wait. A feasible schedule of train r then corresponds to a path $P \subseteq G_T^r$ from the first to the last station. The set $\mathscr{P}^r$ denotes all the potential paths.

In order to formulate the model as an integer program (`IP`), we associate a binary variable $x_a^r \in \{0,1\}$ with each arc $a \in A_T^r$ in each time expanded network, where $x_a^r = 1$ if and only if a is contained in the timetable or track allocation of train r. The headway restrictions impose that two arcs $a = ((u,t_u),(v,t_v)) \in A_T^r$ and $a' = ((u,t_u'),(v,t_v')) \in A_T^{r'}$ with $t_u' - t_u < h_{(u,v)}(r,r')$ must not be used both in a timetable. Therefore, we add the following *packing constraints*

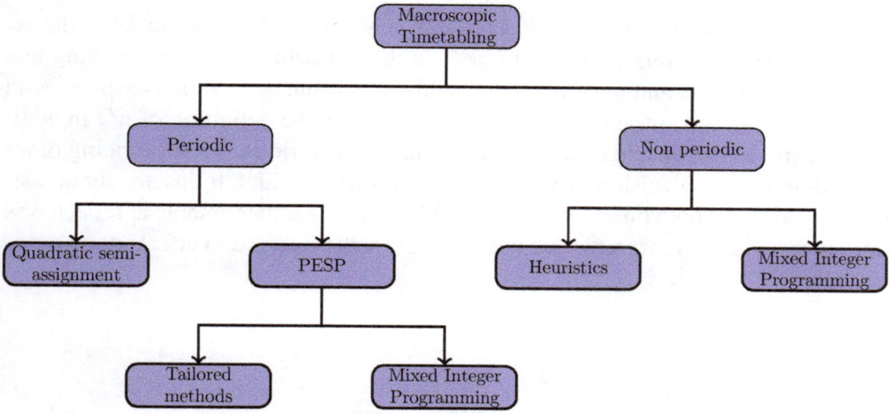

Fig. 7.2: Principal methods in the literature for macroscopic timetabling by Caimi [5]

$$x_a^r + x_{a'}^{r'} \leq 1, \qquad \{(r,a),(r',a')\} \in H, \tag{7.1}$$

where H is the set of pairs of incompatible train arcs. Let

$$\mathscr{H} := \left\{ x = (x^r)_{r \in R} \colon x^r \in \{0,1\}^{|A_T^r|}, x \text{ satisfies (7.1)} \right\}$$

denote the set of integer vectors satisfying all packing constraints. The capacity constraints in the nodes mean that at each time instance $t \in T$ at most $c_u \in \mathbb{N}$ trains may be in $u \in V^I$ at the same time. Hence, the sum over all arcs representing a train being in u at time t

$$K(u,t) := \left\{ (r,a) \colon a = ((u',t'),(u,t)) \in A_T^r, r \in R \right\}$$

must be at most c_u

$$\sum_{(r,a) \in K(u,t)} x_a^r \leq c_u, \qquad u \in V^I, t \in T.$$

Putting all together, the basic track allocation problem (TAP) can be formulated as IP as follows:

Given arc weights $w^r \in \mathbb{R}^{|A_T^r|}$, $r \in R$,

$$\begin{aligned} \text{maximize} \quad & \sum_{r \in R} \langle w^r, x^r \rangle \\ \text{subject to} \quad & x_a^r + x_{a'}^{r'} \leq 1, && \{(r,a),(r',a')\} \in H, \\ & \sum_{(r,a) \in K(u,t)} x_a^r \leq c_u, && u \in V^I, t \in T, \\ & x^r \in X^r, && r \in R, \end{aligned} \tag{7.2}$$

i.e., we select for each train $r \in R$ a feasible schedule $x^r \in X^r$ where X^r is the set of arc incidence vectors of the paths $\mathscr{P}^r$, so that all paths satisfy the packing and capacity constraints and maximize a general weight function. Various aspects with respect to time and space can be directly handled by the definition of w^r. In addition, the model can be extended to tackle further real world aspects as running times depending on stop decision or single-track sections on which trains are simultaneously operated in both directions, see [11, 24]. In [16] another notable approach was introduced where hypergraphs are used to capture the interaction effects of resource transitions.

7.3.1 Clique Separation

The basic model (7.2) is quite general and allows very flexible definitions of conflicts in terms of the set of pairwise conflicting arcs H (we defined conflicts only by violated headway conditions, but other operational restrictions could be formulated as well). However, a disadvantage of the "packing-like constraints" (7.1) is the "lost structure" which leads to weak linear programming relaxations of (7.2). We will examine this issue for the case of headway conflicts.

For the given set of conflicting arcs H we create a conflict graph $G = (V,E)$ with node set $V = A^r_{\text{run}}$ of all running time-arcs. In order to strengthen the linear relaxation of the formulation, the pairwise conflict sets may be enlarged to inclusion-maximal ones which correspond to maximal cliques in G. Certainly, constraints (7.1) can be strengthened to constraints

$$\sum_{a \in C} x_a \leq 1, \qquad C \text{ maximal clique in } G.$$

Although these constraints lead to much stronger relaxations, the disadvantage is the potentially large number of maximal cliques in G, and thus the large number of constraints, and hence the problem of detecting violated constraints. In the following we will collect some basic facts about detection and occurrence of maximum cliques in special graph classes.

In the general case the separation of the maximal clique constraints is hard. This is because the minimal headway times are in general different for each train type combination and for each stopping behavior combination. Lukac [21] gives an extensive analysis of the structure of clique constraints arising from triangle-linear and quadrangle-linear headway matrices and proves that the time window of interest is bounded by twice the maximum headway time.

The arguably simplest case is "full block occupation", that is, the headway time is set to the running time of the preceding train. In this setting headway times are completely independent from the type of the successor train and the graph G becomes an interval graph. Figure 7.3 illustrates the construction of G and the maximal cliques in that case. On the left hand side six trains are shown on one track from left to right. Consider the time is going from bottom to the top. Thus, in the middle of Fig. 7.3 the projected time intervals of the block occupation of each train are shown. Consequently, overlapping intervals are in conflict and correspond to edges in the conflict graph G on the right hand side. The gray area highlights the maximal clique in G. Note there are two other maximal cliques of size 3. The maximal cliques of the conflict graph are collections of compact real intervals. By Helly's Theorem, see [17], the intervals of each such clique contains a common point. It follows that the conflict graph G has $O(V)$ inclusion maximal cliques, which can be enumerated in polynomial time, see [24].

The authors of [6] provide an efficient algorithm to determine the maximal conflict cliques on the microscopic level. Moreover, they develop an alternative model using the sequence of resources that each train path passes, encoded in a resource tree. In tests with real-world data from the Swiss Federal Railways, the resource tree conflict graph model was able to reduce the computation time by roughly two orders of magnitude when compared to previous approaches.

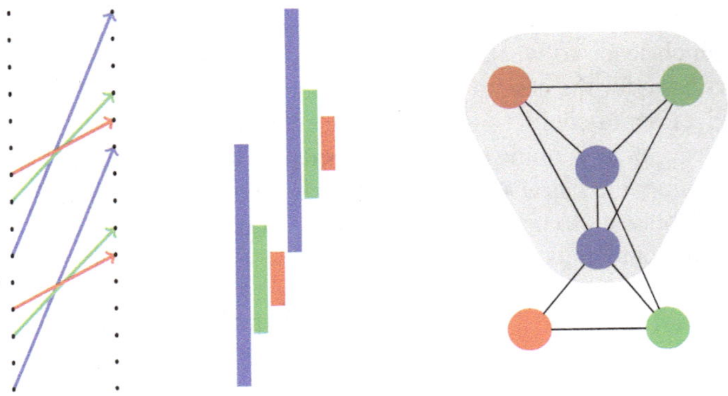

Fig. 7.3: Example for maximum cliques for block occupation conflicts

7.3.2 *Configuration Networks*

An alternative formulation for the track allocation problem that guarantees a conflict free routing by allowing only feasible orderings was proposed by Borndörfer and Schlechte [2]. Instead of excluding conflicting arc sets as described in Sect. 7.3.1, the formulation is based on the concept of feasible arc *configurations*, i.e., sets of arcs on a track without headway conflicts which correspond one-to-one to paths in so called *Configuration Networks*. In [24] was shown that this model is an extended formulation of model `TAP` which can be directly solved by column generation techniques. Furthermore, the corresponding LP-relaxation of the extended formulation based on configuration networks provides at least the same lower bound as the packing formulation. Weaknesses of those formulations are the larger size and poor convergence properties, see [14]. However, a major advantage of the configuration models are the direct control of train orders and the possibility to use those for branching or primal heuristics in order to construct high quality primal solutions, see [12].

7.4 Algorithmic Techniques for the `TAP`

Real world railway networks are notoriously very large and hence lead to large optimization models. For instance, in track allocation models the infrastructure networks comprise thousands of nodes (stations, switches) and arcs (tracks connecting the nodes) as well as thousands of trains running on a single day in the network. For these reasons advanced algorithmic techniques have been developed. We will present some of them in this section.

The basic concept, which all those techniques built on, is decomposition of the complete model into smaller subproblems. Each subproblem is solved independently several times and an overall algorithm coordinates the subproblem evaluation in such a way, that a global solution can be derived. The standard decomposition approach is *Lagrangian relaxation*, which will be described shortly in the context of track allocation problems below. In order to solve the decomposed model, several methods have been proposed. Very successful ones are *bundle methods* which are examined in Sect. 7.4.2. Finally, we describe the *Dynamic Graph Generation* technique, whose development has been motivated by large scale track allocation problems and which aims at managing the model size dynamically during the solution process, so that the overall memory requirements are reduced dramatically.

7.4.1 Lagrangian Relaxation

Formally, the track allocation problem can be written as a mixed-integer programming problem in the following compact way

$$\begin{aligned} \text{maximize} \quad & \sum_{r \in R} \langle w^r, x^r \rangle, \\ \text{subject to} \quad & x^r \in X^r, \\ & \sum_{r \in R} M^r x^r \leq b, \end{aligned} \tag{7.3}$$

by collecting all coupling constraints in the inequalities $\sum_{r \in R} M^r x^r \leq b$, see Sect. 7.3. An important observation is that omitting constraints (7.3) decomposes the problem in several subproblems of the form

$$\max\{\langle w^r, x^r \rangle : x^r \in X^r\},$$

one for each train $r \in R$. These subproblems turn out to be simple shortest-path problems in the time-expanded train graph. Therefore, the idea of Lagrangian relaxation is to penalize the violation of the complicating coupling constraints (7.3) instead of enforcing their validity. In fact, for a given *Lagrangian multiplier* vector $y \in \mathbb{R}^m_+$, the relaxation reads

$$\begin{aligned} \text{maximize} \quad & \langle b - \sum_{r \in R} M^r x^r, y \rangle + \sum_{r \in R} \langle w^r, x^r \rangle, \\ \text{subject to} \quad & x^r \in X^r. \end{aligned}$$

It is easy to see that for any vector $y \in \mathbb{R}^m_+$ the optimal value of the relaxation is a lower bound on the real optimal value. Therefore, the natural goal is to find the Lagrangian multiplier $\tilde{y} \in \mathbb{R}^m_+$ that gives the best bound. This leads to the optimization problem

$$\tilde{y} \in \operatorname*{arg\,min}_{y \in \mathbb{R}^m_+} \left[\langle b, y \rangle - \sum_{r \in R} \max\left\{ \langle w^r - (M^r)^T y, x^r \rangle : x^r \in X^r \right\} \right]. \tag{7.4}$$

This is a non-smooth, convex optimization problem. Algorithmically, it can be solved using subgradient methods or bundle methods to be described in the next section.

7.4.2 Bundle Methods

Bundle methods and subgradient methods are iterative algorithms to solve non-smooth, convex optimization problems of the form $\min\{f(y)\colon y \in Y\}$ with f a convex function and $Y \subseteq \mathbb{R}^m$ a "simple" convex set. Both algorithms assume that f is given via a *function oracle* that, for a given point $y \in \mathbb{R}^m$ computes the function value $f(y)$ and a subgradient $g \in \partial f(y)$. Both types of algorithms are iterative methods that evaluate the function f at certain trail points y_k, $k \in \mathbb{N}$, and require the function value $f_k = f(y_k)$ and a subgradient $g_k \in \partial f(y_k)$. The algorithms use this information to compute the next trail point y_{k+1} such that the sequence of trail points converges to an optimal solution $\tilde{y} := \lim_{k\in\mathbb{N}} y_k$. In practice, bundle methods turned out to have better convergence in practical applications, in particular for problems of the form (7.4), hence we focus on them in this section.

The idea of Bundle methods (see [18, 19]) is to use the subgradient information obtained by evaluating the function to build a cutting plane model (or "bundle") $\hat{f}_k(y) := \max\{f_i + \langle g_i, y - y_i\rangle : i \in I\} \leq f(y)$ of f with I a finite set and $g_i \in \partial f(y_i)$, $f_i \leq f(y_i)$ for $i \in I$ (see Fig. 7.4 for an illustration of the cutting plane model). This model is then used to determine the next candidate $\bar{y}_{k+1}$ and the function is evaluated at $\bar{y}_{k+1}$. If the function value $f(\bar{y}_{k+1})$ is reasonably close to the model value $\hat{f}_k(\bar{y}_{k+1})$ the function accepts the trial point as new iteration point $y_{k+1} \leftarrow \bar{y}_{k+1}$. Otherwise, the model seems to be a bad approximation of f at $\bar{y}_{k+1}$. Then the bundle method stays at the old point $y_{k+1} \leftarrow y_k$ and uses the subgradient information $g_{k+1} \in \partial f(\bar{y}_{k+1})$ to improve the model.

The Lagrangian relaxation of the track allocation problem is equivalent to

$$\min\{f(y)\colon y \geq 0\},$$

with

$$f(y) := \langle b, y\rangle + \sum_{r\in R} \max_{x^r \in X^r} \langle w^r - (M^r)^T y, x^r\rangle,$$

see [11]. The feasible set is $Y := \{y \in \mathbb{R}^m\colon y \geq 0\}$. The function $f(y)$ is convex as sum of maximums of linear functions. For a given point y_k, each optimal solution $x(y_k) = (x^r(y_k))_{r\in R} \in \bigtimes_{r\in R} X^r$ of the inner problems defines a subgradient $g_k := b - \sum_{r\in R} M^r x^r \in \partial f(y_k)$. Hence, one can apply subgradient or bundle methods to solve the Lagrangian relaxation problem. In addition to an (approximate) optimal solution $\tilde{y}$ of the Lagrangian relaxation, the algorithm also generates a sequence $(\bar{x}_k^r)_{k\in\mathbb{N}}$ of primal solutions $\bar{x} \in \bigtimes_{r\in R} \operatorname{conv} X^r$, such that each accumulation point $\tilde{x} = (\tilde{x}^r)_{r\in R}$ of this sequence satisfies the coupling constraints (7.3). In particular, such an accumulation point is an optimal solution of

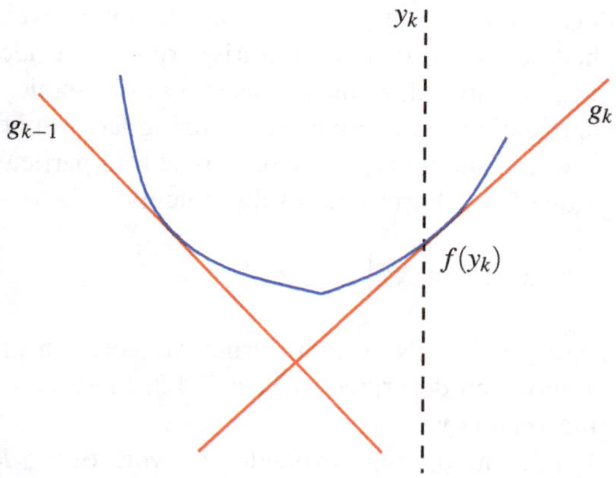

Fig. 7.4: The blue function f has to be minimized. The two red subgradients g_k and g_{k-1} define a lower cutting plane model of f

$$\begin{aligned} \text{maximize} \quad & \sum_{r\in R} \langle w^r, x^r \rangle, \\ \text{subject to} \quad & x^r \in \operatorname{conv} X^r, \\ & \sum_{r\in R} M^r x^r \leq b. \end{aligned}$$

Note that $\bar{x}^r \in \operatorname{conv} X^r \supsetneq X^r$ is, in general, not necessarily an integral schedule but a convex combination of feasible schedules for train $r \in R$. Although this (fractional) solution is not an optimal solution of the IP formulation for the track allocation problem, it provides an upper bound on the optimal value and is often used as a starting point for rounding heuristics.

7.4.3 Dynamic Graph Generation

One major disadvantage of time expanded models is their enormous size if the number of time steps increases, e.g., due to a finer discretization. In track allocation the time horizon typically consists of several hours up to several days and the discretization step is often 1 min or less. Hence, the time expanded network of a single train can have many thousands nodes and arcs, which leads to large memory requirements.

In practice there are often conditions that allow reducing the size of these networks. For instance, often a train may not be shifted arbitrarily in time but its schedule may only deviate from an "ideal" schedule by a few minutes. However, this is not always the case and further techniques are required to handle these huge models.

Dynamic Graph Generation, see [11, 13], is a technique developed for the case where the train schedules may deviate arbitrarily from some ideal timetable, but the general objective is to have all trains run as early as possible. This is a reasonable goal in many applications, because a train running late implies a high delay in the train's schedule, which should typically be avoided. In particular, the algorithm focuses on the solution of the shortest path subproblems

$$\max\left\{\langle w^r - (M^r)^T y, x^r\rangle : x^r \in X^r\right\}, r \in R, \tag{7.5}$$

for some given vector $y \in \mathbb{R}^m_+$. Note that during the solution process of the Lagrangian relaxation approach described in Sect. 7.4.2, these subproblems must be solved w.r.t. changing vectors y.

Let $G^r_T = (V^r_T, A^r_T)$ denote the time expanded network of $r \in R$. In general, this network is very large but because of the objective to run as early as possible, the train schedules will mainly use the early time steps. However, it is important to note that the model does not enforce this. It may be the case that some trains do get a large delay and the corresponding schedules will use late time steps. But we do not know in advance which trains are delayed. Hence, we have to prepare for the case that each train might receive a large delay. The idea of Dynamic Graph Generation can be described as follows. Instead of keeping the complete time expanded graph G^r_T in memory, only a small subgraph $G^r_{\text{mem}} = (V^r_{\text{mem}}, A^r_{\text{mem}}) \subseteq G^r_T$ with $|V^r_{\text{mem}}| \ll |V^r_T|$ and $|A^r_{\text{mem}}| \ll |A^r_T|$ is stored. Let $X^r_{\text{mem}} \subseteq X^r$ denote the set of all schedules of train r that correspond to paths in G^r_{mem}. The goal is to solve problem (7.5), but only

$$\max\left\{\langle w^r_{\text{mem}}, x^r\rangle : x^r \in X^r_{\text{mem}}\right\}, \tag{7.6}$$

is solved, which is a shortest path problem only in the stored subgraph w.r.t. some special objective function w^r_{mem}. Given an optimal solution $\tilde{P} \in X^r_{\text{mem}}$ the question is to decide whether $\tilde{P}$ is also an optimal solution of (7.5). In general, this is not possible without solving (7.5) itself, but then the approach would be pointless.

Fortunately, the special structure of the optimization problem enables us not only to detect whether $\tilde{P}$ is an optimal solution but also gives a clue how to enlarge G^r_{mem} if this is not the case. The idea is to split the stored subgraph G^r_{mem} into two parts, the *active subgraph* $G^r_{\text{act}} = (V^r_{\text{act}}, A^r_{\text{act}}) \subset G^r_{\text{mem}}$ and the rest. The graph G^r_{act} is the union of all paths that have been computed in some earlier iteration. The objective function $w^r_{\text{mem}} : A^r_{\text{mem}} \to \mathbb{R}$ is defined as follows

$$w^r_{\text{mem}}(a) := \begin{cases} w^r(a) - (M^r_a)^T y, & a \in A^r_{\text{act}}, \\ w^r(a), & a \in A^r_{\text{mem}} \setminus A^r_{\text{act}}. \end{cases}$$

In [11] was shown that if, for a given active subgraph G^r_{act}, the stored subgraph G^r_{mem} is large enough the following property holds:

$$\tilde{P} \text{ is optimal for (7.6) and } \tilde{P} \subseteq G^r_{\text{act}} \Rightarrow \tilde{P} \text{ is an optimal solution of (7.5).} \tag{7.7}$$

Furthermore it was shown that under reasonable technical conditions a stored subgraph G^r_{mem} satisfying (7.7) can be constructed efficiently. Thus, the algorithm can test whether $\tilde{P} \subseteq G^r_{\text{act}}$. If $\tilde{P} \subseteq G^r_{\text{act}}$ then $\tilde{P}$ is an optimal solution of (7.5). Otherwise, the algorithms adds $\tilde{P}$ to G^r_{act}, possibly enlarges G^r_{mem} appropriately (so that (7.7) is satisfied) and solve the shortest path problem w.r.t. the enlarged graph again, see Algorithm 7.1.

Algorithm 7.1: Dynamic graph generation shortest path

```
Data: The active subgraph G^r_act and the multiplier vector y
Result: optimal solution P̃ ∈ X^r
loop
    Determine appropriate G^r_mem ⊃ G^r_act satisfying (7.7);
    Compute w^r_mem;
    Solve (7.6) ⇝ P̃ ∈ X^r_mem;
    if P̃ ⊆ G^r_act then
        return P̃
    else
        G^r_act ← G^r_act ∪ P̃
    end
end
```

In fact, it was shown that under mild assumptions this algorithm always returns after a finite number of iterations and the stored subgraph G^r_{mem} is only slightly larger than the union of all computed paths G^r_{act}. Furthermore, the dynamic graph generation approach allows dealing with models that do not have an a priori bound on the number of time steps. Instead, the graphs grow dynamically as much as needed to contain an optimal solution, see [11].

7.5 Status Quo and Future Opportunities

Research papers demonstrated that the presented mathematical models are able to solve instances for corridors or regional areas (Fig. 7.6) in several case studies. In [3] optimal track allocations for approximately 40 km from Brig to Domodossola and back for a single day are computed and verified by simulations. In Fig. 7.5 the microscopic infrastructure of the Simplon corridor based on the simulation tool [23], is shown. The microscopic network consists of 1154 nodes and 1831 edges where roughly 200 trains can be scheduled a day.

To the best knowledge of the authors, there are no commercial software solutions providing fully automated timetabling by using the discussed optimization models and methods. Railway infrastructure manager use mainly tailor-made decision support tools provided by their in-house development teams.

railML.org is a major initiative towards a standardization of data for the railway industry which started 2002, see www.railml.org. Since then, there are some

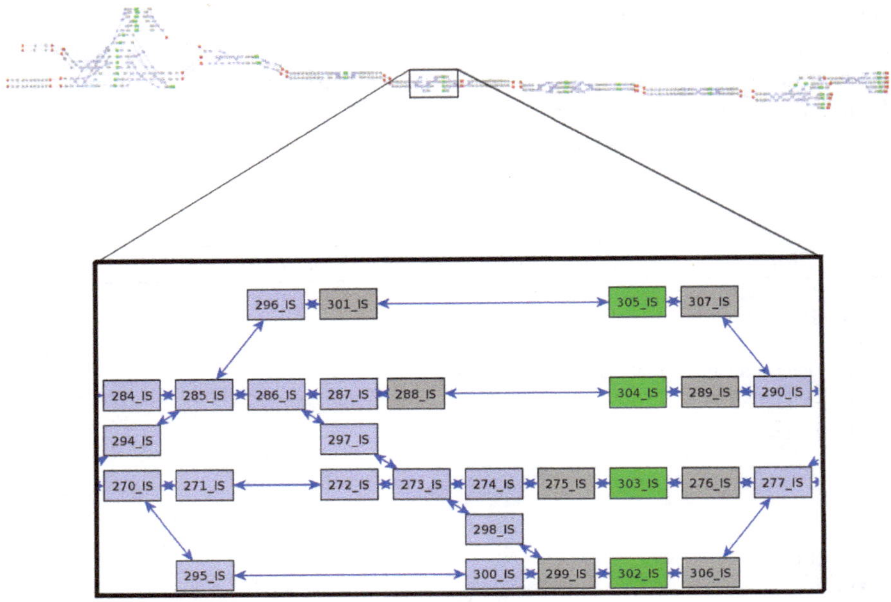

Fig. 7.5: Microscopic network of the Simplon corridor in Switzerland and detailed representation of station Iselle

large and promising undertakings from the industry and the research community to bridge the gap between theoretic case studies and real-world applications. Nevertheless, we believe there is a huge opportunity to fill this gap between theory and practice in this particular application. On the scientific side, there are powerful techniques to increase the tractability of those models. But, on the practical side due to technical details on several layers, peculiarities of the dedicated railway system, and a complex human interaction the technology transfer is very challenging. The research community together with railway companies must work on defining standard processes such that the work will be then aligned and synergies can be created. This will give software vendors a real possibility to offer a standardized software on the market.

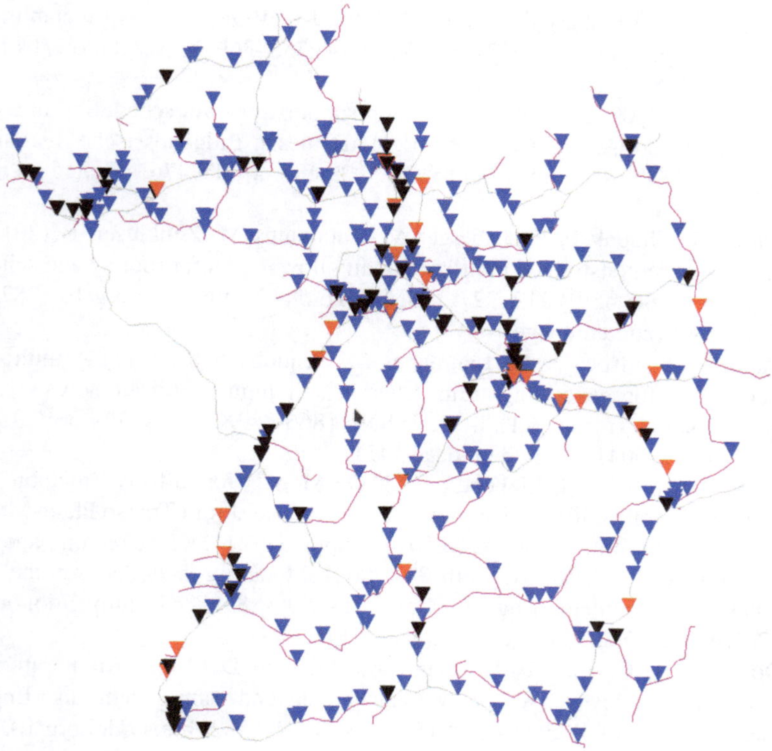

Fig. 7.6: Macroscopic network of Baden-Württemberg (roughly 10% of the German railway network) from an instance with 6 h planning horizon and about 2000 trains. Each triangle denotes the current position of one train (long distance (red), short distance (blue), freight (black))

References

1. Blanco M, Schlechte T (2013) Analysis of micro-macro transformations of railway networks. In: Huisman D, Louwerse I, Wagelmans APM (eds) Operations research proceedings 2013. Springer, Berlin, pp 37–42. ISBN: 978-3-319-07000-1. https://doi.org/10.1007/978-3-319-07001-8_6 (cited on page 146)
2. Borndörfer R, Schlechte T (2007) Models for railway track allocation. In: Liebchen C, Ahuja RK, Mesa JA (eds) Proceedings of the 7th workshop on algorithmic approaches for transportation modeling, optimization, and systems (ATMOS'07). OpenAccess series in informatics (OASIcs), vol 7. Schloss Dagstuhl, Wadern. ISBN: 978-3-939897-04-0. https://doi.org/10.4230/OASIcs.ATMOS.2007.1170 (cited on page 150)
3. Borndörfer R, Erol B, Graffagnino T, Schlechte T, Swarat E (2014) Optimizing the Simplon railway corridor. Ann Oper Res 218(1):93–106. https://doi.org/10.1007/s10479-012-1260-9 (cited on page 155)

4. Brännlund U, Lindberg PO, Nou A, Nilsson J-E (1998) Railway timetabling using Lagrangian relaxation. Transp Sci 32(4):358–369. https://doi.org/10.1287/trsc.32.4.358 (cited on page 146)
5. Caimi GC (2009) Algorithmic decision support for train scheduling in a large and highly utilised railway network. PhD thesis, Eidgenössische Technische Hochschule Zürich. https://doi.org/10.3929/ethz-a-005947637 (cited on pages 145, 146, 147)
6. Caimi G, Chudak F, Fuchsberger M, Laumanns M, Zenklusen R (2011) A new resource-constrained multicommodity flow model for routing and scheduling. Transp Sci 45(2):212–227. ISSN: 1526-5447. https://doi.org/10.1287/trsc.1100.0349 (cited on page 149)
7. Caimi G, Fuchsberger M, Laumanns M, Schüpbach K (2011) A multi-level framework for generating train schedules in highly utilised networks. English. Public Transp 3(1):3–24. ISSN: 1866-749X. https://doi.org/10.1007/s12469-011-0041-1 (cited on page 145)
8. Caimi G, Kroon LG, Liebchen C (2017) Models for railway timetable optimization: applicability and applications in practice. J Rail Transp Plann Manage 6(4):285–312. https://doi.org/10.1016/j.jrtpm.2016.11.002 (cited on page 146)
9. Caprara A, Fischetti M, Toth P (2002) Modeling and solving the train timetabling problem. English. Oper Res 50(5):851–861. https://doi.org/10.1287/opre.50.5.851.362 (cited on pages 146, 147)
10. Dollevoet T, Corman F, D'Ariano A, Huisman D (2014) An iterative optimization framework for delay management and train scheduling. English. Flex Serv Manuf J 26(4):490–515. ISSN: 1936-6582. https://doi.org/10.1007/s10696-013-9187-2 (cited on page 145)
11. Fischer F (2013) Dynamic graph generation and an asynchronous parallel bundle method motivated by train timetabling. PhD thesis, Chemnitz University of Technology. URN: urn:nbn:de:bsz:ch1-qucosa-118358 (cited on pages 148, 152, 154, 155)
12. Fischer F (2015) Ordering constraints in time expanded networks for train timetabling problems. In: Italiano GF, Schmidt M (eds) 15th workshop on algorithmic approaches for transportation modelling, optimization, and systems (ATMOS 2015). OpenAccess Series in Informatics (OASIcs), vol 48. Schloss Dagstuhl, Wadern, pp 97–110. ISBN: 978-3-939897-99-6. https://doi.org/10.4230/OASIcs.ATMOS.2015.97 (cited on page 150)
13. Fischer F, Helmberg C (2014) Dynamic graph generation for the shortest path problem in time expanded networks. English. Math Program A 143(1–2):257–297. ISSN: 0025-5610. https://doi.org/10.1007/s10107-012-0610-3 (cited on page 154)
14. Fischer F, Schlechte T (2015) Comparing two dual relaxations of large scale train timetabling problems. In: Proceedings of conference on advanced systems in public transport (CASPT 2015) (URN links to ZIB report). URN: urn:nbn:de:0297-zib-56068 (cited on page 150)

15. Goverde R, Besinovic N, Binder A, Cacchiani V, Quaglietta E, Roberti R, Toth P (2015) A three-level framework for performance-based railway timetabling. In: 6th international conference on railway operations modelling and analysis, RailTokyo. UUID: b095c1cd-36dd-4fb3-98a4-ef96c2fff7d7 (cited on page 146)
16. Harrod S (2010) Modeling network transition constraints with hypergraphs. English. Transp Sci 10:293–310. https://doi.org/10.1287/trsc.1100.0337 (cited on page 148)
17. Helly E (1923) Über Mengen konvexer Körper mit gemeinschaftlichen Punkten. German. Jahresbericht der Deutschen Mathematiker-Vereinigung 32:175–176 (cited on page 149)
18. Hiriart-Urruty J-B, Lemaréchal C (1993) Convex analysis and minimization algorithms I. A series of comprehensive studies in mathematics, vol 305. Springer, Berlin. https://doi.org/10.1007/978-3-662-02796-7 (cited on page 152)
19. Hiriart-Urruty J-B, Lemaréchal C (1993) Convex analysis and minimization algorithms II. A series of comprehensive studies in mathematics, vol 306. Springer, Berlin. https://doi.org/10.1007/978-3-662-06409-2 (cited on page 152)
20. Liebchen C (2006) Periodic timetable optimization in public transport. PhD thesis, TU Berlin. dissertation.de. ISBN: 978-3-86624-150-3 (cited on page 146)
21. Lukac SG (2004) Holes, antiholes and maximal cliques in a railway model for a single track. English. Technical Report, ZIB Report 04-18. Zuse-Institut Berlin. URN: urn:nbn:de:0297-zib-7939 (cited on page 149)
22. Lusby RM, Larsen J, Ehrgott M, Ryan D (2011) Railway track allocation: models and methods. OR Spectr 33(4):843–883. https://doi.org/10.1007/s00291-009-0189-0 (cited on page 146)
23. OpenTrack (2010) ETH Zürich. www.opentrack.ch (cited on page 155)
24. Schlechte T (2012) Railway track allocation - models and algorithms. PhD thesis, TU Berlin. https://doi.org/10.14279/depositonce-3124 (cited on pages 145, 148, 149, 150)
25. Schlechte T, Borndörfer R, Erol B, Graffagnino T, Swarat E (2011) Micro-macro transformation of railway networks. English. J Rail Transp Plann Manage 1(1):38–48. https://doi.org/10.1016/j.jrtpm.2011.09.001 (cited on page 146)
26. Sewcyk B (2004) Makroskopische Abbildung des Eisenbahnbetriebs in Modellen zur langfristigen Infrastrukturplanung. Deutsch. PhD thesis. Universität Hannover. Eurailpress. ISBN: 978-3-7771-0323-5. http://d-nb.info/973559918 (cited on page 144)

Chapter 8
Use of Optimization Tools for Routing in Rail Freight Transport

Armin Fügenschuh, Henning Homfeld, Marc Johann, Hanno Schülldorf, and Anke Stieber

A. Fügenschuh · A. Stieber (✉)
Helmut Schmidt University/University of the Federal Armed Forces Hamburg, Holstenhofweg 85, 22043 Hamburg, Germany
e-mail: fuegenschuh@hsu-hh.de; anke.stieber@hsu-hh.de

H. Homfeld
Deutsche Bahn AG, DB Management Consulting, Gallusanlage 8, 60329 Frankfurt am Main, Germany
e-mail: henning.homfeld@deutschebahn.com

M. Johann · H. Schülldorf
Deutsche Bahn AG, DB Analytics, Poststr. 20, 60329 Frankfurt am Main, Germany
e-mail: marc.johann@deutschebahn.com; hanno.schuelldorf@deutschebahn.com

R. Borndörfer et al. (eds.), *Handbook of Optimization in the Railway Industry*, International Series in Operations Research & Management Science 268,
https://doi.org/10.1007/978-3-319-72153-8_8

Abstract Deutsche Bahn, one of the largest European railway companies, offers mainly two products to commercial and industrial customers for freight transportation. Customers with high demand order unit trains, that are pulled by one or two locomotives from their respective origins to their destinations. In contrast, customers with less demand order a limited amount of single cars, that are first pulled to a classification yard. There they are grouped together with single cars from other customers into a train unit. On the way from their respective origins via intermediate yards to their destinations, the cars are reclassified several times, which is a time-consuming and personnel-intensive procedure. To support the strategic long-term planning process of the single car freight routing, a mathematical optimization tool based on mixed-integer nonlinear programming was developed and is in practice use since 2011. However, real-world constraints have changed over the last years. For example, unit trains and single cars are no longer strictly separated products, but they are more and more integrated: In some unit trains there are still residual capacities that can be used for single cars. For some of these additional new requirements, the existing optimization tool has to be extended slightly by formulating new additional mathematical constraints. For some other requirements, a substantial re-development will be necessary in the future. The purpose of this chapter is to review the existing single car routing model, to discuss how it is used in real-life, and to demonstrate how it can be extended to meet the new requirements in the present and future.

8.1 Introduction

At the major German railway company Deutsche Bahn (DB), the rail freight transport is separated into single car transportation and transportation of unit trains. Customers with high shipment demand (e.g., 20–30 cars) order unit trains to transport their goods via rail. The unit trains are pulled by a set of locomotives from their origin to their destination. In contrast, customers with less demand order only a limited amount of single cars (e.g., 1–5 cars). To route these single cars through the network, a locomotive pulls the cars to the next classification yard, where they are grouped together with single cars from other customers to be transported again to the next yard. On the way to their final destination the cars are reclassified several times. A route is defined by the sequence of yards cars of a certain order pass, beginning with their origin and finishing with their destination.

The problem of routing single freight cars was modeled as a mixed-integer nonlinear program, see Fügenschuh et al. [11] and Homfeld [13]. This model is used in practice for strategic long-term planning and simulation purposes at DB since 2011. Input is a predicted or real demand matrix and the German rail freight network including all yards and tracks that can be used by freight cars and trains. As the whole German network is too large to be solved in acceptable time, the network is aggregated. To do this, DB uses a hierarchy of yards that basically depends on the size and handling capacity of a yard—*Rangierbahnhof (Rbf)* is the largest, *Knotenbahnhof (Kbf)* and *Satellit mit Rangiermitteln (SmR)* follow. The *Güterverkehrsstelle (Gvst)* is the smallest hierarchy class. There is no exact definition to this classification but it helps defining a hub structure that is typical for this kind of problem. There are several algorithms for aggregation and disaggregation of the network.

There are many applications for this model already. The main application is finding optimal routes for future rail freight cars to predict the load of the railway network and identify bottlenecks in yards and tracks. For this, the yard constraints are relaxed to review the actual yard usage of the optimal routing solution for supporting yard re-dimensioning decisions. Other applications include decision support in call for proposals when new transportation networks are to be designed from scratch.

Furthermore, real-world constraints have changed since 2011. For example, unit trains and single cars as two separated products are merged step by step: Some unit trains contain residual capacities, that can be used for single cars in a first step. Both entities are connected to build a full unit train.

For a company with a huge variety of projects, it would be obviously too expensive and too time-consuming to develop an optimization model and algorithm for every new application or changed constraint. From this point of view, flexible models are desired that can help a project team within days rather than months.

From a mathematical point of view, the existing model, which is introduced in Sect. 8.3 on page 164 is already quite powerful. It is a multicommodity-flow problem with both linear and fixed costs, with node capacities, out-degree-constraints and path restrictions.

We show how a smart parametrization and small changes (if any) to the model provide solutions for a variety of different practical applications.

8.2 Survey of the Literature

The single car rail freight transportation problem first emerged in the literature in the 1960s. For a survey of the approaches that were published till 1980 we refer to [2]. Different model formulations were presented by Bodin et al. [6], Assad [3], and Crainic et al. [9]. Keaton [16] presented a model that focuses on train building instead of car routing.

Newton [17] developed a model for a more general network budget design problem (BDP). The author formulated the railroad blocking problem (RBP) as a BDP with additional constraints. The single car routing problem is similar to the railroad blocking problem, but there are differences. One difference results from the fact, that German trains are restricted to a maximum length, usually of 700 m due to the length of bypass tracks in the German railway network. Freight trains are parked on the bypass tracks when faster passenger trains pass by. The consequence of this fact is, that a block of cars with a length of 710 m results in much higher train costs than a block with a length of 690 m, because the first one needs two trains. Another difference compared to the RBP is, that in our model shipment is not routed over the shortest path, instead we minimize train costs as a main component and have to add additional constraints that model the unique successor rule to prohibit distinct ways between a pair of origin-destination (OD) nodes. For a more detailed comparison between European and North American railway systems, we refer to [8].

An arc-based model was presented by Ahuja et al. [1]. Despite the special application context, the single car routing problem has similarities to the broader class of network design problems, see Balakrishnan et al. [4] for a general survey. Similar structures occur, for example, in the design of telecommunication networks, see Bley [5].

8.3 Model Formulation

A model for the single car railway freight routing problem was presented in [11]. Deutsche Bahn based its strategic analysis and simulation of the single car routing on this model since 2011. For the reader's convenience we give the model formulation in a concise way as a general basis, so that we can concentrate on present and future adaptations of the model to different application cases.

The model described in [11] is a nonlinear mixed-integer programming problem, which is linearized, so that linear programming based branch-and-cut (MILP) solvers, such as [14], can be used for the numerical solution of given instances. For an overview of the symbols used in the model formulation we refer to Table 8.1. Some of the model parameters refer to a certain time period, usually assumed to be one "model day", which is an abstract 24 h period of weighted averaged transportation demands, that were aggregated over a longer period (usually 3–6 months).

To formulate the model, three families of integer variables are introduced. The routes of the cars are described by decision variables $x^p_{i,j} \in \{0,1\}$ for all arcs $(i,j) \in A$ and orders $p \in K$. It is $x^p_{i,j} = 1$, if and only if station j is the successor of station i in the route for the cars belonging to order p. The second decision of the model is the number of trains from station i to j, for which we use the integer variables $n_{i,j} \in \mathbb{Z}_+$. Finally, we introduce the integer variables $y_{i,j} \in \mathbb{Z}_+$, that decide on the number of sorting tracks at station i on which trains are assembled exclusively in direction of station j.

The objective is to find economically optimal car routes, where the most crucial cost component is the number of train kilometers. The other two components are of less importance: the amount of used infrastructure (number of sorting tracks) and the number of car kilometers.

$$\min \quad \alpha_1 \cdot \sum_{(i,j)\in A} \delta_{i,j} \cdot n_{i,j} + \alpha_2 \cdot \sum_{(i,j)\in A} y_{i,j} + \alpha_3 \cdot \sum_{p\in K,(i,j)\in A} \delta_{i,j} \cdot x^p_{i,j}, \tag{8.1}$$

where $\alpha_1, \alpha_2, \alpha_3$ are user-defined weight parameters to give a human planner some control on the global shape of the solution.

The necessary constraints can be assigned to three major groups: constraints for the orders, for the trains and for the yards. The central constraints of the first group are the multi-commodity flow conservation equations:

$$\forall i \in V, p \in K: \sum_{j:(i,j)\in A} x^p_{i,j} - \sum_{j:(j,i)\in A} x^p_{j,i} = \begin{cases} 1 \text{ if } i = o_p, \\ -1 \text{ if } i = d_p, \\ 0 \text{ otherwise.} \end{cases} \tag{8.2}$$

Table 8.1: Symbols used in the model formulation

Type	Symbol	Meaning
Sets	V	Stations (yards and terminals)
	A	Links between stations
	K	Customer orders
Indices	i, j, k	Stations ($\in V$)
	p, q	Orders ($\in K$)
Parameters	o_p	Origin of order p
	d_p	Destination of order p
	v_p	Number of cars of order p
	ℓ_p	Length of cars belonging to order p
	w_p	Weight of cars belonging to order p
	T_p	Maximal travel time for order p
	u_i	(Constant) waiting time at yard i
	H_i	Maximal car capacity at yard i
	N_i	Maximal number of trains on sorting track at yard i
	Y_i	Number of available sorting tracks at yard i
	$\delta_{i,j}$	The distance between stations i and j
	$t_{i,j}$	Travel time from i to j including constant coupling times
	$L_{i,j}$	Maximum length of trains between i and j
	$W_{i,j}$	maximum weight of trains between i and j
Variables	$x^p_{i,j} \in \{0,1\}$	Decision variable, $= 1$ iff order p goes from i to j
	$n_{i,j} \in \mathbb{Z}_+$	Decision variable on the number of trains from i to j
	$y_{i,j} \in \mathbb{Z}_+$	Decision variable on the number of sorting tracks from i to j

Typically, the operating rail freight company offers different services of transportation such as express, premium, or standard (delivery within 24, 48, and 72 h respectively). The given input parameter $t_{i,j}$ denotes the travel time from station i to j, including all constant times for coupling in yard i and decoupling in yard j. By $w^p_{i,j}$ we denote the waiting time of the cars of order p spent in station i in the direction of station j. The time limit constraints ensure feasibility of the subsequent steps in the planning process:

$$\forall p \in K: \sum_{(i,j)\in A} w^p_{i,j} + t_{i,j} \cdot x^p_{i,j} \leq T_p. \tag{8.3}$$

The waiting time is a variable, that depends on the number of assembled trains on the respective sorting track. As soon as enough cars have been assembled on the sorting track, they are connected to a new train and re-enter the network to travel to the next station. Consequently, the number of trains assembled per time period on a particular sorting track $n_{i,j}$ and the waiting time $w^p_{i,j}$ of the cars of order p are related in the following way:

$$w^p_{i,j} = \frac{T}{n_{i,j}} \cdot x^p_{i,j}, \tag{8.4}$$

where T denotes the maximum waiting time for cars in a classification yard if only one train departs per day. Equation (8.4) introduces a nonlinear constraint into the model formulation. There are several options to deal with this nonlinear constraint, that is, to linearize it. For the details we refer to [11].

Each yard i has a maximum capacity of cars H_i that it can handle per time period. For large yards this value refers to the respective hump capacity, while for flat classification yards it refers to their shunting capacity. We may formulate the capacity restrictions by the following constraints:

$$\forall i \in V: \sum_{p\in K, j:(i,j)\in A} v_p \cdot x^p_{i,j} \leq H_i. \tag{8.5}$$

Another station dependent parameter is the number of trains that can be assembled per time period on a single sorting track N_i. Thus we get the constraints:

$$\forall (i,j) \in A: n_{i,j} \leq N_i \cdot y_{i,j}. \tag{8.6}$$

Each yard i has a total number of sorting tracks Y_i, which leads to the following capacity limit of assigned sorting tracks:

$$\forall i \in V: \sum_{j:(i,j)\in A} y_{i,j} \leq Y_i. \tag{8.7}$$

The trains have an upper bound on the total length and weight of the cars due to the length of bypass and sorting tracks and due to the strength of Deutsche Bahn locomotives. If $L_{i,j}$ denotes the maximum length of trains between i and j and $W_{i,j}$ the maximum weight respectively, the following constraints have to be satisfied:

$$\forall (i,j) \in A: \sum_{p \in K} \ell_p \cdot x^p_{i,j} \leq L_{i,j} \cdot n_{i,j}. \tag{8.8}$$

$$\forall (i,j) \in A: \sum_{p \in K} w_p \cdot x^p_{i,j} \leq W_{i,j} \cdot n_{i,j}. \tag{8.9}$$

The unique successor constraints model the way of operating the classification yards. More precisely, if two different orders that share the same destination meet at some station in the network, they are not allowed to split up their paths from then on and therefore enter the same yards till the destination:

$$\forall p,q \in K, p \neq q, d_p = d_q, i \in V, (i,j_1),(i,j_2) \in A, j_1 \neq j_2 : x^p_{i,j_1} + x^q_{i,j_2} \leq 1. \tag{8.10}$$

If we do *not* include constraints (8.10) into the formulation, we will obtain a model for the blocking of cars, a similar operational mode for classification yards.

The following constraints refer to the structure of the car routes with the aim of achieving a model formulation that a MILP solver is able to solve faster. The general idea for these kind of constraints was introduced in [10] under the term *heuristic cuts*. Heuristic cuts are not exact in the way that they do not cut off parts of the feasible solution. To impose a certain structure of the routes, a hierarchy level h_i for each station $i \in V$ is introduced. The larger the station, the smaller the value of h_i. Cars first ascend in the hierarchy from small to large yards, then stay for a while on that level, and finally descend from large to small again. Not all shipments have to go to the highest level. In particular, when the transport distances are short, the routes may stay more often on the lower levels. However, for long-distance transports the routes typically go to higher levels as quickly as possible, because there are enough capacities to reassign many cars to new trains while in lower levels you can reclassify only few trains. For any car it is not allowed to ascend in the hierarchy level after a descent. This behavior is ensured by the following hierarchy constraints:

$$\forall j \in V, p \in K: \sum_{i:(i,j) \in A, h_i \leq h_j} x^p_{i,j} + \sum_{k:(j,k) \in A, h_j \geq h_k} x^p_{j,k} \leq 1. \tag{8.11}$$

For a better understanding of these constraints we refer to a family's holiday journey on the roads. Here a vehicle may start on small roads in a residential area, then uses streets to finally enter the motorway. Usually the vehicle is on the motorway for the main part of the journey. Then the other way round it takes streets and small roads to get to his destination.

8.4 Model Adaptation

The presented model was successfully deployed for DB's strategic long-term planning of single car rail freight transport. A screenshot of the graphical user interface of the planning system with a national scenario running is shown on page 169 in Fig. 8.1. Over the years, some requirements changed and new ones arose. We describe the upcoming situation referring to different application cases and present how to cope with the new requirements through adaptations to the optimization model.

8.4.1 Interlocking of Unit Trains and Single Cars

Customers of DB Schenker Rail (DBSR) have to specify the number of cars to be transported as well as the corresponding times in the new business model in much more detail and in advance, see Novak [18]. This gives DBSR the ability to provide a more reliable transport plan for their customers. An early capacity checking is possible while taking into account the entire network and concentrating on traffic flows. This results in the identification of block trains. Block trains are unit trains that have residual capacities in train length and weight. The unit trains are filled with a sufficient number of single cars that have a similar destination nearby, so that the train then is maximal in length or weight. These unit trains can be transported over long distances without reclassification at yards. The new blocking concept provides more reliable planning, more stable and often faster transportation, saves production costs and aims to an optimal use of existing resources. First parts of the new concept have been in use since the end of 2012. The new business model is going to be implemented gradually. For more information we refer to [12].

According to the blocking concept, unit trains as long as they provide free capacities do no longer use a direct trip from their origin to their destination, but have to stop at intermediate yards to take and unload single cars. Since unit trains often are imposed by tight time restrictions, they do not need to be classified in these intermediate yards. For the model adaptation, this means we have to consider two sets of orders. K^S is the set of customer orders for single cars and K^U is the set of customer orders for unit trains with residual capacities, where the set of all customer orders is $K = K^S \cup K^U$. Then we can use the basic model formulation as follows.

Flow conservation (8.2), time limit constraints (8.3) and (8.4), capacity constraints (8.5), (8.6), (8.7), (8.8), and (8.9) as well as the unique successor rule (8.10) have to hold for the set of orders K. The hierarchy constraints only have to be satisfied for the set of single car orders:

$$\forall j \in V, p \in K^S: \quad \sum_{i:(i,j)\in A, h_i \le h_j} x^p_{i,j} + \sum_{k:(j,k)\in A, h_j \ge h_k} x^p_{j,k} \le 1. \tag{8.12}$$

Since the unit train orders are built for long-distance transport without being reclassified very often, the hierarchy constraints do not have to be fulfilled.

Unit trains are not classified at intermediate yards, even if they stop at a hump yard, they will not be pushed over the hump for classification. However, unit trains occupy classification tracks and shunting resources for coupling and decoupling of single cars at the respective yards, they have to satisfy the yard capacity restriction (8.5).

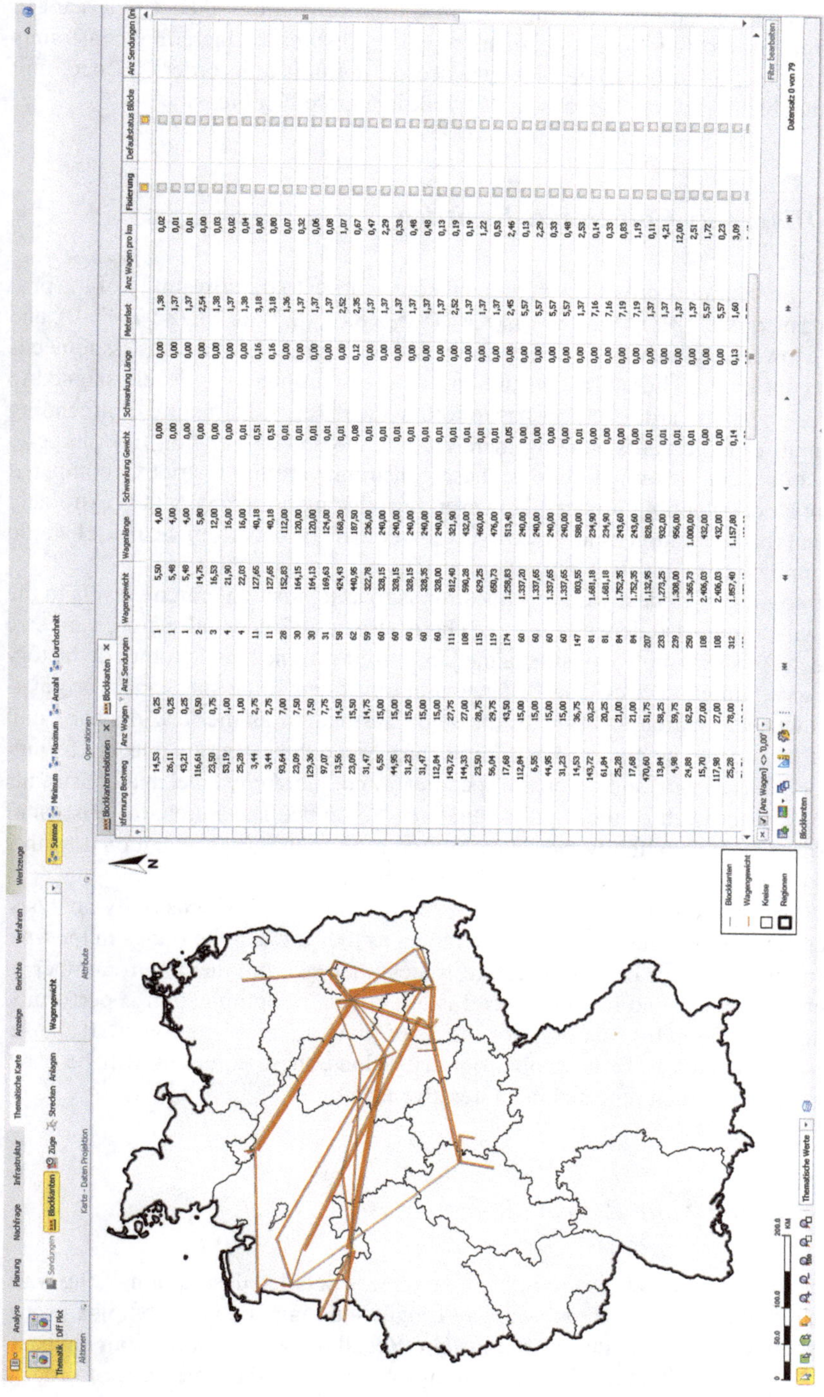

Fig. 8.1: Graphical user interface of the planning system at DB. Rail freight transport within Germany

In general, unit trains hold more narrow time restrictions than single cars to travel from their origin to their destination. Due to this reason hierarchy constraints (heuristic cuts) as in (8.12) are not imposed on the unit train orders. Hence, unit trains are able to traverse the railway network in a more flexible way.

8.4.2 Hierarchy Constraints Revisited

Hierarchy constraints provide a significant contribution to the solvability of the proposed basic model (8.1), (8.2), (8.3), (8.4), (8.5), (8.6), (8.7), (8.8), (8.9), (8.10), and (8.11). However, planning processes in recent years unfolded that this issue needs to be reconsidered. Hierarchy cuts are based on certain hierarchy level assigned to each classification yard. Each car has to follow the described structure of ascending from small to large yards, staying there for a while and descending from large to small yards again. It turned out that these constraints are too restrictive compared to the practical planning process. However, introducing hierarchy cuts is extremely beneficial for calculation time. To completely remove the cuts from the model would result in practically unsolvable problem instances.

One possibility to deal with this question is to assign new hierarchy levels to the yards, which may also depend on some characteristics of the local region, yard size and order. As an example, consider a medium-sized yard near an automobile factory having a certain number of long-distance unit trains. For some single cars, originating in that local region and having the same destination as the long-distance unit trains, the yard can be regarded as a large yard. Since the hierarchy cuts are formulated for a pair of yard and order, the specification of hierarchy level and constraints can be done for each of those pairs separately to embed special characteristics of the yards. This gives slightly more flexibility for the solution space, but on the other hand does not lead to entirely unsolvable instances.

Another way to deal with this issue is to perform a kind of sensitivity analysis. All hierarchy constraints are kept in the model as they are and the routes in the final solution are analyzed. In case there are routes, that take the maximum number of hierarchy levels (i.e., no hierarchy level is left out), a re-optimization is performed removing the hierarchy cuts only for the respective orders of the routes. If the re-optimization results in better global optimal values, these solutions will be used. This procedure is then repeated in an iterative fashion.

8.4.3 Waiting Time Revisited

The goal of this adaptation is to implement a more detailed description of the waiting time in classification yards. Railway freight shunting is a very complex procedure, where Chap. 9 provides a good insight. Recall the waiting time from the basic model, $w^p_{i,j}$ is described in (8.4). This number represents the time, cars of order p

have to wait on a sorting track (i, j) to be assembled to a train, that then re-enters the network. However, if the number of cars, that enter yard i for classification, reaches a certain yard specific threshold (which is below the hump capacity H_i), congestion starts and classification of cars needs additional time due to the high car volume in the yard. This threshold is denoted by S_i. As soon as the number of cars goes further beyond S_i the average delay for each car will increase even faster. A common way to model the connection between the number of cars using a certain infrastructure and the resulting delay is to apply a *capacity restraint (CR) function*. These functions were first developed for describing jam situations in road traffic [15, 20], and [7] applied it to railway freight transport. Likewise we apply the adjusted and simplified version of the CR function to model the average delay

$$\tau \left(\frac{m_i}{\kappa_i} \right)^{\beta},$$

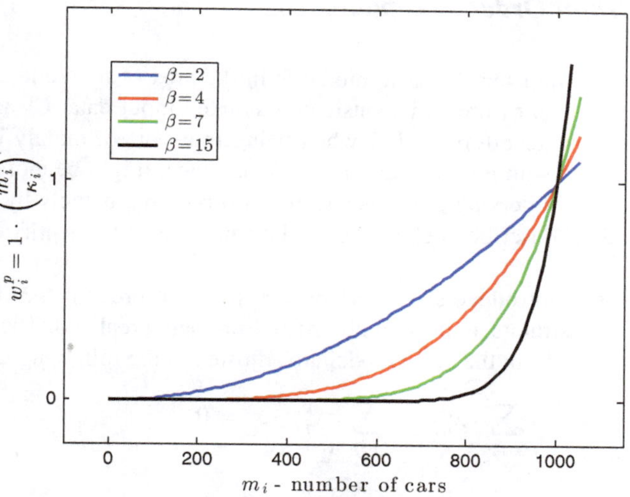

Fig. 8.2: CR functions with $\tau = 1$, $\kappa_i = 1000$ and $\beta \in \{2, 4, 7, 15\}$

where τ denotes the running time, κ_i the capacity limit of yard i, m_i the number of cars at yard i and β controls the growth of the penalization. Since we have no running time for traversing a yard, we use τ as a scaling factor. Assuming there is a maximum capacity limit $H_i > S_i$ for yard i, we have $\kappa_i = H_i - S_i$. Congestion only occurs, when the number of cars is between S_i and H_i. We denote the new additional waiting time by w_i and the number of cars, that are processed in yard i during a certain time period by

$$m_i = \sum_{p \in K, j:(i,j) \in A} v_p \cdot x_{i,j}^p.$$

With this, w_i is defined in the following way:

$$w_i = \begin{cases} 0 & \text{if } m_i < S_i, \\ \tau \cdot \left(\frac{m_i - S_i}{H_i - S_i} \right)^{\beta} & \text{if } S_i \leq m_i \leq H_i. \end{cases} \tag{8.13}$$

A visualization of the nonlinear CR function is given in Fig. 8.2. The respective time limit constraints of the basic model (8.3) change to

$$\forall p \in K: \sum_{(i,j) \in A} (w^p_{i,j} + w_i \cdot x^p_{i,j} + t_{i,j} \cdot x^p_{i,j}) \leq T_p. \tag{8.14}$$

Techniques for a linear approximation of (8.13) are described in [7]. The term $w_i \cdot x^p_{i,j}$ in (8.14) can be linearized using a straight-forward big-M formulation.

8.4.4 Restraint Order Acceptance

A further application for the basic model is in the area of marketing. So far, it is assumed that the set of orders is considered as fixed input data. Clearly, some orders require a higher effort for DB when being transported, mainly because they emerge in regions with little traffic. In such situations it might be more economical to transport these orders by other means of transportation, namely by trucks. With the modifications described below, the model can be used to identify these critical orders.

The idea is to introduce slack variables s^p for each order $p \in K$ into the flow conservation constraints, to prevent all orders from being realized. The flow conservation constraints from the basic model are adjusted in the following way:

$$\forall p \in K, i = o_p: \sum_{j:(i,j) \in A} x^p_{i,j} - \sum_{j:(j,i) \in A} x^p_{j,i} - s^p = 0,$$

$$\forall p \in K, i = d_p: \sum_{j:(i,j) \in A} x^p_{i,j} - \sum_{j:(j,i) \in A} x^p_{j,i} + s^p = 0,$$

$$\forall p \in K, i \in V, i \neq o_p, i \neq d_p: \sum_{j:(i,j) \in A} x^p_{i,j} - \sum_{j:(j,i) \in A} x^p_{j,i} = 0.$$

Additionally, a penalization term is inserted into the objective function. That means, we extend the objective function (8.1) from our basic model as follows:

$$\min \quad \alpha_1 \cdot \sum_{(i,j) \in A} \delta_{i,j} \cdot n_{i,j} + \alpha_2 \cdot \sum_{(i,j) \in A} y_{i,j} + \alpha_3 \cdot \sum_{p \in K, (i,j) \in A} \delta_{i,j} \cdot x^p_{i,j} + \alpha_4 \cdot \sum_{p \in K} s^p,$$

where α_4 is a scaling factor. It prevents all orders resulting in higher scaled costs than the value of α_4 from being scheduled. The orders that were not accepted are analyzed step by step according to their costs. The information that the realization of an order is highly expensive is then used by the planning department to enter into negotiations with the corresponding customer.

8.4.5 A Multi-Period Approach Towards Robust Planning

Planning is done for a certain period of time, usually it is referred to one abstract model day (24 h). However, the demand (set of orders) is not constant over a week, it usually changes from one day to the other. In order to plan trains, locomotives

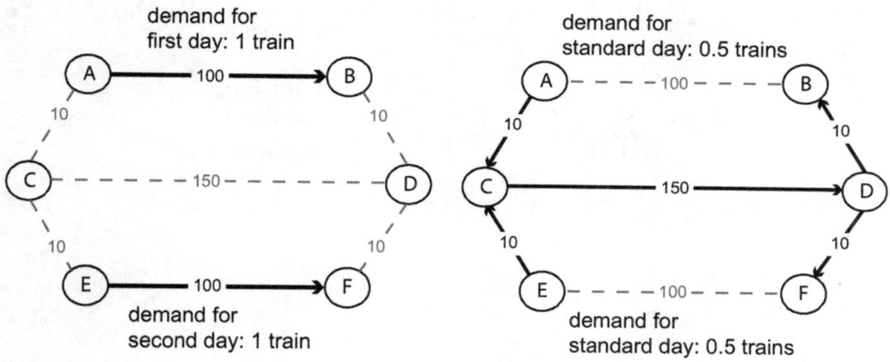

Fig. 8.3: The left figure shows a network with different demands for a period of 2 days. The right figure shows the average demand on a standard day. The respective routes are visualized by black solid lines

and staff currently a single train schedule is computed from the proposed model. Therefore, the given demand values are accumulated over a certain period of time (e.g., a season of 3–6 months), and average values for one standard day are used as input demand. Based on these average values a basic schedule is computed and rolled out to every single day. This schedule will only be suitable if transportation demand varies slightly. In case the demand significantly differs from one day to the other, the quality of the resulting optimal solution may be weaker, as the example in Fig. 8.3 demonstrates. Here, a network with six nodes A,B,C,D,E and F is given. Train operation cost for one train traversing an arc is given on the respective arc. In the left picture, there is a demand of one train from node A to node B and from node E to node F on the second day. In case that average demand values are used as input data, we obtain a demand of 0.5 trains for each of both days between nodes $A-B$ and $E-F$, as depicted in the right figure. This leads to scheduled trains from A to C, E to C, C to D, D to B and D to F on both days, resulting in an objective function value of $2 \cdot 190 = 380$. In case orders of different days are not combined for scheduling as described in the left figure, we obtain an objective function value of $2 \cdot 100 = 200$.

There is a need for a schedule, that is robust against errors induced by averaging orders over a certain period of time. One way to achieve this, is to compute a schedule for every single day of the period. Let us consider the set of days P that have to be differentiated. Then we introduce a set of links $B \subseteq K \times P$ between orders and their specified days. The tuple $(p,r) \in B$ denotes an order $p \in K$ of a customer with the specified day $r \in P$ the order p refers to. The decision variables $x^p_{i,j}$ then change to $x^{p,r}_{i,j}$ and $n_{i,j}$ to $n^r_{i,j}$ respectively. We have that $x^{p,r}_{i,j} = 1$ if and only if order k that is specified for day r goes from i to j, and $n^r_{i,j} \in \mathbb{Z}_+$ represents the number of trains from i to j on day r. The basic model may be adjusted as follows to fit to the described situation:

$$\min \quad \alpha_1 \cdot \sum_{r\in P}\sum_{(i,j)\in A} \delta_{i,j}\cdot n^r_{i,j} + \alpha_2 \cdot \sum_{(i,j)\in A} y_{i,j} + \alpha_3 \cdot \sum_{(p,r)\in B,(i,j)\in A} \delta_{i,j}\cdot x^{p,r}_{i,j}, \qquad (8.15)$$

$$\forall i \in V, (p,r) \in B: \sum_{j:(i,j)\in A} x^{p,r}_{i,j} - \sum_{j:(j,i)\in A} x^{p,r}_{j,i} = \begin{cases} 1 \text{ if } i = o_p, \\ -1 \text{ if } i = d_p, \\ 0 \text{ otherwise,} \end{cases}$$

$$\forall (p,r) \in B: \sum_{(i,j)\in A} w^{p,r}_{i,j} + t_{i,j}\cdot x^{p,r}_{i,j} \le T_p,$$

$$w^{p,r}_{i,j} = \frac{T}{n^r_{i,j}} \cdot x^{p,r}_{i,j},$$

$$\forall i \in V: \sum_{(p,r)\in B, j:(i,j)\in A} v_p \cdot x^{p,r}_{i,j} \le H_i,$$

$$\forall (i,j) \in A, r \in P: n^r_{i,j} \le N_i \cdot y_{i,j},$$

$$\forall i \in V: \sum_{j:(i,j)\in A} y_{i,j} \le Y_i,$$

$$\forall (i,j) \in A: \sum_{(p,r)\in B} \ell_k \cdot x^{p,r}_{i,j} \le L_{i,j}\cdot n^r_{i,j},$$

$$\forall (i,j) \in A: \sum_{(p,r)\in B} w_k \cdot x^{p,r}_{i,j} \le W_{i,j}\cdot n^r_{i,j},$$

$$\forall (p,r),(q,r) \in B, p \ne q, d_p = d_q, i \in V, (i,j_1),(i,j_2) \in A, j_1 \ne j_2: \quad x^{p,r}_{i,j_1} + x^{q,r}_{i,j_2} \le 1,$$

$$\forall j \in V, (p,r) \in B: \sum_{i:(i,j)\in A, h_i \le h_j} x^{p,r}_{i,j} + \sum_{k:(j,k)\in A, h_j \ge h_k} x^{p,r}_{j,k} \le 1.$$

We mention that this model is very similar to the approach of [19] with the difference that in our model the number of trains can differ on each day, whereas Voll suggested that also the number of trains have to be the same on each day for reducing planning efforts and operational advantages.

Due to the increased number of variables and constraints, the proposed model may not easily being solved by standard optimization software. However, an optimal solution to (8.15)–(8.25) will lead to a robust schedule, since each day is considered separately.

8.4.6 Moving Horizon

In Sect. 8.4.5 we discussed how the model can be adapted to produce robust plans in the sense that for different OD-matrices the assignment of the classification tracks is the same, while the number of trains may differ. Now we present another possibility how our model can be adapted to optimize different OD-matrices for each day of 1 week.

Note that—while optimizing one particular day—the travel times of orders might be much higher than 1 day, i.e., the orders are affecting the next day. This was no problem, because the OD-matrix was the same every day. However, if we dismiss our assumption of a recurrent OD-matrix, there is the consequence that we have to consider orders that remained from the previous day.

In Sect. 8.4.4 we showed a way how to model that an order can be declined, if its production cost are too high. We can use this idea with another intention: instead of generating an additional train, we leave an order for the next day. If we apply our model for consecutive days, we will add the remaining orders to the following OD-matrix, while prohibiting that it is left on two consecutive days.

This idea does not work properly though. In the formulation of Sect. 8.4.4 a declined order is always left at its origin, while in the new interpretation we would like to ship an order as far as possible towards its destination, unless there is no space in the train left on the current day.

This can be modeled with the following extension of the formulation. We use the formulation of Sect. 8.4.4 but we add an additional index i to the slack variable s^p with $s_i^p = 1$, if order p is left at station i and 0 otherwise. Thus we change the flow conservation constraints to:

$$\forall p \in K, i = o_p : \sum_{j:(i,j)\in A} x_{i,j}^p - \sum_{j:(j,i)\in A} x_{j,i}^p - s_i^p = 0,$$

$$\forall p \in K, i \neq d_p : \sum_{j:(i,j)\in A} x_{i,j}^p - \sum_{j:(j,i)\in A} x_{j,i}^p + s_i^p = 0.$$

Note that we do not need a particular constraint for the destination since the meaning of s_i^p that order p is left in station i also applies for the destination.

Now to model that the order is shipped towards its destination we have to set the right costs. Let $D_p(i)$ denote the distance of node i to the destination of order p. The coefficient of the slack variable s_i^p depends on the distance to the destination: the closer the station i is to the destination, the lower the costs.

$$\min \quad \alpha_1 \cdot \sum_{(i,j)\in A} \delta_{i,j} \cdot n_{i,j} + \alpha_2 \cdot \sum_{(i,j)\in A} y_{i,j} + \alpha_3 \cdot \sum_{p\in K,(i,j)\in A} \delta_{i,j} \cdot x^p_{i,j} + \alpha_4 \cdot \sum_{p\in K} D_p(i) \cdot s^p.$$

Note that, since $D_p(d_p) = 0$, there is no penalty term if an order reaches its destination. This formulation can be regarded as a generalization of the formulation in Sect. 8.4.4 in the sense that it is possible to fix $s^p_i = 0$ for all $i \neq d_p$.

With this extension we can now set up a moving horizon. We define our horizon as a time span—1 day, for example—and reduce the maximum transportation times of all shipments to this time span. If an order has a maximum transportation time less than the time span, it will not be changed. Now we apply our model to this first time window, obtaining positions of each order at the end of that time span evaluating $s^p_i = 1$. For the second time span, i.e., the next day, we add all orders from the previous day to the OD-matrix that are not at their respective destination. For those orders, we reduce the maximum transportation time by the time span, i.e., 1 day.

With this framework, it is easy to set up more sophisticated techniques such as adding "urgency" by increasing penalty costs for orders that have a high D_p/T_p ratio, meaning that an order has to travel a rather large distance in a short time. Also, it is possible to use overlapping time windows in a way that the routing of an order from the previous time span can be revised in order to obtain a better overall solution.

By this formulation, the model size would increase by considerably $|V| \cdot |K|$ variables. However, if the time span is sufficiently small, such as 6 or 12 h, the number of orders will be reduced significantly.

8.4.7 Less-Than-Truckload Optimization

Besides rail freight traffic, DB Schenker Logistics is also one of the five biggest road freight transportation company. Hence, our methods for modeling and solving the car routing problem were also re-used to solve routing problems for road traffics. This refers to the class of less-than-truckload (LTL) problems. The main difference is the following. In rail freight transportation, we had a bilevel structure of entities: the customers have wagons, which are assembled into trains. Now, in road freight transportation, we have in fact a trilevel structure. The customers have loads on pallets. The pallets are grouped into swapbodies (containers), and the containers are transported by trucks, where each truck can carry one or two swapbodies. Swapbodies can be moved between trucks without being opened. In a terminal, a gate is assigned to another terminal, i.e., the swapbody at that gate will be moved to the respective terminal where it will be unloaded. It is possible that the pallets then are assigned to another gate for a further terminal until they finally reach their respective destination. Also, it is possible that after departure of one swapbody, another swapbody is used for the same destination. The overall objective is to deliver all pallets to their destination with minimal costs.

We can solve this problem heuristically by applying our model twice. First, we consider the assignment of pallets to swapbodies. To apply our model, we regard a terminal as station, the gates as classification tracks, and swapbodies as trains. The pallets can be seen as the orders in our model. By this interpretation, we can set the costs and parameters like capacities of the swapbodies and handling costs in the terminal. In the first run we optimize the assignment of pallets to swapbodies and obtain the number of swapbodies between the terminals as a result. For the second run, we consider the swapbodies as orders where origin and destination of the order correspond with origin and destination terminal of the swapbody. The volume of the order is the number of swapbodies on the particular OD-pair. Now we have trucks—regarded as trains in our model—carrying the swapbodies. A truck has the capacity of two swapbodies. The objective for the second run is to move all swapbodies from their origin to their destination at minimal transport costs that are linear in truck kilometers.

We applied this method in several projects for DB Schenker Logistics. It turned out to be very effective, and the solutions of this heuristic were regarded as very plausible and accepted by the planners. Note that we used the original model described in Sect. 8.3 and did not change a single line of code for solving this LTL problem.

8.5 Conclusion

We presented a model for freight train car routing, that is in use at DB since 2011 for strategic planning and simulation purposes. Over the years, several demands and new application scenarios emerged, that require an adaptation of the model by adding new constraints, so that it is able to cope with these new scenarios. In most cases, the modifications can easily be carried out, because the modifications are just modest. However, the request of a robustification of the model is severe, so that it would require to think about the solution process for a longer time in a future research project.

The situation at DB shows that today the railway freight market is such a dynamic environment, that once established optimization tool have a very short lifetime in practice, and require a constant and continuous modification and extension to newly emerging demands. For DB (or any large railway company) it is thus vital to have a team of highly specialized experts in Operations Research, Mathematics, and Computer Science at hand, that are able to understand the algebraic models, formulate new requests as constraints, and work on numerical solution algorithms, and, equally important, are able to speak the same language as railroad practitioners, in order to implement the solutions back in the company's planning process, and demonstrate a value added over the still dominating manual planning.

References

1. Ahuja RK, Jha KC, Liu J (2007) Solving real-life railroad blocking problems. Interfaces 37(5):404–419. ISSN: 0092-2102. https://doi.org/10.1287/inte.1070.0295 (cited on page 164)
2. Assad AA (1981) Analytical models in rail transportation: an annotated bibliography. Inf Syst Oper Res 19(1):59–80. https://doi.org/10.1080/03155986.1981.11731807 (cited on page 163)
3. Assad AA (1983) Analysis of rail classification policies. Inf Syst Oper Res 21(4):293–314. https://doi.org/10.1080/03155986.1983.11731905 (cited on page 163)
4. Balakrishnan A, Magnanti TL, Mirchandani P (1997) Network design. In: Dell'Amico M, Maffioli F, Martello S (eds) Annotated bibliographies in combinatorial optimization. Wiley, London, pp 311–334. ISBN: 978-0-471-96574-9 (cited on page 164)
5. Bley A (2007) Routing and capacity optimization for IP networks. PhD thesis, Technische Universität Berlin. https://doi.org/10.14279/depositonce-1588 (cited on page 164)
6. Bodin LD, Golden BL, Schuster AD, Romig W (1980) A model for the blocking of trains. Transp Res B 14(1):115–120. https://doi.org/10.1016/0191-2615(80)90037-5 (cited on page 163)
7. Borndörfer R, Klug T, Schlechte T, Fügenschuh A, Schang T, Schülldorf H (2016) The freight train routing problem for congested railway networks with mixed traffic. Transp Sci 50(2):408–423. https://doi.org/10.1287/trsc.2015.0656 (cited on pages 171, 172)
8. Clausen U, Voll R (2013) A comparison of North American and European railway systems. Eur Transp Res Rev 5(3):129–133. https://doi.org/10.1007/s12544-013-0090-4 (cited on page 164)
9. Crainic T, Ferland J-A, Rousseau J-M (1984) A tactical planning model for rail freight transportation. Transp Sci 18(2):165–184. https://doi.org/10.1287/trsc.18.2.165 (cited on page 163)
10. Eisenblätter A, Fledderus ER, Fügenschuh A, Geerdes H-F, Heideck B, Junglas D, Koch T, Kürner T, Martin A (2003) Mathematical methods for automatic optimization of UMTS radio networks. Tech. rep. Zuse Institut Berlin (ZIB), Berlin. `momentum.zib.de` (cited on page 167)
11. Fügenschuh A, Homfeld H, Schülldorf H (2015) Single-car routing in rail freight transport. Transp Sci 49(1):130–148. https://doi.org/10.1287/trsc.2013.0486 (cited on pages 162, 164, 166)
12. Heinrici T (June 2017) Platzkarten für den Güterverkehr (internet article of DVZ, Logistik auf der Schiene). `dvz.de` (cited on page 168)
13. Homfeld H (2012) Consolidating car routes in rail freight service by discrete optimization. Verlag Dr. Hut, München. `urn:nbn:de:101:1-20120618285` (cited on page 162)
14. IBM ILOG. CPLEX. Apr. 2016. `ibm.com` (cited on page 164)

15. Irwing N, Cube HV (1962) Capacity restraint in multi-travel mode assignment programs. Highw Res Board Bull 347:258–287 (cited on page 171)
16. Keaton MH (1989) Designing optimal railroad operating plans: Lagrangian relaxation and heuristic approaches. Transp Res B 23(6):415–431. https://doi.org/10.1016/0191-2615(89)90042-8 (cited on page 163)
17. Newton H (Jan. 1997) Network design under budget constraints with application to the railroad blocking problem. PhD thesis, Auburn University, Auburn, AL (cited on page 164)
18. Novak A (2013) Innovation: full speed ahead for Netzwerkbahn. Railways 2:17–22 (cited on page 168)
19. Voll R (2014) Methoden der mathematischen Optimierung zur Planung taktischer Wagenrouten im Einzelwagenverkehr. Verlag Dr. Hut, München. https://doi.org/10.17877/DE290R-191 (cited on page 174)
20. Wohl M (1968) Notes on transient queing behavior, capacity restraint functions, and their relationship to travel forecasting. Pap Reg Sci 21(1):191–202. https://doi.org/10.1007/bf01952729 (cited on page 171)

Chapter 9
Optimization of Railway Freight Shunting

Markus Bohlin, Ronny Hansmann, and Uwe T. Zimmermann

Abstract Railway freight shunting is the process of forming departing trains from arriving freight trains. The process is continuously performed at rail yards. The shunting procedure is complex and rail yards constitute bottlenecks in the rail freight network, often causing delays to individual shipments. One of the problems is that planning for the allocation of tracks at rail yards is difficult, given that the planner has limited resources (tracks, shunting engines, etc.) and needs to foresee the consequences of committed actions for the current inbound trains.

M. Bohlin
KTH Royal Institute of Technology, Stockholm, Sweden
e-mail: mbohl@kth.se

R. Hansmann (✉)
Ostfalia – University of Applied Sciences, Salzgitter, Germany
e-mail: r.hansmann@ostfalia.de

U. T. Zimmermann
Institute of Mathematical Optimization, TU Braunschweig, Germany
e-mail: u.zimmermann@tu-bs.de

R. Borndörfer et al. (eds.), *Handbook of Optimization in the Railway Industry*, International Series in Operations Research & Management Science 268,
https://doi.org/10.1007/978-3-319-72153-8_9

The required schedules highly depend on the particular infrastructure of the rail yard, on the configuration of inbound and outbound trains, and on the business objectives. Thus, new optimization tools as active decision support for the dispatchers are closely tailored to the actual processes. Due to its practical relevance, a broad range of variants has been discussed and solved by the scientific community in recent years. For selected relevant variants, we describe their fruitful relation to scientific research topics such as graph coloring, sequence partitioning, and scheduling, we discuss their computational complexity and approximability, and we outline efficient optimization procedures.

In particular, we consider a set of models and algorithms which are applicable in practice, and discuss their application to the shunting yards in Ludwigshafen, Germany and in Hallsberg, Sweden. We also discuss similarities and differences between the different approaches and outline the need for future research.

9.1 Introduction

Railway freight transportation is a classical application area of operations research; a recent overview is given by Nemani and Ahuja [61]. In the literature, much of the focus has been on railway freight in the U.S., where the infrastructure is typically owned by the operator, freight trains are longer and heavier and the market share for rail freight is higher than in Europe. For freight operations in particular, the *railroad blocking problem*, which considers the formation of a cost-minimal freight transportation plan, has been studied. In this problem, freight yards are considered on a macroscopic level, and operations within the yard are not relevant.

However, this chapter is concerned with operations research approaches for freight yard planning on a microscopic level, where the actual car movements and shunting operations are considered. Consequently, the various solution methods for the railroad blocking problem (cf. [1, 5, 7, 38, 46, 51, 62, 69]) are not directly applicable.

9.1.1 Classification Problems in Classification Yards

Railway freight transportation offers two distinct types of services. In full train load service, all cars of a single train have the same origin and destination, while in car load service, cars can have different origins and destinations. In the latter case, freight trains are typically composed of individual cars, which share some legs of the full journey. Transportation in a car load service involves one or more break-up and formation steps, which are performed at specialized rail yards called classification yards (also shunting yards, marshalling yards and hump yards). The services offered at a classification yard include the following:

- Arrival and departure: car registration.
- Arrival and departure: inspections, wherein cars are checked for damage.
- Corrective maintenance of cars which fail inspections.
- Decoupling of arriving trains (and coupling of departing trains).
- Train formation, i.e., assigning outbound cars to departing trains according to destination.
- Train sorting, i.e., arranging the cars of a departing train according to a preferred car order.
- Parking empty freight cars.

To enable these services, shunting yards are quite complex, and contain several side tracks and smaller yards. The layouts of three shunting yards in Sweden are shown in Fig. 9.1. Planning all above-mentioned services and operations in the entire yard results in crucial, challenging tasks for the dispatchers. Uncertain arrival times, ad hoc changing orders of inbound cars, tight capacities, and financial constraints complicate the process and offer large potential for optimization.

In this chapter we focus on *classification problems* (CP) modelling train formation and train sorting as performed in a part of the shunting yard commonly named as *classification bowl*. A solution of a CP assigns any inbound car of an arriving train to a departing train. In more detail, a solution describes a schedule of the car movements into and out of the so-called *sorting tracks* of the classification bowl. Optimal solutions describe most efficient *classification schedules* for transforming a fixed sequence of inbound cars into one or many departing trains with regard to prescribed requirements. Typical goals and measures for the efficiency of a schedule are discussed in Sects. 9.3 and 9.4.

Profit maximizing selections of inbound cars which can feasibly be shunted to departing trains are studied in [15]. Other real-world applications with underlying theoretical problems similar to CP have also been tackled by mathematical optimization approaches: dispatching buses in parking depots, see Gallo and Di Miele [35] and Hamdouni et al. [41]; stowage planning of containers in container ships, see Avriel et al. [4]; storage planning of steel slabs in integrated steel production, see König et al. [55], König and Lübbecke [53], and König [52]; job sequencing on conveyor or automated storage systems, see Han [42] and Demange et al. [26].

9.1.2 Short Historical Review of Classification Methods

Rearranging cars remains one of the biggest challenges in operating freight railways. In the literature, first methods for rearranging cars are called *sorting schemes*. Such schemes construct a classification schedule for a requested outbound car sequence without considering the inbound car order. Futhner's method is one of the oldest sorting schemes. According to [48], Harry Futhner was the first to apply it in practice in 1880, i.e., to the parallel dead-ended tracks at Liverpool station.

Over the course of time, similar rule-based methods for rearranging cars were developed and applied in practice. Among the most famous are the *simultaneous*,

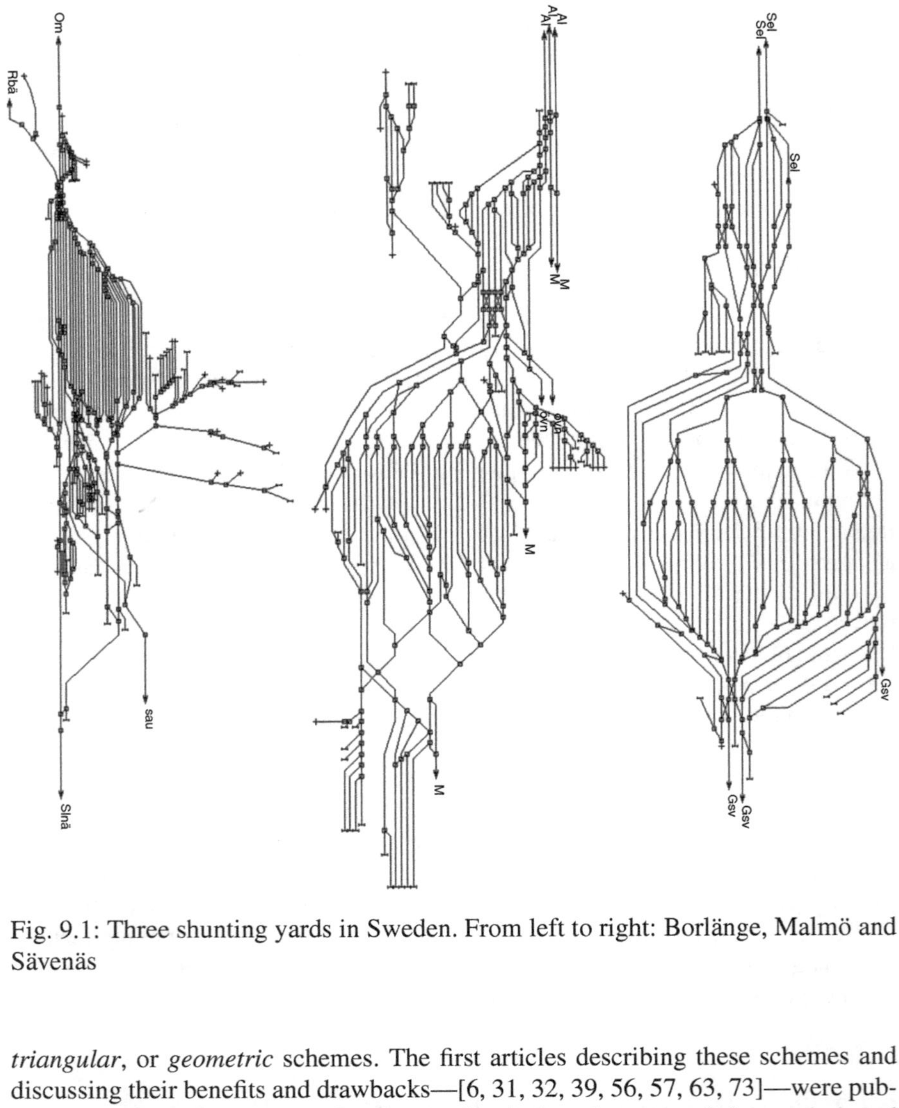

Fig. 9.1: Three shunting yards in Sweden. From left to right: Borlänge, Malmö and Sävenäs

triangular, or *geometric* schemes. The first articles describing these schemes and discussing their benefits and drawbacks—[6, 31, 32, 39, 56, 57, 63, 73]—were published in railway magazines. On the one hand, the advantage of these rule-based heuristical methods is their simplicity and transparency; experienced staff exactly knows what to do, no matter how the inbound cars are actually ordered. On the other hand, the required sorting may of course finish faster, with fewer tracks and less shunting operations if it implements an optimal scheme computed by an exact classification method. In the literature, cf. [54, 66], the oldest such strategy was introduced by the Operations Research group at the Schweizer Bundesbahnen in the late sixties of the last century. Under the direction of the mathematician Peter Schaltegger, they developed and implemented a mathematical optimization algorithm in FORTRAN IV. Subject to certain assumptions, they could generate an optimal simultaneous schedule for the predicted order of inbound cars on a Univac 1107 within seconds. However at that time implementation of computed schedules in daily operation was due to fail. One reason is that there were no adequate, fast information management systems for the many employees involved in the process. Alas, optimization methods did not replace rule-based schemes in practice. The next three decades produced little methodological progress for train classification. On the theoretical side, the effectiveness of rule-based strategies for different scenarios was analysed, e.g., in [2, 3, 20–23, 64, 65, 67, 68].

In the course of time, technical advances in rail yards created the prerequisites for a convenient use of automatically generated efficient schedules, e.g., automatic switches and brakes replaced mechanical ones. Moreover, at least in principle, information management systems became available. Nowadays a dispatcher may monitor and control the classification process from a control center, and the few people who are still physically involved are connected via fast communication networks.

As a consequence, on the practitioners side there is increasing interest for supporting optimization tools. Before the turn of the millennium, practitioners and researchers rarely worked together in this field. However, in recent years, quite a few joint projects for rail operators and universities were launched. Luckily for the engaged researchers, the underlying theoretical problems turned out to be challenging as well as beautiful.

For the last decade, publications on shunting and classification problems have in fact shown to address a hot topic from both the theoretical as well as the practical perspectives. Recent introductory surveys and literature reviews are given by Boysen et al. [14], Hansmann [43], and Gatto et al. [36]. Relevant references to details and results are included in Sects. 9.3 and 9.4.

This chapter deals with offline variants of CP where all relevant information is known in advance. Some online variants of CP are discussed in Hansmann [43]. Since robustness is not a particular issue in this chapter, we just briefly mention some related literature. For example, robust approaches may allow solutions to be valid even under disturbances such as delays and deviations in the predicted car order. In particular, recoverable robustness, in which well-defined recovery actions can be taken to correct the effects of a disturbance, has been considered for freight shunting problems, cf. [59]. In general, recoverable robustness models have to be simplified substantially in order to be tractable, and are therefore mostly of theoretical value. The concept was first considered for shunting by Cicerone et al. [18] for two disruption types (a car being in the wrong place, and a new car being introduced) and different recovery strategies. Büsing and Maue [16] uses the same concepts and presents a general algorithm for finding a classification schedule that allows a recovery strategy where k additional sorting steps can be inserted after the p first steps while minimizing the total number of re-classification steps in the original schedule.

At the present time, despite all the mentioned research and developments, there still exists no active optimization tool as decision support for the dispatchers in most railway operating companies. One reason may be that it is hard to come up with a standard approach. The schedules that need to be generated highly depend on the particular infrastructure of the rail yard, the configuration of arriving and departing trains, and the requested objective. Thus, algorithms for computing high quality schedules still have to be tailor-made to the actual yard situation.

9.2 Classification Scheme of Classification Problem Variants

There is a huge number of quite different variants of CP; these result from alternative goals, from different assumptions on the track topology, on the necessary chronology of car movements, or on the requirements for the sequence of outbound cars in departing trains. In this chapter, the classification scheme and the necessary notations from [43] are slightly adapted and extended.

First of all, we describe a generic framework including all CP variants. In all CP variants, we consider a fixed sequence of inbound cars that approaches the classification bowl on a virtual inbound track, cf. Fig. 9.2. We do not allow that cars arrive at the same time. If car u arrives earlier than car v, then u has to leave the inbound track before v. In any variant and for any car, we allow a movement from the inbound track to a sorting track in the classification bowl (*i-t-move*) as well as a movement from a sorting track to a virtual outbound track (*t-o-move*).

Based on the above-mentioned generic framework, we list further properties and requirements specifying a particular CP variant.

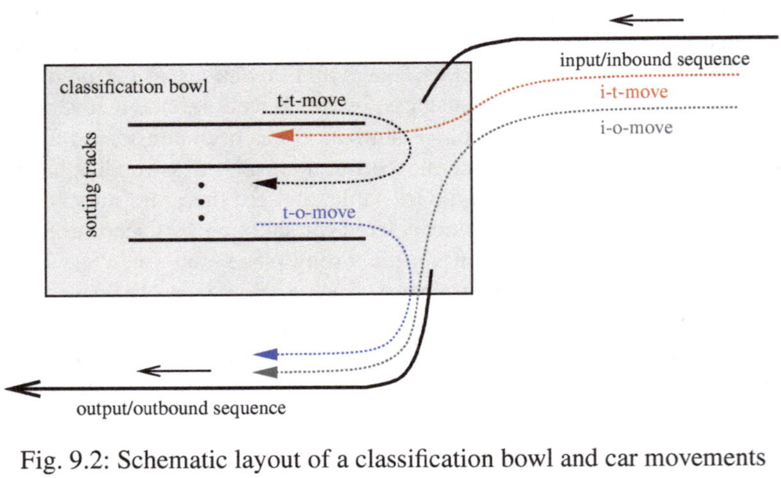

Fig. 9.2: Schematic layout of a classification bowl and car movements

9.2.1 Track Topology

There are various topologies of the classification bowl depending on the design and the length of the sorting tracks. In the following, if we speak of *tracks* we refer only to these sorting tracks.

9.2.1.1 Design

If the tracks are dead-ends and hence may be accessed only from one side, we speak of **stacks**. Note that any two cars parked on the same stack will change their order from arrival to departure if they are not additionally rearranged or shunted. Under the same assumption, any two cars preserve their order from arrival to departure when placed on a queue (**queues**), which is a one way track where the units arrive at one end and leave at the opposite side. In the case denoted as **stacks/queues**, one may freely decide whether a track is used as a queue or a stack track. Since this is an unnecessary restriction from practical point of view, this rather is a theoretical assumption. In the above three cases, both the entrance and exit are only on one—possibly differing—side of the track, which is known in the literature as *siso* (single in single out). In the following track designs, units may arrive at or depart from both sides of the tracks: **sido** (single in double out), i.e., entrance is on one side, exit is on both sides; **diso** (double in single out), i.e., entrance is on both sides, exit is on one side; **dido** (double in double out), i.e., entrance and exit are on both sides. With respect to the number t of tracks in the rail yard we denote the above-mentioned track designs by t-**stacks**, t-**queues**, and so on.

9.2.1.2 Length

Of course, in real rail yards, tracks are bounded in length. In the general **bounded** case, we consider sorting tracks with arbitrary (nonuniform) lengths. In the special case b-**bounded**, at most b cars may be placed on each track. Though **unbounded** track lengths are seemingly a rather theoretical issue, they may still well be reasonable from a practical point of view. For a discussion on this, see Sect. 9.5.

9.2.2 Sorting Mode

9.2.2.1 Shunting

Remember, in any CP variant we allow i-t-moves and t-o-moves of the cars. The infrastructure and topology of some rail yards may enable further movements from the virtual inbound track to the virtual outbound track (i-o-moves) or from the sorting track to another sorting track (t-t-moves), cf. Fig. 9.2. If no t-t-moves are allowed, we speak of **single-stage** variants. Otherwise, we speak of **multi-stage** variants. We refer to **single-stage** variants where i-o-moves are not allowed as **ito-shunting** variants (in [43] called "no shunting" variants). A particular **multi-stage** variant allowing i-o-moves is the so-called **hump-shunting** variant, practically motivated by common hump yards, see Sect. 9.4.

9.2.2.2 Timing of t-o-Moves

In **sequential** variants, t-o-moves of the cars are not allowed before the *parking process* is finished, that is, not before the i-o- or i-t-movement of the last incoming car. In contrary, in **concurrent** or **anytime** variants, the *parking process* is not separated in time from t-o-moves. However, in the **concurrent** case, cars of group g may be t-o-moved only after the last incoming car of group g is either i-t- or i-o-moved. In **anytime** variants, i-t-moves of cars are allowed at any point in time.

9.2.2.3 Splitting

The inbound cars are characterized by a particular distinctive criterion, e.g., their destination, their designated departing train, or their construction type. We say that cars satisfying the same criterion form a *group*.

Let us consider a **sequential, ito-shunting** variant with the additional requirement that the cars depart group by group. Then, it might be reasonable to require **0-split**, that is, that all cars of one group are placed (en bloc) on the same sorting track. Because, then, a shunting engine could carry the cars of one group out of the classification bowl with a single pull-out-operation. For example, the **0-split** assignment of seven incoming cars to two sorting tracks as follows, $|| - 3 - 1 - 1 \rightarrow$ and $|| - 4 - 4 - 2 - 2 \rightarrow$, allows a groupwise departure with four pull-out-operations. In general, the s-**split** condition requires that the cars of one group may be distributed to at most $s+1$ sorting tracks with $s \geq 0$.

The wish of reducing the number of pull-out-operations even further leads to the particularly restrictive splitting condition **chain-split**, also reasonable for **sequential, ito-shunting** variants. In **chain-split** variants, cars may be distributed over all tracks such that a single pull-out-operation per track leads to the required sequence of the outbound cars. For example, the **chain-split** assignment of seven incoming cars to two tracks as follows, $|| - 2 - 1 - 1 \rightarrow$ and $|| - 4 - 4 - 3 - 2 \rightarrow$, yields the desired output sequence $-4-4-3-2-2-1-1 \rightarrow$ with two pull-out-operations.

Finally, in **split** variants, we may arbitrarily distribute the cars of one group to the sorting tracks.

9.2.3 Requirement for Outbound Sequence

Of course, in order to obtain a particular CP variant, we need to define which outbound sequence of the cars is feasible and which is infeasible. Avoiding a lengthy, summarizing description of various practically motivated assumptions at this point, we distinguish later between usual requirements for **single-stage** classification, cf. Sect. 9.3, and those which are common for **multi-stage** classification, Sect. 9.4.

9.2.4 Goal

So far, we stated that CP aims at schedules that contain information about car movements that "efficiently" lead to a sequence of outbound cars as desired. What "efficiently" means in practice depends on the particular CP variant, on the amount of cars, on the capacities of the yard, as well as on the intention of the dispatchers. Thus, CP variants with various objective functions were studied in the literature. We discuss goals for **single-stage** variants in Sect.9.3, and later, in Sect. 9.4, other objective functions tailored to **multi-stage** variants.

Table 9.1: Parameter values for CP variants

Track topology (α)		Sorting mode (β)			Outbound sequence (γ)	Goal (δ)
Capacity	Design	t-o-moves	Shunting	Splitting		
unbounded	**(t-)stacks**	**sequential**	**ito-shunting**	**0-split**	**(o-)ordered blocks**	**t-min**
(b-)bounded	**(t-)queues**	**concurrent**	**(h-)hump-shunting**	**split**	**(o-)free blocks**	**h-min**
	(t-)stacks/queues	**anytime**		**chain-split**	**ordered pattern**	**c-min**
	(t-)sido	**scheduled**			**free pattern**	
	(t-)diso				**time windows**	
	(t-)dido				**sotwoc**	

We classify CP variants by an $\alpha|\beta|\gamma|\delta$ notation, where α specifies the track topology, β the sorting mode, γ the requirements for the outbound sequence, and δ the goal, see Table 9.1 for a summary of studied parameter values. The parameter values in the last two columns regarding γ and δ are discussed at appropriate points in the following two sections.

9.3 Single-Stage Classification

In this section, we focus on **ito-shunting** variants. Firstly, we further classify these CP variants by various requirements for the outbound sequence of cars.

Requirements for Outbound Sequence

Remember, the incoming cars are classified by a distinctive criterion, that is, by groups. We label the groups by consecutive natural numbers beginning from 1. In many practically relevant variants, a block pattern of the outbound sequence of cars is required that may be described by $(a,\dots,a,b,\dots,b,c,\dots,c,\dots)$ for short where a, b, c, ... are different group labels. In other words, the cars have to depart group by group. In practice, the groups are often associated with departing trains. In this case, the block pattern enables a train by train departure. However, if a departing train delivers its cars to several unloading stations, then a block pattern for the train itself might be reasonable. Because, then, the car groups may be decoupled at the rear of the departing train at each unloading station without additional shunting. In that case, the groups are associated with the unloading stations.

In a **free blocks** variant, we allow any of the $g!$ possible block patterns of the outbound sequence, where g is number of different groups of the inbound cars. In other words, the cars have to depart group by group and the departing order of the groups does not matter. On the contrary, in an **ordered blocks** variant, we only allow the block pattern $(1,\dots,1,2,\dots,2,3,\dots,3,\dots)$ of the outbound sequence. That is, we require that all inbound cars of group 1 depart consecutively at first, followed by all cars of group 2, followed by all cars of group 3, and so on.

Allowing arbitrary patterns—not necessarily block patterns—for the outbound sequence leads to **free pattern** and **ordered pattern** variants analogously. For example, if the sequence of inbound cars consists of three cars of group 1, of two cars of group 2, and of three cars of group 3, then the desired **free pattern** (a,b,c,b,b,c,a,a) allows the two particular patterns $(1,3,2,3,3,2,1,1)$ (either assign group 1 to positions a) and $(3,1,2,1,1,2,3,3)$ (or assign group 3 to positions a) of the outbound sequence.

Now let us consider a particular ordered blocks variant which additionally requires that the groups depart at predefined times. In these so-called **time windows** variants, the arrival time and departure time for any car are given in advance, and any feasible schedule to compute has to comply with these times strictly. For the sake of a clear inbound sequence we assume that no two cars arrive at the same time. Another assumption is that any two groups depart at different times. In other words, a group is determined by all cars with an identical departure time.

Note that, we do not require a particular order of the departing cars within one group.

Goal

For the **ito-shunting** variants discussed in this section, our goal is to compute a feasible schedule that "occupies" only a minimum number of sorting tracks in the classification bowl. This objective, abbreviated by ***t*-min**, is motivated by the following reasons. First, the less tracks used, the larger the number of unused sorting tracks, and, thus, the less the maintenance costs. Second, a large number of free sorting tracks may be very useful in case of an unpredictable change in the inbound sequence or of operational disturbances. In this sense, ***t*-min** contributes to the robustness of the schedules.

For many **single-stage** variants we present complexity results that mainly stem from connections to either minimum sequence partitioning or to graph coloring. The following subsection provides concepts and notations that are relevant for particular sequence partitioning problems. For definitions regarding graph coloring and graph classes we refer to a standard book on Graph Theory, for example, to Golumbic [37].

9.3.1 Notations

The inbound sequence of cars can be formulated as an *integer sequence* $S = (s_1,\ldots,s_n)$ that lists n natural numbers s_i, where $1 \le s_i \le g$ and g is the number of groups, in a particular ordering. The i-th incoming car (car i) thus belongs to the s_i-th group. Thinning S out results in a subsequence of S. That is, a *subsequence* of S is either an empty sequence () or an integer sequence $(\tilde{s}_1,\tilde{s}_2,\ldots,\tilde{s}_m)$ with $\tilde{s}_i = s_{j_i}$ for $i = 1,\ldots,m$ and $j_1 < j_2 < \cdots < j_m$. By this definition, S is a subsequence of S.

Concatenating two integer sequences $S = (s_1,s_2,\ldots,s_n)$ and $T = (t_1,t_2,\ldots,t_m)$ results in the sequence $S \oplus T = (u_1,\ldots,u_{n+m})$ with $u_i := s_i$ for $i = 1,\ldots,n$ and $u_{n+i} := t_i$ for $i = 1,\ldots,m$.

$S = (s_1,\ldots,s_n)$ is called *increasing* if $n = 1$ or if $s_i < s_j$ for $1 \le i < j \le n$; if $n = 1$ or if $s_i > s_j$ for $1 \le i < j \le n$, then S is called *decreasing*. We say a sequence S is *upper unimodal* if it is either increasing, or decreasing, or if there is an index j with $1 < j < n$ such that $s_1 < s_2 < \cdots < s_j$ and $s_j > s_{j+1} > \cdots > s_n$. The sequence $S = (s_1,\ldots,s_n)$ is said to be an *arrow sequence* if there is a partition of $(s_2,\ldots,s_n)$ into two subsequences R and T such that R is either empty or increasing with elements larger than s_1 and T is either empty or decreasing with elements smaller than s_1. Examples of upper unimodal sequences and of arrow sequences are shown in Fig. 9.3.

Finally, $S = (s_1,\ldots,s_n)$ is said to be *unimodec* if there is partition of $(s_2,\ldots,s_n)$ into two subsequences R and T (possibly empty) such that $(s_1) \oplus R$ is upper unimodal and $(s_1) \oplus T$ is decreasing. For example, the sequence $(4,\mathbf{6},2,\mathbf{5},1,\mathbf{3})$ is unimodec, since $(4,\mathbf{6},\mathbf{5},\mathbf{3})$ is upper unimodal and $(4,2,1)$ is decreasing.

9.3.2 Complexity Results

In the following we list the results for **ito-shunting**, ***t*-min** variants from the literature. We start with relevant results for **unbounded** variants, followed by the results for the ***b*-bounded** variants.

9.3.2.1 Unbounded Variants

Table 9.2 summarizes the knowledge about special CP variants in which no two incoming cars belong to the same group. In other words, the number n of inbound cars equals the number g of different groups. In particular, the second last and the last column of Table 9.2 list complexity and approximability results, respectively. If c is the number in a cell of the last column, then a "fast" method (with polynomial running time) is known for the corresponding variant which determines a schedule that occupies at most c-times more tracks than the optimal schedule.

Table 9.2: Results for **unbounded, ito-shunting, *t*-min** variants with the following properties: $n = g$ (**0-split=split**), **sequential, ordered (blocks)**

No.	Track design	Equivalent to min partition of inbound sequence into subsequences that are	Complexity	Approximable within factor
1	**stacks** or **queues**	Isotone	**O(n log n)** (well-known)	
2	**stacks/queues**	Monotone	**NP-hard**[a]	**1.71**[b]
3	**sido**	Upper unimodal	**NP-hard**[c]	**2**[d]
4	**diso**	Arrow("-like")	**NP-hard**[c]	**2**[d]
5	**dido**	unimodec	**NP-hard**[c,d]	**3**[d]

[a] Wagner [70]
[b] Fomin et al. [33]
[c] Di Stefano and Koči [27]
[d] Di Stefano et al. [28]

All variants listed in Table 9.2 are equivalent to the problem of finding a minimum partition of the inbound sequence into subsequences with particular properties. A partition into *isotone* subsequences means that all subsequences are increasing or all subsequences are decreasing. If we speak of a partition into *monotone* subsequences, then any subsequence is either increasing or decreasing. Figure 9.3 gives first insights into the connection to sequence partitioning for two example variants.

Further complexity results for **dido,unbounded|ito-shunting,split|time windows|*t*-min** variants where the cars enter and leave the sorting tracks in pre-defined directions are discussed in [19].

Of course, the hardness results listed in Table 9.2 carry over to the corresponding variants with the common, more general assumption $g \leq n$. Table 9.3 summarizes further knowledge about **unbounded, ito-shunting, *t*-min** variants that mainly stem from a fruitful connection to another well-known problem, the minimum vertex coloring problem (MVC).

For several variants, one can construct geometrical objects that help to answer the question whether any two groups (in **0-split** variants) or any two cars (in **split** variants) may be placed on the same sorting track or not. In particular, we may not assign two cars or two groups to the same track in a feasible schedule if and only if the two corresponding geometrical objects intersect. For a detailed construction of theses objects—lines (1,2,3), intervals (4,5), triangles (6,7,8), chords (9,10,11),

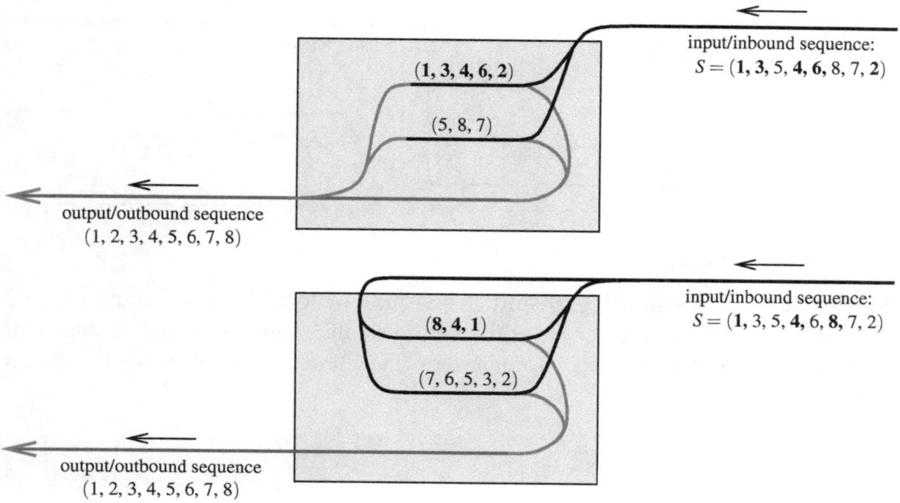

Fig. 9.3: **Unbounded, ito-shunting, sequential, ordered (blocks)**, $n = g$ variants: feasible partition of inbound sequence into two upper unimodal subsequences for **sido** (above) or into two arrow subsequences for **diso** (below)

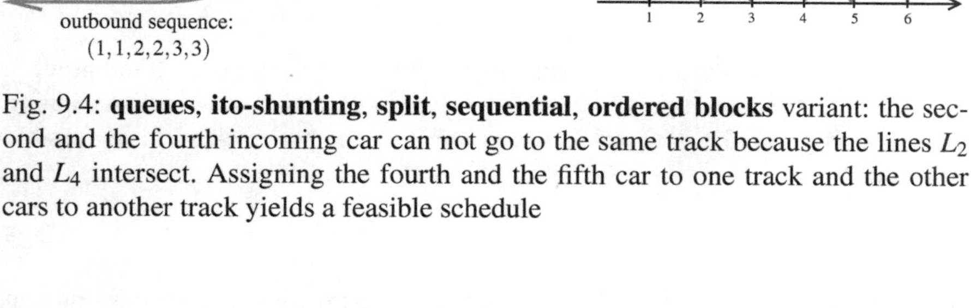

Fig. 9.4: **queues, ito-shunting, split, sequential, ordered blocks** variant: the second and the fourth incoming car can not go to the same track because the lines L_2 and L_4 intersect. Assigning the fourth and the fifth car to one track and the other cars to another track yields a feasible schedule

or polygons with vertices on a circle line (12)—we refer to Hansmann [43] and Di Stefano and Koči [27]; the numbers in the parentheses relate to the number of the respective variants listed in Table 9.3. Though, for the interested reader, Figs. 9.4, 9.5, and 9.6 might reveal how to construct the objects for three different variants.

Graphs whose vertices correspond either to the cars or to the groups, and in which an edge between two vertices exits if an assignment of the corresponding cars/groups to the same track is infeasible (i.e., if the corresponding geometrical

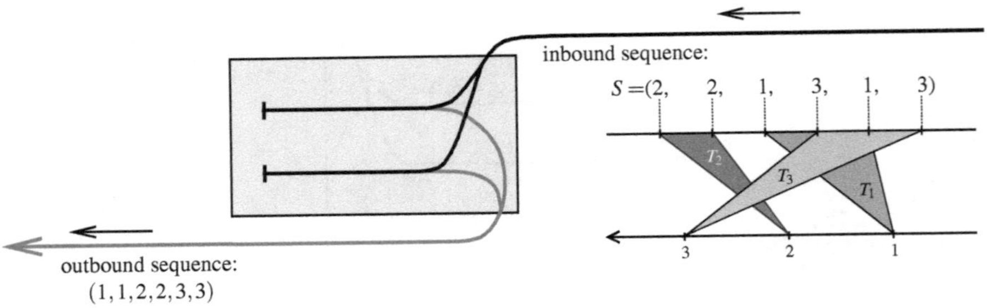

Fig. 9.5: **stacks, ito-shunting, 0-split, sequential, ordered blocks** variant: group 2 and group 3 can not go to the same track because the triangles T_2 and T_3 intersect. Assigning group 2 and group 1 to one track and group 3 to another track yields a feasible schedule

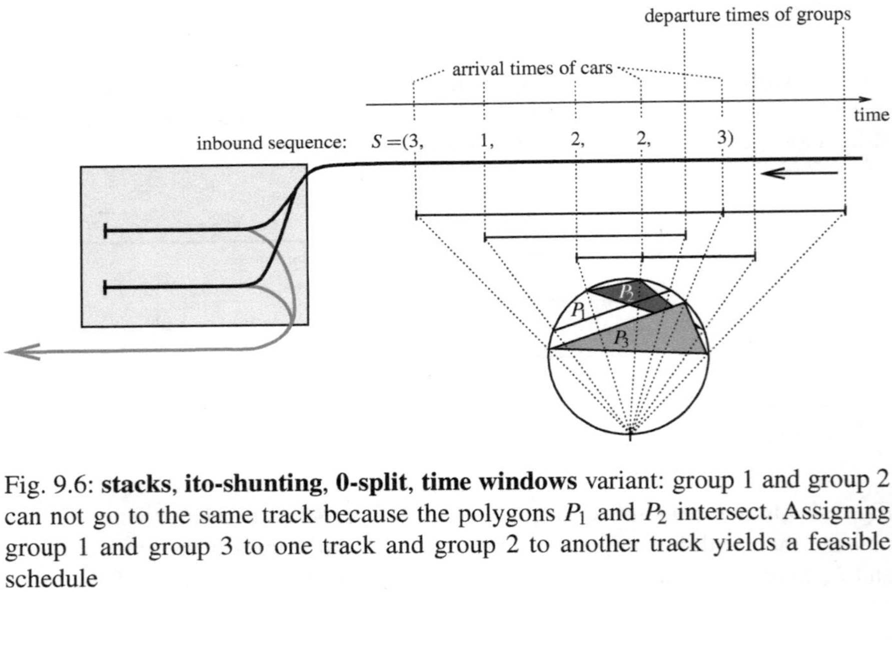

Fig. 9.6: **stacks, ito-shunting, 0-split, time windows** variant: group 1 and group 2 can not go to the same track because the polygons P_1 and P_2 intersect. Assigning group 1 and group 3 to one track and group 2 to another track yields a feasible schedule

objects intersect) belong to well-known graph classes: permutation graphs, interval graphs, point-interval graphs, circle-graphs, polygon-circle graphs. Feasible vertex colorings of these graphs with minimum number of colors (result of MVC) match feasible schedules with the least number of tracks used. Based on this connection complexity results and solution methods can be derived, cf. Table 9.3. In particular, for the variants in rows 1–8 optimal schedules can be computed by fast greedy methods. The hard variants in rows 9–12 can be solved to optimality by any exact method for MVC, and in particular, for MVC of polygon-circle graphs, since any circle graph is a polygon-circle graph. Currently, it is an open question whether MVC of circle graphs, and, thus, the variants in rows 9–12 are constant factor approximable. Computational experience with LP- and Lagrangian-based branch-and-bound implementations for MVC of polygon-circle graphs is discussed in [43]. Accordingly, one can compute optimal schedules for the variants in rows 9–12 within reasonable time for a moderate number of inbound cars.

Table 9.3: Results for **unbounded, ito-shunting, *t*-min** variants

No.	Track design	Sorting mode		Outbound condition	Equivalent to MVC of	Complexity
		t-o-moves	splitting			
1	**queues**	**sequential** or **concurrent** or **anytime**	**split**	**ordered blocks**	Permutation graphs	**O(n log n)**[a]
2	**queues**	**concurrent**	**split**	**time windows**		
3	**Stacks**	**sequential**	**split**	**ordered blocks**		
4	**queues**	**sequential** or **concurrent** or **anytime**	**0-split**	**free blocks**	Interval graphs	**O(n log n)** (folklore)
5	**Stacks**	**sequential**	**0-split**	**free blocks**		
6	**queues**	**sequential** or **concurrent** or **anytime**	**0-split**	**ordered blocks**	Point-interval graphs	**O(n log n)**[a]
7	**queues**	**concurrent**	**0-split**	**time windows**		
8	**Stacks**	**sequential**	**0-split**	**ordered blocks**		
9	**Stacks**	**concurrent** or **anytime**	**split**	**ordered blocks**	Subclass of circle graphs	NP-hard[a,e]
10	**Stacks**	**anytime**	**0-split**	**ordered blocks**		
11	**Stacks**	**concurrent**	**split**	**time windows**	Circle graphs	**NP-hard**[b]
12	**Stacks**	**concurrent**	**0-split**	**ordered blocks** or **free blocks** or **time windows**	Polygon-circle graphs	**NP-hard**[a]
13	**Stacks** or **queues**	**sequential**	**chain split**	**ordered blocks**		**O(n)**[c]
14	**Stacks** or **queues**	**sequential**	**chain split**	**free blocks**		**NP-hard**[d] **(2-approx.)**[a]

[a] Hansmann [43]
[b] Di Stefano and Koči [27]
[c] Dahlhaus et al. [25]
[d] Dahlhaus et al. [24]
[e] Hansmann [44]

9.3.2.2 *b*-Bounded Variants

Table 9.4 summarizes some knowledge for *b*-**bounded, ito-shunting, split**, *t*-**min** variants. This table and the following explanations revise some incorrect statements in Table 4.2 presented in [43].

For the **anytime** variants in rows 1 and 2 of Table 9.4, the bounded track length is not restricting, because any car can depart immediately after it arrived on the sorting track. In other words, the optimal "**unbounded**" schedules are also feasible (and optimal) for the *b*-**bounded** case. This is also true for the **concurrent, 0-split** variants in rows 1 and 2 if no group contains more than *b* cars (otherwise, there is no feasible schedule at all).

Table 9.4: Results for *b*-**bounded, ito-shunting**, *t*-**min** variants

No.	Track design	Sorting mode		Outbound condition	Complexity	
		t-o-moves	Splitting		"easy" for	NP-hard for
1	**queues**	**anytime**	**split**	**ordered blocks**	Any b $[O(n\log n)]$	
2	**queues**	**concurrent** or **anytime**	**0-split**	**ordered blocks** or **free blocks**	Any b $[O(n\log n)]$	
3	**stacks** or **queues**	**sequential**	**0-split**	**free blocks**	$b \leq 3$[b]	Fixed $b \geq 8$ $(7/3$-approx.$)$[a]
4	**stacks** or **queues**	**sequential**	**split**	**ordered blocks**	$b \leq 2$	Fixed $b \geq 6$ (2.5-approx.)[a]
5	**stacks** or **queues**	**sequential**	**0-split**	**ordered blocks**		
6	**stacks** or **queues**	**concurrent**	**split**	**time windows**	$b = 1$[b]	Fixed $b \geq 6$
7	**stacks** or **queues**	**concurrent**	**0-split**	**time windows**		
8	**stacks**	**concurrent** or **anytime**	**split** or **0-split**	**ordered blocks**	$b = 1$[b]	Fixed $b \geq 6$[b]
9	**stacks**	**concurrent**	**0-split**	**free blocks**	$b \leq 2$[b]	General b
10	**stacks**	**sequential**	**split**	**ordered pattern**	$b = 1$	Fixed $b \geq 3$[c,d,e]
11	**stacks** or **queues**	**sequential**	**chain-split**	**ordered blocks**	Any b $[O(n)]$[f]	

[a] Epstein and Levin [30]
[b] Hansmann [44]
[c] Winter [71]
[d] Winter and Zimmermann [72]
[e] Eggermont et al. [29]
[f] Hansmann [43]

The variants in row 3 where any group consists of exactly two cars are equivalent to $b/2$-MES of interval graphs, which is shown to be NP-hard even for fixed $b \geq 4$ in [8]. Thus, variants in row 3 are NP-hard for fixed $b \geq 8$. Besides that, the optimal schedule for any instance of the variants in row 3 corresponds to an optimal solution for an instance of the bin packing problem with conflicts (BPC) on interval graphs (as conflict graphs). BPC seeks for a minimum vertex coloring in the conflict graph whose vertices are assigned with natural weights complying with the additional requirement that the weights of the vertices colored with the same color sum up to at most b. Thus, BPC generalizes MVC and b-MES. Since, BPC is $7/3$-approximable on interval graphs [30], this is also true for the variants in row 3.

The variants in row 4 are equivalent to the mutual exclusion scheduling problem (b-MES) of permutation graphs, which is shown to be NP-hard even for fixed $b \geq 6$ in [50] and to be 2.5-approximable in [30]. b-MES can be stated as follows: find a minimum vertex coloring in a graph such that at most b nodes are colored with the same color.

The special cases of variants in row 5 with $g = n$ are equivalent to b-MES of permutation graphs. Thus, by previously mentioned results, variants in row 5 are NP-hard for fixed $b \geq 6$. The optimal schedule for any instance of the variants in row 5 corresponds to an optimal solution of an instance of BPC on point-interval graphs. Since, BPC is 2.5-approximable on point-interval graphs [30], any point-interval graph is a perfect graph), this is also true for the variants in row 5.

Note that, for the variants in rows 6–9, there is no direct connection to graph coloring as for its **unbounded** counterparts. Nevertheless, the NP-hardness carries over from variants in row 4 to variants in row 6 as well as from variants in row 5 to variants in row 7. The reason is, that for any instance of the **sequential** case we obtain an equivalent instance of the corresponding (**concurrent**,) **time windows** case by setting the arrival time and the departure time for the i-th incoming car to i and to $n + s_i$, respectively.

That the variants in row 8 are NP-hard for fixed $b \geq 6$ can be shown by a simple reduction from their **sequential** counterparts. The idea is to add a car of an extra group 0 at the end of the inbound sequence that has to depart before any other cars.

If an **unbounded** variant is NP-hard, this is also true for the corresponding b-**bounded** variant for general b. Hence, variant 9 is NP-hard for general b.

After all, there still are interesting open theoretical challenges. For example, for any NP-hard variant listed in Table 9.4, the threshold value b for which the b-**bounded** case is polynomially solvable and for which the $b+1$-**bounded** case is NP-hard is currently unknown.

The variants in row 3 are solvable in polynomial time for fixed $b \leq 3$: for $b \leq 2$ this is trivial; for $b = 3$ the ideas are as follows. There is a feasible schedule if and only if the inbound sequence contains no group with more than three cars. In case of feasibility, an optimal schedule results from assigning any group of three cars to one track, from a maximum (feasible) matching of the groups containing two cars to groups of a single car, and from arbitrarily filling other tracks with the unmatched groups of one car [44].

The variants in rows 4 and 5 are solvable in polynomial time for fixed $b = 2$: for the variants in row 4 this is due to its equivalence to 2-MES (of permutation graphs), which is polynomially solvable; and, for the variants in row 5 the reasoning is as follows. An incoming sequence containing a group with more than two cars does not admit a feasible **0-split** schedule. Each group with two cars is assigned to exactly one track. The assignment of the remaining cars to a minimum number of tracks can be computed with a method for a variant in row 4, and therefore with the known polynomial time method for 2-MES.

Finally, the optimal schedules of the variants in rows 6–8 with $b = 1$ and of the variant in row 9 with $b = 2$ can be determined by optimal interval colorings [44].

9.4 Multi-Stage Classification

In this section, we focus on **multi-stage** variants that relate to *hump yards*, the largest class of rail yards. Other **multi-stage** variants with the characteristics *t*-**stacks**, *b*-**bounded**, **sequential**, **split**, **ordered pattern**, with no allowed i-o-moves, with $n = t \cdot b$ inbound cars and with an objective function relating to the number of necessary t-t-moves are discussed in [17] and [13].

Figure 9.7 shows the schematic layout of the main part of a typical hump yard.

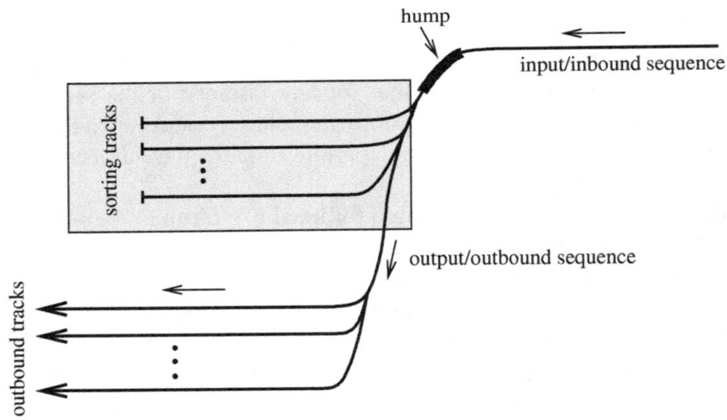

Fig. 9.7: Layout of a typical hump yard

Trains arrive at an *arrival yard* (prior to our input) where the cars are inspected and decoupled. Then, the inbound cars are pushed over the hump by a shunting engine before rolling down one by one onto either appropriately chosen sorting tracks (i-t-moves) or an outbound track (i-o-moves) by means of gravity. This procedure reduces the number of time-consuming pushing/pulling operations of cars by shunting engines significantly. In the same convenient manner the hump is used for t-t-moves and t-o-moves. At each *humping step* all cars placed on one sorting track are pulled back over the hump. Then, these cars are again pushed over the hump either to an outbound-track (t-o-move) or again to some sorting tracks (t-t-move).

If we allow arbitrary many or h humping steps, we speak of **hump-shunting** or h-**hump-shunting** variants of CP, respectively.

Not allowing humping steps before the last inbound car is pushed over the hump corresponds to our **sequential** requirement (first t-o-move after last i-t-move and last i-o-move). In contrary, in **concurrent** variants, humping steps are allowed at any point in time during the roll-in process of the inbound cars. Note that we slightly changed the definition of **sequential** and **concurrent** presented in [43] to enable a broader classification scheme. Thus, some h-**hump-shunting** variants classified as **concurrent** in [43] are classified as **sequential** in this chapter.

The following alternative requirement regarding the chronology of the humping steps is denoted by **scheduled**. In this case, the points in time when humping steps are performed are pre-defined in advance. Note that we assume instantaneous car movements. In other words, in the **concurrent** and **scheduled** case, the roll-in process of the sequence of inbound cars is instantaneously interrupted by the humping steps.

Because the **0-split** requirement does not make sense for **hump-shunting**, we only consider **split** variants in this section.

9.4.1 Requirements for the Outbound Sequence

The **ordered blocks** and **free blocks** requirements are suited for the **hump-shunting** variant when forming one departing train (on one outbound track) with an **ordered blocks** or a **free blocks** ordering of the cars.

Now, assume that o departing trains can be formed simultaneously (on o parallel outbound tracks) and that the groups $g_{<i}+1,\dots,g_{<i}+g_i$ have to leave in departing train i where g_i is the number of groups for departing train i and $g_{<i}$ is the total number of groups for the trains $1,\dots,i-1$, that is, $g_{<i}=\sum_{k=1}^{i-1} g_k$. If we require **free blocks** (or **ordered blocks**) for any of the departing trains, then the outbound sequence is feasible if it decomposes into o subsequences $S_1,\dots,S_o$ such that S_i only contains cars for departing train i and that S_i has the structure **free blocks** (or **ordered blocks**) for $i=1,\dots,n$. These requirements are abbreviated by o-**free blocks** and o-**ordered blocks**, respectively.

For instance, consider the incoming sequence $(2,4,1,5,3,3,2,3,5,4)$ of ten cars for three outbound trains. The two $(g_1=2, g_{<1}=0)$ groups 1 and 2 have to be assigned to train 1, group 3 $(g_2=1, g_{<2}=2)$ to train 2, and the two $(g_3=2, g_{<3}=3)$ groups 4 and 5 to train 3. Then, the outbound sequence $(3,4,1,3,2,4,5,3,5,2)$ has the 3-**ordered blocks** property, since it decomposes into the three subsequences $(1,2,2)$, $(3,3,3)$, and $(4,4,5,5)$ with **ordered blocks**.

By the previous requirements, we do not pay attention to the number of outbound tracks available as well as to their capacities. However, these capacities are a limiting factor when creating the overall classification schedule for simultaneous train formation at hump yards, as too high traffic volumes can make it impossible to reserve an outbound track for the full time interval between the first roll-in of a car for this train and the final departure of this train. Using the sorting tracks as a buffer area, where cars of different departing trains may be temporarily stored, makes it possible to simultaneously build more departing trains than there are outbound tracks.

In the **soft time windows and outbound capacity** variant abbreviated **sotwoc** we consider a fixed number of available outbound tracks. Each outbound track ("formation track") f with length n_f is assumed to be free at the very beginning. For each incoming car i its length, the arrival time a_i as well the latest departure time d_i are known. In fact, d_i is the time when car i's dedicated departing train leaves the outbound track. We assume, that no two trains leave the outbound tracks at the same time. Thus, we recognize all cars of one departing train o by an identical latest departure time. In a schedule to compute, let t_o denote the point in time when a car of departing train o is t-o-moved or i-o-moved for the first time. Then, a schedule is feasible for this **sotwoc** variant if any car i is t-o-moved or i-o-moved not later than d_i and if there is a free outbound track f (not blocked by other departing trains) at time t_o that is long enough for all cars of departing train o. Thus, the assignment of the departing trains to the outbound tracks has to be an implicit part of the solution. For an example of a **sotwoc** variant, we refer to Fig. 9.9 in Sect. 9.4.5.

9.4.2 Goals

The size of the hump yard and the volume of traffic determine whether it is more efficient to form the departing trains with a minimum number of used sorting tracks for a given upper bound on the number of humping steps (t-**min**) or rather with as few humping steps as possible for a given number of sorting tracks. We refer to the latter objective as h-**minimizing** or h-**min** for short.

Because the pullback and roll-in operations of cars wear down switches and tracks, it is also reasonable to relate the objective function to the number of car movements over the hump. Thus, c-**minimizing** (c-**min**) variants aim at schedules with a minimum number c of t-t-moves and t-o-moves of the cars over all humping steps.

The goals h-**min** and c-**min** aim in a similar direction, still they are not equivalent for any variant. For example, there always exists an optimal schedule for the **unbounded,stacks|sequential,hump-shunting,split|ordered blocks|**h-**min** variant which also is optimal for the c-**minimizing** counterpart variant, cf. Sect. 9.4.3. However, there are trivial examples (with three incoming cars of two groups) that show the following for the same variants. There are "c-**minimal**" schedules that are not "h-**minimal**", and there other schedules which are "h-**minimal**" but not "c-**minimal**".

9.4.3 Complexity Results

Table 9.5 summarizes complexity results for **stacks**, h-**hump-shunting**, **split** variants. Furthermore, the h-**minimizing** variant in row 2 is shown to be 2-approximable in [43]; which is also true for the h-**minimizing** variant in row 3 with $g = n$, cf. [49] and [45].

Note that the track length and track design listed in the second and third column only refer to the sorting tracks in the classification bowl. Assuming that each outbound track is long enough for any departing train, the 0-**hump-shunting** (= 0-**stacks**) **sotwoc** variant in row 4 with no humping steps (no sorting tracks are used) is equivalent to MVC of interval graphs, and therefore solvable with a fast greedy method. Under the same assumption, the **unbounded** variant in row 5 is also easily solvable, cf. [9].

Table 9.5: Results for **stacks**, h-**hump-shunting**, **split** variants

No.	Track design	Track length	Sorting mode t-o-moves	Outbound condition	Objective	Complexity
1	t-**stacks**	**unbounded**	**sequential**	(o) **ordered blocks**	t-, h-, or c-**min**	Linear[a,b]
2	t-**stacks**	**unbounded**	**sequential**	**free blocks**	t-, h-, or c-**min**	NP-hard[a]
3	t-**stacks**	b-**Bounded**	**sequential**	**ordered blocks**	t- or h-**min**	NP-hard[b]
4	**0-stacks** $h = 0$	(Irrelevant)	**Scheduled**	**sotwoc**	c-**min**	NP-hard[c]
5	**1-stack**	**unbounded** or b-**bounded**	**Scheduled**	**sotwoc**	c-**min**	NP-hard[c]

[a] Hansmann [43]
[b] Jacob et al. [49]
[c] Bohlin et al. [12]

In the following we sketch the main ideas behind the algorithm presented in [43] that solves the h-**minimizing** variant in row 1 of Table 9.5. It can be implemented such that it runs in linear time in the size of the schedule to compute. In [49] seemingly the same algorithm is described in different terminology for the special case where no cars belong to the same group ($g = n$). Figure 9.8 shows the optimal car moves for an example instance of this **unbounded**,2-**stacks**|**sequential**,**hump-shunting**,**split**|**ordered blocks**|h-**min** variant. The list of observations and results justifying Algorithm 9.1 reads as follows; for detailed proofs we refer to [43].

- For each car, we consider its *path* through the sorting tracks, i.e., the integer sequence of the track numbers of visited sorting tracks. In our example, the first inbound car (of group 7) takes the path $(1, 2, 1)$, i.e., firstly to the upper sorting track 1, secondly to track 2, and thirdly to track 1 again.

- Cars taking the same path correspond to a subsequence of the inbound sequence. In our example, the cars of group 1 as well as the last inbound car are i-o-moved with their first roll-in. These three cars constitute the subsequence $(1,1,2)$ of the inbound sequence $(7,6,5,4,3,4,1,2,1,2)$.
- A path corresponds to a subsequence of the *track execution order* $E = (e_1,\ldots,e_h)$ where e_i is the number of the sorting track "executed" at humping step i. In our example, the path $(1,1)$ of the car of group 5 is a subsequence of $E = (1,2,1)$.
- For fixed numbers of humping steps and of sorting tracks, the *cyclic track execution order* $E = (1,2,\ldots,t,1,2,\ldots,t,1,\ldots)$ enables the largest number of different paths. This number can recursively be computed in linear time. For example, $E = (1,2,1)$ enables seven different paths, $E = (1,1,1)$ only four.

Step 1 in Algorithm 9.1 is equivalent to the variant in row 10 of Table 9.3, and therefore executable in linear time in the number of inbound cars. This is also true for Step 2 by a recursive computation. Finally, Step 3 and hence the complete Algorithm 9.1 can be realized in linear time in the size of the schedule to compute, that is, in $O(nh^*)$ time where n is the number of incoming cars and h^* is the desired minimum number of humping steps.

Algorithm 9.1: Greedy for the CP variant:
unbounded,t-stacks|sequential,hump-shunting,split|ordered blocks|h-min

---

Step 1: Compute a minimum partition of the input sequence S into subsequences $S_1,\ldots,S_{p^*}$ such that $S_1 \oplus S_2 \oplus \cdots \oplus S_{p^*}$ has the structure **ordered** g**-blocks**.

(*In the example in Fig. 9.8, the minimum partition of the input sequence* $(7,6,5,4,3,4,1,2,1,2)$ *with seven subsequences reads:* $(1,1,2)$, (2), $(3,4)$, (4), (5), (6), (7). *Their concatenation (from left to right) yields the desired* ***ordered blocks*** *output sequence. There is no partition into less subsequences with the same property.*)

Step 2: Determine the minimum number h^* of humping steps enabling (with cyclic track execution) at least p^* different realizable car paths.

(*Our example cont.: The cyclic track execution* $E = (1,2,1)$ *with* $h^* = 3$ *humping steps enables seven different paths, namely,* $()$, (1), (2), $(1,2)$, $(1,1)$, $(2,1)$, $(1,2,1)$, *which is not possible with less than three humping steps.*)

Step 3: For all $p = 1,\ldots,p^*$, assign a suitable path p regarding the cyclic track execution order with h^* humping steps to all cars in subsequence S_p.

(*Our example cont.: We obtain an optimal schedule by assigning all cars in* $S_1 = (1,1,2)$ *to the "fastest" path* $()$, *by assigning the car in* $S_2 = (2)$ *to the second "fastest" path* (1), *by assigning the cars in* $S_3 = (3,4)$ *to the third "fastest" path* (2), *and so on.*)

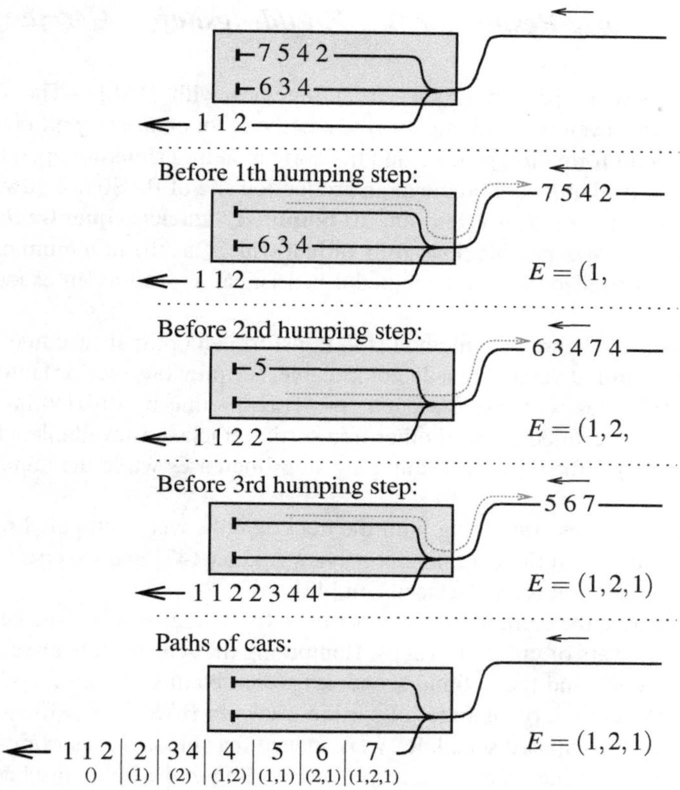

Fig. 9.8: **unbounded,2-stacks|sequential,hump-shunting,split|ordered blocks**|h-**min**: optimal car moves and paths of the cars for the instance $(7,6,5,4,3,4,1,2,1,2)$

It is easily seen that the schedule computed by Algorithm 9.1 is also optimal for the corresponding c-**minimizing** variant. Besides that, we get an optimal schedule for the corresponding o-**ordered blocks** variant for forming o departing trains simultaneously by "gluing" together the o optimal schedules obtained by considering the departing trains separately.

In the following two subsections we focus on real-world applications in rail yards in Germany and Sweden. We describe the main ideas of the solution approaches and summarize computational experiences. Similar approaches for other rail yards in Europe were presented in the literature, for example in the Netherlands, cf. Freling et al. [34], Kroon et al. [58] or in Switzerland, cf. Márton et al. [60].

9.4.4 Real-World Results for BASF, Ludwigshafen, Germany

Within our 3-years project (2004–2007) together with BASF—The Chemical Company—we developed and implemented methods for computing efficient schedules for the train formation process and the corresponding shunting operations performed at a large hump yard on the main production site of BASF in Ludwigshafen, Germany. For the practically relevant 30-**bounded**,t-**stacks**|**sequential**,h^*-**hump-shunting**,**split**|o-**ordered blocks**|c-**min** variant with h^* as the minimum number of humping steps needed, we implemented a heuristical as well as an exact solution approach.

The idea of the heuristical method is to transform an optimal schedule of the respective **unbounded** version (easily computable, see previous section) into a schedule which complies with the track lengths. This is done by distributing the cars assigned to an overfilled track to other free sorting tracks—if available. Of course, by this "splitting", the number of humping steps increases while the number of car moves stays the same.

Optimal schedules complying with the track lengths were computed by solving Binary Programs with the commercial solvers `Cplex` [47] and `Gurobi` [40]. We omit these models and refer to Hansmann [43].

We summarize the computational experience for 30 real-world instances with up to 400 inbound cars of up to 71 groups. Computing the schedules heuristically took less than a second and the optimal schedules were obtained within a few minutes. Due to a sufficiently large number of sorting tracks in BASF's classification bowl, the heuristically computed schedules were indeed feasible and no sorting track had to be emptied twice (i.e., $E = (1, 2, \ldots, h^*)$) according to the optimal schedules. On the average over the 30 daily instances, the cars roll 1.44 times down the hump according to the heuristically computed schedules, instead of 1.41 times as in the optimal case. Moreover, the heuristical schedules use 0.85 humping steps more than the optimal ones on average.

After all, our project partner preferred the schedules that are optimal for the **unbounded** case, due to the following reasons.

- Schedules complying with the track lengths are still often not directly applicable, since cars rolling down to tracks are slowed down by automatic brakes which may lead to gaps between the cars on the sorting track. Thus, it seems to be more reasonable to compute optimal "*unbounded*" schedules and leave it to the dispatcher to handle "*full*" tracks adequately.
- Already the flexible realization of the optimal "*unbounded*" schedules offers major potential for dropping the overall processing time in comparison to the rule-based schedules in daily action.
- The fast computation (within a second) of optimal "*unbounded*" schedules makes it possible to react quickly on any disruptions or real time changes in the order of the incoming cars.
- Optimal "*unbounded*" schedules can be computed without the aid of commercial solvers.

9.4.5 Real-World Results for Hallsberg, Sweden

The variant **bounded**,1-**stack**|**scheduled**,h-**hump-shunting**,**split**|**sotwoc**|c-**min**, which corresponds closely to the operation of the largest Swedish hump yards in Hallsberg and Sävenäs, was studied in a 4-year research effort started in 2010. The initiative was performed together with the Swedish infrastructure manager Trafikverket, and with support from Sweden's largest freight operator Green Cargo AB (as measured in number of freight trains). The variant is illustrated in Fig. 9.9.

The core of the problem variant is the allocation of formation tracks to outbound trains, subject to a set of realistic constraints on train scheduling, arrival and departure timeliness, and track capacity. In this variant, formation tracks have unique lengths, and departing trains are required to strictly follow a fixed timetable. Furthermore, the problem formulation allows the temporary storage of freight cars on a single dedicated mixed-usage track, which is treated like a stack. This variant was inspired by a real-world practice at the Hallsberg yard, and increases the number of simultaneous trains that can be managed on the yard. In the real world practice, it

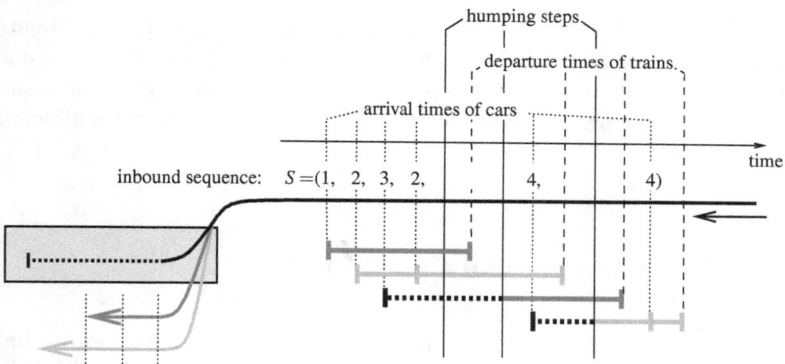

Fig. 9.9: **unbounded**,1-**stack**|**scheduled**,3-**hump-shunting**,**split**|**sotwoc**|c-**min**—an example with six inbound cars for two outbound trains, two available outbound tracks, and three humping steps predefined in time. All cars have length 1, one outbound track has length 1, the other length 2. A schedule with the minimum number of 3 additional car moves (t-t or t-o) is depicted by the thick intervals: Trains 1 and 3 are assigned to the short outbound track, trains 2 and 4 to the longer one. The car of train 3 rolls down at the first and second humping step; the first incoming car of train 4 at the third humping step

is also possible that several shunting tracks are allocated for temporary car storage during certain time intervals of the week only. However, the variant described here, where a single track is employed, was seen as a reasonable compromise between solvability and degree of detail. Further, the variant easily generalizes to the case where multiple tracks are used as long as all tracks are emptied at the same fixed time points. For this problem, we have proposed and evaluated several optimization models. The models include the following.

1. A series of mixed-integer linear programming (MILP) formulations using the Big-M method, cf. [9],
2. a pure integer reformulation of the above, cf. [10],
3. a column-based integer programming model which is solved using branch-and-price, cf. [11], and
4. a simplified reformulation of the first model as an arc-indexed integer linear program, which has the same LP relaxation as the column-based model, cf. [12].

All models have been evaluated on real-world data from the Hallsberg yard. Further, the last two models, being the ones which performed best in the experiments, have been adapted for rolling horizon planning and evaluated on a 5-month historical data set. From this data set, 784 instances of different types and lengths, spanning from 2 to 5 days, were created.

In the experiments, the arc-indexed model proved optimality on average twice as fast as the column-based model for the independent instances, and three times faster for the rolling horizon instances. For the arc-indexed model, the average solution time for a reasonably sized planning horizon of 3 days was 16 s. Regardless of size, no instance took longer than 8 min to be solved. The results indicate that optimization approaches are a suitable alternative for scheduling and track allocation at classification yards.

9.5 Practical Relevance and Conclusions

From a more theoretical perspective, in recent research one tries to draw a line between "easy" and "hard" (NP-hard) variants. "Easy" means that a fast, i.e., polynomial running time algorithm is known that computes optimal classification schedules. However, this map of computational complexity is still rather incomplete. For example, the complexity for the following two variants is open:

- **queues,unbounded**|**ito-shunting,concurrent,split**|**free blocks**|t-**min**,
- 1-**stack,bounded**|h-**hump-shunting,scheduled,split**|**sotwoc**|c-**min** under the assumption that each outbound track is long enough for any departing train.

At first sight, some variants seem to be rather restricted from a practical point of view, but we can provide some reasoning why corresponding methods might in fact be very helpful in order to improve classification processes in practice.

Ito-shunting variants answer the question whether it is possible to form a departing train using only i-t-moves and t-o-moves. In the positive case the corresponding solution method computes a classification schedule that does not require additional time-consuming shunting operations, as, e.g., couplings, decouplings, and engine moves. In the negative case t-t-moves are unavoidable and one will apply

multi-stage methods. In the latter case, **single-stage** methods may help (as subroutines) to compute efficient classification schedules for the **multi-stage** problem, cf. Sect. 9.4.3.

Although tracks in rail yards are bounded in length, methods for the **unbounded** case may propose classification schedules that are quite useful for the dispatcher in practice, cf. Sect. 9.4.4. In general, when dropping such "hard" requirements as real track lengths, optimal classification schedules are much easier to compute. Once a track gets filled to its real capacity during actual operations, the dispatcher may redirect cars scheduled for the filled track to another unfilled track in the classification bowl or elsewhere in the yard.

Another benefit is that minimal "unbounded" schedules provide benchmarks for "bounded" schedules. In particular, the minimum objective value of an optimal "unbounded" schedule defines a lower bound on the minimum objective value for the respective **bounded** variant. Whenever a lower bound matches the objective value of a "bounded" schedule, this schedule obviously is optimal.

To the best of our knowledge, the problem of finding an optimal layout of a new or rebuilt rail yard by means of mathematical optimization approaches has not been studied explicitly in the literature so far. An analysis of schedules obtained with the proposed CP methods for scenario sets of instances modeling various traffic volumes could help to decide on the future yard layout. In particular, it could give hints how many sorting or outbound tracks are necessary or sufficient.

For real-world data from rail yards in Hallsberg, Sweden, and in Ludwigshafen, Germany, we could compute efficient classification schedules complying with the track lengths. In particular, we obtained very good solutions by fast heuristical methods or optimal schedules (under certain assumptions) in reasonable time, using either commercial solvers, or tailor-made branch-and-bound-implementations, or column generation approaches. In theory, the computed schedules indicate that considerable improvements are possible in daily action. For future research and projects, we recommend to analyze and to quantify the real cost reductions that result from supporting human experts by the knowledge of automatically optimized schedules. Real significant savings would convince further practitioners of (mathematical) optimization tools and would justify the effort for (re-)developing tailor-made methods whenever a new or different requirement in the rail yard has to be considered.

References

1. Ahuja RK, Jha KC, Liu J (2007) Solving real-life railroad blocking problems. Interfaces 37(5):404–419. ISSN: 0092-2102. https://doi.org/10.1287/inte.1070.0295 (cited on page 182)
2. Assad AA (1981) Analytical models in rail transportation: an annotated bibliography. Inf Syst Oper Res 19(1):59–80. https://doi.org/10.1080/03155986.1981.11731807 (cited on page 185)

3. Assad AA (1983) Analysis of rail classification policies. Inf Syst Oper Res 21(4):293–314. https://doi.org/10.1080/03155986.1983.11731905 (cited on page 185)
4. Avriel M, Penn M, Shpirer N (2000) Container ship stowage problem: complexity and connection to the coloring of circle graphs. Discret Appl Math 103(1):271–179. https://doi.org/10.1016/S0166-218X(99)00245-0 (cited on page 183)
5. Barnhart C, Jin H, Vance PH (2000) Railroad blocking: a network design application. Oper Res 48(4):603–614. https://doi.org/10.1287/opre.48.4.603.12416 (cited on page 182)
6. Baumann O (1959) Die Planung der Simultanformation von Nahgüterzügen für den Rangierbahnhof Zürich-Limmattal. Rangiertechnik 19:25–35 (cited on page 184)
7. Bodin LD, Golden BL, Schuster AD, Romig W (1980) A model for the blocking of trains. Transp Res B 14(1):115–120. https://doi.org/10.1016/0191-2615(80)90037-5 (cited on page 182)
8. Bodlaender HL, Jansen K (1995) Restrictions of graph partition problems. Part I. Theor Comput Sci 148(1):93–109. https://doi.org/10.1016/0304-3975(95)00057-4 (cited on page 196)
9. Bohlin M, Flier H, Maue J, Mihalák M (2010) Hump yard track allocation with temporary car storage. Tech. rep. T2010:09. Swedish Institute of Computer Science. HDL: 20.500.11850/99142 (cited on pages 201, 206)
10. Bohlin M, Flier H, Maue J, Mihalák M (2011) Track allocation in freight-train classification with mixed tracks. In: Proceedings of the 11th workshop on algorithmic approaches for transportation modelling, optimization, and systems (ATMOS 2011), Schloss Dagstuhl, Wadern, Sept 2011, vol 20, pp 38–51. https://doi.org/10.4230/OASIcs.ATMOS.2011.38 (cited on page 206)
11. Bohlin M, Dahms F, Flier H, Gestrelius S (2012) Optimal freight train classification using column generation. In: Proceedings of the 12th workshop on algorithmic approaches for transportation modelling, optimization, and systems (ATMOS 2012), Schloss Dagstuhl, Wadern, vol 25, pp 10–22. https://doi.org/10.4230/OASIcs.ATMOS.2012.10 (cited on page 206)
12. Bohlin M, Gestrelius S, Dahms F, Mihalák M, Flier H (2016) Optimization methods for multistage freight train formation. Transp Sci 50(3):823–840. https://doi.org/10.1287/trsc.2014.0580 (cited on pages 201, 206)
13. Borndörfer R, Cardonha C (2009) A set partitioning approach to shunting. Electron Notes Discret Math 35:359–364. https://doi.org/10.1016/j.dam.2011.06.009 (cited on page 198)
14. Boysen N, Fliedner M, Jaehn F, Pesch E (2012) Shunting yard operations: theoretical aspects and applications. Eur J Oper Res 220(1):1–14. https://doi.org/10.1016/j.ejor.2012.01.043 (cited on page 185)
15. Boysen N, Emde S, Fliedner M (2015) The basic train makeup problem in shunting yards. OR Spectr 38(1):207–233. https://doi.org/10.1007/s00291-015-0412-0 (cited on page 183)

16. Büsing C, Maue J (2010) Robust algorithms for sorting railway cars. In: Proceedings of 18th annual European conference on algorithms: Part I (ESA'10). Springer, Berlin, pp 350–361. ISBN: 978-3-642-15774-5. https://doi.org/10.1007/978-3-642-15775-2_30 (cited on page 185)
17. Cardonha, C (2012) Applied methods for the vehicle positioning problem. Ph.D. thesis, TU Berlin. https://doi.org/10.14279/depositonce-3141 (cited on page 198)
18. Cicerone S, D'Angelo G, Di Stefano G, Frigioni D, Navarra A (2009) Recoverable robustness for train shunting problems. Algorithmic Oper Res 4(2):102–116. ISSN: 1718-3235 (cited on page 185)
19. Cornelsen S, Di Stefano G (2007) Track assignment. J Discret Algorithms 5(2):250–261. https://doi.org/10.1016/j.jda.2006.05.001 (cited on page 192)
20. Daganzo CF (1986) Static blocking at railyards: sorting implications and track requirements. Transp Sci 20(3):189–199. https://doi.org/10.1287/trsc.20.3.189 (cited on page 185)
21. Daganzo CF (1987) Static blocking at railyards: part I. Homogeneous traffic. Transp Res B 21(1):1–27. https://doi.org/10.1016/0191-2615(87)90018-X (Not cited.)
22. Daganzo CF (1987) Static blocking at railyards: part II. Heterogeneous traffic. Transp Res B 21(1):29–40. https://doi.org/10.1016/0191-2615(87)90019-1 (Not cited.)
23. Daganzo CF, Dowling RG, Hall RW (1983) Railroad classification yard throughput: the case of multistage triangular sorting. Transp Res A 17(2):95–106. https://doi.org/10.1016/0191-2607(83)90063-8 (cited on page 185)
24. Dahlhaus E, Horák P, Miller M, Ryan JF (2000) The train marshalling problem. Discret Appl Math 103(1–3):41–54. https://doi.org/10.1016/S0166-218X(99)00219-X (cited on page 195)
25. Dahlhaus E, Manne F, Miller M, Ryan JF (2000) Algorithms for combinatorial problems related to train marshalling. In: Proceedings of AWOCA 2000, Hunter Valley, NSW, pp 7–16 (cited on page 195)
26. Demange M, Ekim T, de Werra D (2009) A tutorial on the use of graph coloring for some problems in robotics. Eur J Oper Res 192(1):41–55. https://doi.org/10.1016/j.ejor.2007.09.018 (cited on page 183)
27. Di Stefano G, Koči ML (2004) A graph theoretical approach to the shunting problem. Electron Notes Theor Comput Sci 92:16–33. https://doi.org/10.1016/j.entcs.2003.12.020 (cited on pages 192, 193, 195)
28. Di Stefano G, Krause S, Lübbecke ME, Zimmermann UT (2008) On minimum k-modal partitions of permutations. J Discret Algorithms 6(3):381–392. https://doi.org/10.1016/j.jda.2008.01.002 (cited on page 192)
29. Eggermont C, Hurkens CAJ, Modelski M, Woeginger GJ (2009) The hardness of train rearrangements. Oper Res Lett 37(2):80–82. https://doi.org/10.1016/j.orl.2008.12.005 (cited on page 196)
30. Epstein L, Levin A (2008) On bin packing with conflicts. J Optim 19(3):1270–1298. https://doi.org/10.1137/060666329 (cited on pages 196, 197)

31. Fertig HR (1927) System classification plan improves yard efficiency. Railway Age 515–522, February 19, 1927 (cited on page 184)
32. Flandorffer H (1953) Vereinfachte Güterzugbildung. Rangiertechnik 13:114–118 (cited on page 184)
33. Fomin FV, Kratsch D, Novelli J-C (2002) Approximating minimum cocolourings. Inf Process Lett 84(5):285–290. https://doi.org/10.1016/S0020-0190(02)00288-0 (cited on page 192)
34. Freling T, Lentink RM, Kroon LG, Huisman D (2005) Shunting of passenger train units in a railway station. Transp Sci 39(2):261–272. https://doi.org/10.1287/trsc.1030.0076 (cited on page 203)
35. Gallo G, Di Miele F (2001) Dispatching buses in parking depots. Transp Sci 35(3):322–330. https://doi.org/10.1287/trsc.35.3.322.10151 (cited on page 183)
36. Gatto M, Maue J, Mihalák M, Widmayer P (2009) Shunting for dummies: an introductory algorithmic survey. In: Ahuja R, Möhring R, Zaroliagis C (eds) Robust and online large-scale optimization: models and techniques for transportation systems. Springer, Berlin, pp 310–337. https://doi.org/10.1007/978-3-642-05465-5_13 (cited on page 185)
37. Golumbic MC (2004) Algorithmic graph theory and perfect graphs. North-Holland, Amsterdam (cited on page 191)
38. Gorman MF (1998) An application of genetic and tabu searches to the freight railroad operating plan problem. Ann Oper Res 78:51–69. ISSN: 0254-5330. https://doi.org/10.1023/A:1018906301828 (cited on page 182)
39. Graßmann E (1952) Zur Geschichte der Rangiertechnik. Rangiertechnik 12:9–18 (cited on page 184)
40. Gurobi Optimization, Inc. (June 2017) Website of Gurobi. http:gurobi.com (cited on page 204)
41. Hamdouni M, Desaulniers G, Soumis F (2007) Parking buses in a depot using block patterns: a benders decomposition approach for minimizing type mismatches. Comput Oper Res 34(11):3362–3379. https://doi.org/10.1016/j.cor.2006.02.002 (cited on page 183)
42. Han Y-H (2004) Dynamic sequencing of jobs on conveyor systems for minimizing changeovers. Ph.D. thesis, Georgia Institute of Technology, School of Industrial and Systems Engineering. https://doi.org/10.1007/s00170-010-2704-5 (cited on page 183)
43. Hansmann RS (2010) Optimal sorting of rolling stock. Ph.D. thesis, Institut für Mathematische Optimierung, Technische Universität Carolo-Wilhelmina zu Braunschweig. Cuvillier, Göttingen. ISBN: 978-3-86955-659-8 (cited on pages 185, 186, 188, 193, 195, 196, 199, 201, 204)
44. Hansmann RS (2018, in preparation) Complexity of single-stage rail car rearrangements (cited on pages 195, 196, 197, 198)
45. Hauser A, Maue J (2010) Experimental evaluation of approximation and heuristic algorithms for sorting railway cars. In: Festa P (ed) Proceedings of the 9th international symposium on experimental algorithms (SEA-10). Lecture notes in computer science, vol 6049. Springer, Berlin, pp. 154–165. https://doi.org/10.1007/978-3-642-13193-6_14 (cited on page 201)

46. Huntley CL, Brown DE, Sappington DE, Markowicz BP (1995) Freight routing and scheduling at CSX transportation. Interfaces 25(3):58–71. https://doi.org/10.1287/inte.25.3.58 (cited on page 182)
47. IBM ILOG (Apr 2016) CPLEX. http:ibm.com (cited on page 204)
48. Ivić M, Marković M, Marković A (2007) Effects of the application of conventional methods in the process of forming the pick-up trains. Yugoslav J Oper Res 17(2):245–256. https://doi.org/10.2298/YJOR0702245I (cited on page 183)
49. Jacob R, Márton P, Maue J, Nunkesser M (2011) Multistage methods for freight train classification. Networks 57(1):87–105. ISSN: 1097–0037. https://doi.org/10.1002/net.20385 (cited on page 201)
50. Jansen K (2003) The mutual exclusion scheduling problem for permutation and comparability graphs. Inf Comput 180(2):71–81. https://doi.org/10.1016/S0890-5401(02)00028-7 (cited on page 197)
51. Keaton MH (1992) Designing railroad operating plans: a dual adjustment method for implementing Lagrangian relaxation. Transp Sci 26(4):263–279. https://doi.org/10.1287/trsc.26.4.263 (cited on page 182)
52. König FG (2009) Sorting with objectives - graph theoretic concepts in industrial optimization. Ph.D. thesis, TU Berlin. https://doi.org/10.14279/depositonce-2349 (cited on page 183)
53. König FG, Lübbecke ME (2008) Sorting with complete networks of stacks. In: Hong S-H, Nagamochi H, Fukunaga T (eds) International symposium on algorithms and computation (ISAAC 2008). Lecture notes in computer science, vol 5369. Springer, Berlin, pp. 896–907. https://doi.org/10.1007/978-3-540-92182-0_78 (cited on page 183)
54. König H, Schaltegger P (1967) Optimale Simultanformation von Nahgüterzügen in Rangierbahnhöfen. Monatsschrift der Internationalen Eisenbahn-Kongress-Vereinigung 4(1):1–18 (cited on page 184)
55. König FG, Lübbecke ME, Möhring RH, Schäfer G, Spenke I (2007) Solutions to real-world instances of PSPACE-complete stacking. In: Arge L, Hoffmann M, Welzl E (eds) Proceedings of the 15th European symposium on algorithms (ESA). Lecture notes in computer science. Springer, Berlin, pp 729–740. https://doi.org/10.1007/978-3-540-75520-3_64 (cited on page 183)
56. Krell K (1962) Grundgedanken des Simultanverfahrens. Rangiertechnik 22:15–23 (cited on page 184)
57. Krell K (1962) Ein Beitrag zur gemeinsamen Bildung von Nahgüterzügen. Rangiertechnik 23:16–25 (cited on page 184)
58. Kroon LG, Lentink RM, Schrijver A (2008) Shunting of passenger train units: an integrated approach. Transp Sci 42(4):436–449. https://doi.org/10.1287/trsc.1080.0243 (cited on page 203)
59. Liebchen C, Lübbecke M, Möhring R, Stiller S (2009) The concept of recoverable robustness, linear programming recovery, and railway applications. In: Ahuja RK, Möhring RH, Zaroliagis CD (eds) Robust and online large-scale optimization. Lecture notes in computer science, vol 5868. Springer, Berlin, pp 1–27. ISBN: 978-3-642-05464-8. https://doi.org/10.1007/978-3-642-05465-5_1 (cited on page 185)

60. Márton P, Maue J, Nunkesser M (2009) An improved classification procedure for the hump yard lausanne triage. In: Clausen J, Di Stefano G (eds) Proceedings of the 9th workshop on algorithmic methods and models for optimization of railways (ATMOS-09). Schloss Dagstuhl, Wadern, pp 1–15. https://doi.org/10.4230/OASIcs.ATMOS.2009.2142 (cited on page 203)
61. Nemani AK, Ahuja RK (2011) OR models in freight railroad industry. In: Cochran JJ, Cox LA, Keskinocak P, Kharoufeh JP, Smith JC (eds) Wiley encyclopedia of operations research and management science. Wiley, London. https://doi.org/10.1002/9780470400531.eorms0622 (cited on page 182)
62. Newton HN, Barnhart C, Vance PH (1998) Constructing railroad blocking plans to minimize handling costs. Transp Sci 32(4):330–345. https://doi.org/10.1287/trsc.32.4.330 (cited on page 182)
63. Pentinga KJ (1959) Teaching simultaneous marshalling. Railw Gaz 110(21):590–593 (cited on page 184)
64. Petersen ER (1977) Railyard modeling: part I. Prediction of put-through time. Transp Sci 11(1):37–49. https://doi.org/10.1287/trsc.11.1.37 (cited on page 185)
65. Petersen ER (1977) Railyard modeling: part II. The effect of yard facilities. Transp Sci 11(1):50–59. https://doi.org/10.1287/trsc.11.1.50 (cited on page 185)
66. Schaltegger P (1967) Optimierung eines Rangierverfahrens. Ablauf- und Planungsforschung 8(4):302–314 (cited on page 184)
67. Siddiqee MW (1972) Investigation of sorting and train formation schemes for a railroad hump yard. In: Newell GF (ed) Proceedings of the 5th international symposium on the theory of traffic flow and transportation. American Elsevier, New York, NY, pp 377–388 (cited on page 185)
68. Tarjan RE (1972) Sorting using networks of queues and stacks. J Assoc Comput Mach 19(2):341–346. https://doi.org/10.1145/321694.321704 (cited on page 185)
69. Van Dyke CD (1986) The automated blocking model: a practical approach to freight railroad blocking plan development. Transp Res Forum 27:116–121 (cited on page 182)
70. Wagner K (1984) Monotonic coverings of finite sets. Elektronische Informationsverarbeitung und Kybernetik 20(12):633–639 (cited on page 192)
71. Winter T (2000) Online and real-time dispatching problems. Ph.D. thesis, Abteilung für Mathematische Optimierung, Technische Universität Carolo-Wilhelmina zu Braunschweig. URN: urn:nbn:de:gbv:084-119482 (cited on page 196)
72. Winter T, Zimmermann UT (2000) Real-time dispatch of trams in storage yards. Ann Oper Res 96:287–315. https://doi.org/10.1023/A:1018907720194 (cited on page 196)
73. Wöckel F (1949) Nahgüterzugbildung im Flachbahnhof durch Ablauf. Eisenbahntechnik 1:5–12 (cited on page 184)

Chapter 10
Optimization of Rolling Stock Rotations

Markus Reuther and Thomas Schlechte

Abstract This chapter shows a successful approach how to model and optimize rolling stock rotations that are required for the operation of a passenger timetable. The underlying mathematical optimization problem is described in detail and solved by Rotation Optimizer for Railways (ROTOR), i.e., a complex optimization algorithm based on linear programming and combinatorial methods. ROTOR is used by DB Fernverkehr AG (DBF) in order to optimize intercity express (ICE) rotations for the European high-speed network. We focus on main modeling and solving components, i.e. a hypergraph model and a coarse-to-fine column generation approach. Finally, the chapter concludes with a complex industrial re-optimization application showing the effectiveness of the approach for real world challenges.

10.1 Introduction

The rolling stock, i.e., railway vehicles, is among the most expensive and limited assets of a railway operator. The rolling stock is required to operate a timetable,

M. Reuther (✉) · T. Schlechte
LBW Optimization GmbH & Zuse Institute Berlin, Berlin, Germany
e-mail: reuther@lbw-optimization.de; schlechte@lbw-optimization.de

R. Borndörfer et al. (eds.), *Handbook of Optimization in the Railway Industry*, International Series in Operations Research & Management Science 268,
https://doi.org/10.1007/978-3-319-72153-8_10

i.e., each trip of the timetable needs to be assigned to a concrete vehicle which offers the proposed passenger capacity and which is able to operate the given trip. In this context a trip is a journey through the railway network with given departure and arrival times and corresponding stations. Similar to the journey of passengers also the concrete physical vehicles are performing sequences of trips which could be even part of different railway lines as well as other activities like maintenance. The implementation of a timetable by a rolling stock fleet, i.e., the assignment of vehicles to trips, must be done in a most efficient way to be in the black. Note that railway operators usually have only a limited fleet with vehicles of different types and thus the assignment of vehicles to trips can be very flexible.

In order to have a master plan in operation, *rolling stock rotations* are created in advance. This is done within the production process of a railway operator. A rolling stock rotation is a cycle that covers timetabled trips. It is built in order to plan the succeeding operation of timetabled trips by railway vehicles.

A railway timetable forms the majority of the input data for rolling stock rotation optimization. The timetable consists of timetabled trips, which were created in order to provide a reliable transportation offer for freight or, in our application, for passengers of intercity express (ICE) vehicles. The creation of a railway timetable is a highly complex and substantial task for its own, see Schlechte [26] for a thesis on timetabling for railways.

Note that we consider rolling stock rotations as cycles. Indeed, this is a constraint of the so called *standard week* (concept). The standard week is a rich planning concept, which imposes the general assumption that everything will repeat after 1 week of operation. By this assumption we do not have to deal with precise calendar dates. Therefore, the standard week is applicable for long term (i.e., strategical considerations) as well as for mid-term (e.g., 1 year before operation), and it is also used for tactical plans, which are executed in, say, 6 weeks. Thus, the standard week is an overarching planning concept for the preparation of operations. Almost each decision made in operation w.r.t. rolling stock is derived from, or at least affected by, a cyclic rolling stock rotation at DB Fernverkehr AG (DBF).

The reason for the huge amount of preparation, which railway operators spend, is simply that railway operation nowadays is and always was complicated. For that matter, our cooperation partner DBF is not an exception since it provides the largest intercity passenger railway service in Europe. In fact, the proper complexity in railway planning is only partially caused by aspects of size (e.g., the number of timetabled trips to be covered by rolling stock rotations). The reason why it is complicated and, moreover, expensive originates from the diversity of different requirements, which need to be respected entirely.

The major requirements can be summarized as *vehicle composition rules*, *maintenance constraints*, *capacity constraints*, and *regularity stipulations*. Each of these requirements is already complex in its own right. Moreover, it can be almost impossible to treat these requirements sequentially, i.e., a step by step approach could lead to infeasibilities or inefficient results. These most important requirements can be informally described as follows:

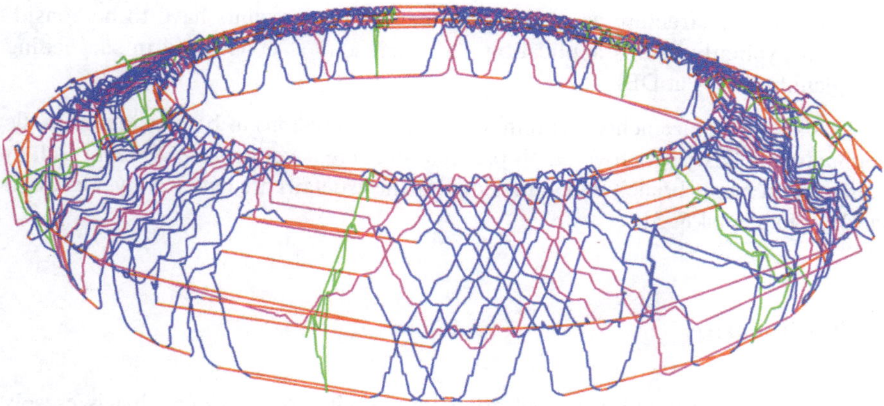

Fig. 10.1: A (small) railway timetable arranged in a torus for a cyclic standard week. The trips of the timetable are covered by two rolling stock rotations: One cycle alternates between blue paths (i.e., the trips of the timetable) and red connections. The second rolling stock rotation is purple. The seven (not six) green graphs indicate the railway infrastructure, which is utilized by railway vehicles on 7 days of operation. The picture has been created using `HyDraw` [10]

- Figure 10.1 shows a cyclic passenger timetable that is valid on 7 days of operation. For each day of operation all given passenger trips are plotted as paths arranged in a torus, in which time proceeds counterclockwise. A profile at a specific time of this torus represents the current location of all vehicles operating the timetable. As one can see in this picture, a common structure of railway timetables is that they are almost periodic, i.e., only a few of the given passenger trips differ from day to day. This implies a first requirement for the rolling stock rotations: They should reflect the periodicity of the timetable. This objective is called *regularity*.
- Another main characteristic of nearly all railway systems is that railway vehicles can be combined to form *vehicle compositions*. Among other things, the expected passenger demand on a timetabled trip leads to an individual and reasonable set of feasible vehicle compositions for each trip of the timetable. This gives many degrees of freedom for the rolling stock rotations, which have to be utilized.
- The rolling stock has to be maintained frequently. This leads to several maintenance constraints with different technical backgrounds. We consider cumulative time and distance resources which are classically constrained by upper bounds. In order to comply to those bounds railway vehicles are required to be maintained in periodic intervals.
- Maintenance and also parking activities usually consume infrastructure and crew capacity. Both types of capacity are limited. Moreover, also the number of railway vehicles of a dedicated type (i.e., of a fleet) is not infinite as is well

known. Infrastructure as well as fleet capacity constraints have to be considered explicitly in this application. Crew capacity is considered in succeeding planning steps at DBF.

This set of requirements is minimal (w.r.t. set inclusion) to be considered in the application at DBF. Moreover, their presentation here is very sparse. In reality, all of these requirements branch into many details. These details are extensively addressed in [25, Chapters 4 & 5].

10.2 Literature

In this section we give a brief but effective overview on literature that is closely related to the rolling stock rotation problem (RSRP) as we consider it. The idea of the section is not to provide an extensive survey, but to arrange the highlights of the chapter in terms of the existing literature on rolling stock optimization.

Vehicle scheduling is extensively discussed in the literature, see Löbel [19] for a survey. Ahuja et al. [1] present a mixed-integer programming (MIP) formulation for a locomotive scheduling problem. The model is solved by a very large-scale neighborhood search technique but does not include any maintenance constraints. In [30] a large-scale non-linear integer programming formulation for the integrated optimization of locomotive schedules including maintenance constraints is developed. Cordeau et al. [8] propose an integer programming model based on a multi-commodity flow formulation for the integrated assignment of locomotives and passenger cars to passenger trips. A three stage heuristic approach to incorporate maintenance tasks in precomputed rolling stock rosters is described in [3]. Furthermore, [21] and [22] present two integer programming formulations and alter optimized rolling stock rosters to incorporate maintenance tasks.

In [11] a MIP model for vehicle composition requirements is given. It considers detailed rules for individual positions of vehicles in compositions. The most important difference to our vehicle composition approach is that almost every timetabled trip has predefined successor trips.

An integer programming model for vehicle composition requirements, but without considering regularity or maintenance requirements, is given in [23]. The results presented in [23] have also been developed in cooperation with our industrial partner DBF.

Constraints that are very similar to the maintenance constraints in the RSRP come up in duty rostering problems, e.g., constraints on the maximal working time per week and the maximal number of successive working days of the drivers. Behrendt [5] introduced several MIP formulations for the duty rostering problem including constraints on cumulative resources such as working time. An adaption of one of these models for the treatment of maintenance constraints for the RSRP is presented in [25, Section 5.2].

The adaptation of the model for the maintenance requirements proposed in [5] is mathematically equivalent to the model developed independently by Giacco et al. [14] for the optimization of rolling stock rosters.[1] Giacco et al. [14] reported computational results for scenarios of the Italian railway company Trenitalia. They assume that the railway timetables are repeated on every day of operation and they consider only the minimization of the number of vehicles as objective function. Moreover, they do not integrate vehicle composition, regularity, and infrastructure capacity in their model.

In [13] a two-step approach is presented for rolling stock rostering and maintenance scheduling by extending the contribution of Giacco et al. [14] by another MIP model that runs on top of the rolling stock rosters. This two-step approach is also part of the thesis by Giacco [12].

Cacchiani et al. [7] present a fast heuristic for the train unit assignment problem (which is similar to the RSRP).

Haahr et al. [16] present a path-based model for timetables involving an acyclic time horizon. They claim that their model can easily take maintenance constraints into account. This model and a branch-and-price algorithm are tested for re-optimization instances of the suburban railway operator in Copenhagen (DSB S-tog) without maintenance constraints. The paper [16] is part of the thesis by Haahr [15].

Wagenaar et al. [29] present a model for railway rolling stock rescheduling. In their approach so called *maintenance appointments* are taken into account. The maintenance appointments are already scheduled for certain railway vehicles and are particularly required to be respected after a disruption in their application.

A comparison of the approaches presented in [29] and [16] is made by Haahr et al. [17]. Another literature comparison in terms of requirements for rolling stock optimization can be found in Thorlacius et al. [27]. A literature overview on re-optimization can be found in [24] and [2].

Conclusion. At this point it must be said that the literature on rolling stock rotation optimization is rather fragmented in the sense that it is hard to identify a significant common line (as we find it in comparison to, e.g., vehicle routing, see Toth and Vigo [28]). One explanation for this issue is that rolling stock rotation optimization is a planning problem whose concrete shape strongly depends on several side conditions. Examples for those side conditions are:

- Which railway operator of which country is involved in the optimization approach?
- What time horizon is to be considered?
- How has the timetable been constructed?
- Which railway vehicles (e.g., locomotives, carriages, or complete ICE vehicles) are to be optimized?

The approach used in RotOR to compute implementable rolling stock rotations by a mathematical optimization algorithm is based on the following modeling particularities:

[1] A rolling stock rotation plan (as we deal with) can be seen as a rolling stock roster.

- A time horizon of seven *individual* days of operation is considered within ROTOR.
- Regularity and handout optimization are not at all investigated in the literature (to the best of our knowledge).
- ROTOR's vehicle composition model is very detailed, i.e., even the orientation of railway vehicles is considered.
- ROTOR's model is fully integrated.

A slightly common line in the literature about rolling stock optimization is that almost all publications agree that the maintenance constraints for railway vehicles are important and, moreover, non-trivial to be tackled. In fact, there seems to be no standard approach to tackle those constraints. This is a particular motivation for [25, Chapter 3].

10.3 Model via Hypergraphs

The RSRP has its roots in a complex railway application. In this application it is not always completely sharp which responsibilities belong to rolling stock rotation planning and which do not. Therefore, we give a mathematical description in this section in order to provide a clearly arranged frame for the responsibilities that we take into account in this chapter. This description is rather short and formal and we refer to [25, Chapters 4 & 5] for all the details of the RSRP.

We will show how hypergraphs can be utilized to model several aspects of rolling stock rotation planning. In many classical works in the literature graphs are used to model the individual activities of vehicles, i.e. via the set of arcs. In order to model activities which several vehicles perform together hypergraphs turn out to be a very elegant and powerful tool. In particular, the fact that railway vehicles can be coupled together to form vehicle compositions is completely captured by the hypergraph.

The input data for the RSRP consists of a timetable that is composed of timetabled trips, so called *services*, a graph-based hypergraph, *maintenance constraints*, and *capacity constraints*:

Timetable. The set of timetabled passenger trips is denoted by T. A trip $t \in T$ has a departure and an arrival date in the standard week as well as a departure and arrival location.

Services. The set of *services* is denoted by S. A service $s \in S$ represents a dedicated activity performed on a vehicle, e.g., refuel and maintain a railway vehicle at Berlin.

Hypergraph. Let V be a set of *nodes* representing departures and arrivals of railway vehicles operating passenger trips of T and let $A \subseteq (V \cup S)^2$ be a set of directed standard arcs. The set of *hyperarcs* $H \subseteq 2^A$ is a subset of the power set of the standard arcs A. Thus, an element of H is a set of standard arcs. The RSRP *hypergraph* is denoted by

$$(V \cup S, A, H).$$

The standard arc $a = (u,v) \in A$ *operates* a trip $t \in T$ if $u \in V$ represents the departure of t and $v \in V$ represents the arrival of T. We say that the hyperarc $h \in H$ *covers* (or *operates*) the trip $t \in T$, if an arc $a \in h \subseteq A$ operates t. We denote the set of hyperarcs that cover the timetabled trip $t \in T$ by $H(t) \subset H$.

Maintenance Constraints. The set of maintenance constraints is denoted by L. A *maintenance constraint* $l \in L$ is represented by a resource function $r_l : S \cup A \mapsto \mathbb{Q}_+$ stating the resource consumption of arcs and maintenance services, a *resource upper bound* $U_l \in \mathbb{Q}_+$, and a set of associated maintenance services $S_l \subseteq S$. A *resource path* $P \subseteq A$ is a simple path starting and ending at nodes of S_l and it is feasible w.r.t. the maintenance constraint l if P fulfills

$$\sum_{a \in P} r_l(a) + \sum_{s \in S(P) \setminus S_l} r_l(s) \leq U_l.$$

Here, $S(P) \subseteq V \cup S$ denotes the set of nodes that is covered by the path P. Note that even a service of $S(P) \setminus S_l$ can consume resources, e.g., the time needed to refuel a vehicle must also be considered in another maintenance constraint that constrains the maximal amount of time between two succeeding maintenance services. We say that each maintenance service of S_l *resets* the maintenance constraint l.

Capacity Constraints. A *capacity constraint* $b \in B$ consists of a resource function $r_b : H \mapsto \mathbb{Q}$ and a *capacity bound* $U_b \in \mathbb{Q}$. Capacity constraints are used to model, e.g., infrastructure capacity, see [25, Section 5.3]. We say that the set of hyperarcs $H^\star \subseteq H$ *fulfills* the capacity constraint $b \in B$ if

$$\sum_{h \in H^\star} r_b(h) \leq U_b. \tag{10.1}$$

The output of the RSRP is a set of *feasible (rolling stock) rotations*:

Rolling Stock Rotation. A cycle $C \subseteq A$ is a *feasible (rolling stock) rotation* w.r.t. the maintenance constraint l if the resource consumption $\sum_{a \in C} r_l(a)$ of the whole cycle is zero or if each node of $V \cup S$ that is covered by C is contained in a resource path that is feasible w.r.t. l. We denote by $V(C) \subseteq V \cup S$ the nodes that the cycle $C \subseteq A$ covers, i.e., $V(C) := \bigcup_{(u,v) \in A} \{u, v\}$.

We denote the RSRP as follows:

Rolling stock rotation problem (RSRP). We are given a set T of timetabled passenger trips and a hypergraph $(V \cup S, A, H)$ with a cost function $c : H \mapsto \mathbb{Q}_+$, a set L of maintenance constraints, and a set B of capacity constraints. The RSRP is

$$\min_{H^\star \subseteq H} \left\{ \sum_{h \in H^\star} c(h) \;\middle|\; \begin{array}{lll} \exists C_1,\dots,C_n \subseteq A, & n \in \mathbb{N}: & \\ V(C_i) \cap V(C_j) = \emptyset & \forall i,j \in \{1,\dots,n\} & \text{and} \\ C_i \text{ is a feasible rotation} & \forall i \in \{1,\dots,n\}, l \in L & \text{and} \\ \bigcup_{h \in H^\star} h = \bigcup_{i=1}^{n} C_i & & \text{and} \\ |H^\star \cap H(t)| = 1 & \forall t \in T & \text{and} \\ \sum_{h \in H^\star} r_b(h) \le U_b & \forall b \in B & \end{array} \right\}.$$

That is, the RSRP is to find a node-disjoint set of rolling stock rotations (which are feasible w.r.t. all maintenance constraints) that form a set of hyperarcs such that each timetabled trip is covered exactly once, all capacity constraints are fulfilled, and the cost function is minimized.

Note that any solution to the RSRP is composed of node-disjoint cycles. Therefore, a hyperarc of the form $\{(u,v),(w,v)\} \subseteq A$ can not be part of a solution because the node v would be covered twice. Therefore, we assume that each hyperarc $h \in H$ defines a perfect matching between its tail nodes $\{u \in V \,|\, (u,v) \in h\}$ and head nodes $\{v \in V \,|\, (u,v) \in h\}$.

Before we proceed with a mathematical program for the RSRP we introduce some notation and definitions.

Notation and Definitions. We denote the subset of all hyperarcs covering the timetabled trip $t \in T$ as $H(t) \subset H$ and we denote by $H(a) \subseteq H$ the subset of hyperarcs that contain the standard arc $a \in A$. W.l.o.g. we assume that $H(a) \neq \emptyset$ for each arc $a \in A$. If $H(a) = \emptyset$, the standard arc a can never be contained in a feasible rotation because we are restricted to select hyperarcs for a solution. For a node $v \in V \cup S$ we denote sets of incoming as well as outgoing hyperarcs (standard arcs) of v as $H(v)^{\text{in}}$ and $H(v)^{\text{out}}$ ($A(v)^{\text{in}}$ and $A(v)^{\text{out}}$), respectively. These notations are formalized by definitions (10.2):

$$\begin{array}{lll} H(t) & := \{h \in H \,|\, h \text{ covers } t\} & \text{for trip } \; t \in T, \\ H(a) & := \{h \in H \,|\, a \in h\} & \text{for arc } \; a \in A, \\ H(v)^{\text{in}} & := \{h \in H \,|\, \exists a \in h : a = (u,v)\} & \text{for node } v \in V \cup S, \\ H(v)^{\text{out}} & := \{h \in H \,|\, \exists a \in h : a = (v,u)\} & \text{for node } v \in V \cup S, \\ A(v)^{\text{in}} & := \{(u,v) \in A\} & \text{for arc } \; a \in A, \\ A(v)^{\text{out}} & := \{(v,u) \in A\} & \text{for arc } \; a \in A. \end{array} \tag{10.2}$$

A Constraint Integer Program as Basic Model. ROTOR's *overall model* for the RSRP is a MIP model, which is presented in [25, Section 5.1] in its full shape. Here, we consider a more compact *basic model* for the RSRP in order to have a convenient reference in the proceeding sections. To this end, we use a constraint programming notation for the maintenance constraints. Obviously, the constraint notation does not help in solving RSRP instances, i.e., it is abstract and can be seen as a placeholder.

The reason why we denote them in this way is that we do not want to anticipate a dedicated modeling idea for these constraints. In fact, Reuther [25] develops two different ways to tackle the maintenance constraints: In ROTOR's industrial application an extra MIP part is used, see Reuther [25, Section 5.2]. This part is sophisticated and customized for the dedicated application at DBF. Additionally, in [25, Chapter 3] a promising alternative, which is also based on a constraint programming formulation, is presented.

Integer programming (IP). Many approaches for combinatorial optimization problems use an IP formulation as basis. The presented approach for the RSRP is not an exception. The major gain that we obtain from an IP model is the direct access to its linear programming (LP) relaxation. This is advantageous for two reasons. First, the solution of LP models of enormous size is a manageable task because a great algorithmic progress achieved during the last 70 years meets fast computers with much memory, nowadays. Second, beside the solution of LP models, the "only" remaining task in solving IP models is to force that all integer variables have to take integer values. This is also a standardized as well as computationally extensively investigated problem.

Basic Model. As explained in Sect. 10.3 an instance of the RSRP is defined through the following data: A hypergraph $(V \cup S, A, H)$ with a cost function $c : H \mapsto \mathbb{Q}_+$, a set of maintenance constraints L with resource functions $r_l : S \cup A \mapsto \mathbb{Q}_+$ and resource upper bounds $U_l \in \mathbb{Q}_+$, and a set of capacity constraints B with resource functions $r_b : H \mapsto \mathbb{Q}_+$ and capacity bounds $U_b \in \mathbb{Q}_+$. We are now ready to present the basic model for the RSRP. We introduce a binary decision variable x_h for each hyperarc $h \in H$ that is equal to one if and only if $h \in H$ is part of the solution and denote the constraint integer programming model (BM):

$$\min \sum_{h \in H} c_h x_h, \tag{BM}$$

$$\sum_{h \in H(t)} x_h = 1 \qquad \forall t \in T, \tag{10.3}$$

$$\sum_{h \in H(v)^{\text{in}}} x_h = \sum_{h \in H(v)^{\text{out}}} x_h \quad \forall v \in V \cup S, \tag{10.4}$$

$$\sum_{h \in H} r_b(h) x_h \leq U_b \qquad \forall b \in B, \tag{10.5}$$

$$\textsc{Maintenance}(l, x) \qquad \forall l \in L,$$

$$x_h \in \{0, 1\} \qquad \forall h \in H. \tag{10.6}$$

Program (BM) is composed of a linear objective function, a *(hyper-) flow part*, i.e., equalities (10.3) and (10.4), capacity constraints (10.5), and maintenance constraints as well as the integrality constraints for the x-variables (10.6).

Linear Objective Function. The objective function of model (BM) minimizes the total cost of the hyperarcs associated with x-variables that take value one. Note that the objective function is simply linear but, however, has still the ability to take distinguished costs for correlated standard arcs into account. For example, we can easily apply the model if a cost structure in which $c(\{a_1\}) + c(\{a_2\}) > c(\{a_1, a_2\})$ for standard arcs $a_1, a_2 \in A$ and hyperarcs $\{a_1\}, \{a_2\}, \{a_1, a_2\} \in H$ is required. This situation is not as easy to handle by using decision variables for standard arcs and has an important industrial application, see Reuther [25, Section 4.4].

Flow Part. The *flow part* (i.e., equalities (10.3) and (10.4)) is a distinguished feature of the basic model. It is responsible to force that the solution is a set of cycles that cover the timetabled trips. Hereby, the *covering constraints* (10.3) assign exactly one hyperarc to each timetabled trip $t \in T$, while the *flow conservation constraints* (10.4) in combination with the integrality constraints (10.6) force the solution to be a set of cycles. Note that the flow part is notationally identical in terms of a standard graph but its column vectors are significantly different, i.e., they belong to hyperarcs instead of standard arcs.

Reformulation by Using Standard Arc Variables. In order to make the modeling of cycles more clearer we reformulate the flow conservation constraints in the following way. We additionally introduce binary decision variables for standard arcs, i.e., the variable y_a takes value one if and only if the standard arc $a \in A$ is part of a solution. We couple these new arc variables to the hyperarc variables by:

$$y_a = \sum_{h \in H(a)} x_h \quad \forall a \in A \tag{10.7}$$

and replace the flow conservation constraints of model (BM) by equalities:

$$\sum_{a \in A(v)^{\text{in}}} y_a = \sum_{a \in A(v)^{\text{out}}} y_a \quad \forall v \in V \cup S. \tag{10.8}$$

By substituting all y-variables in (10.8) according to equality (10.7) we notice that we still deal with exactly the same solution space in terms of the x-variables of the basic model. Obviously, the standard flow conservation constraints (10.8) together with the integrality constraints for the y-variables force the solution to be a set of cycles which is exactly what we wanted to illustrate by the reformulation. In addition, we observe that the reformulation does not promise a gain because we additionally would have to deal with a new type of variables, namely the y-variables, and a large set of additional coupling constraints (10.7).

Inequalities for Capacity Constraints. Inequalities (10.5) are the natural formulation for the capacity constraints (10.1) of the RSRP. They are used to respect the capacity of fleets (i.e., the number of available railway vehicles) as well as the capacity of maintenance and parking facilities, see Reuther [25, Section 5.3].

Maintenance Constraints. The maintenance constraints denoted in the basic model (BM) play a central role in the RSRP. In fact, they are also responsible for the need of graph-based hypergraphs such that the perfect matching between the tail and head nodes of a hyperarc is precisely defined. The constraint notation of a main-

Table 10.1: Industrial requirements of rolling stock rotations for ICE vehicles

Requirement	Section in [25]	Feasibility	Optimality	Modeling
Vehicle configuration	4.1	x	x	Hypergraph
Turn duration rules	4.2	x	x	Hypergraph
Railway topology	4.3	x		Hypergraph
Deadhead trips	4.4	x	x	Hypergraph
Service paths	4.5	x	x	Hypergraph
Vehicle orientation	4.6	x	x	Hypergraph
Vehicle composition	4.7	x		Hypergraph
Trip sequences	4.8	x		Hypergraph
Coupling and decoupling	4.9	x		Hypergraph
Regularity patterns	4.10		x	Hypergraph
Re-optimization templates	4.11		x	Hypergraph
Objective function	4.12		x	Hypergraph
Maintenance constraints	5.2	x		MIP
Capacity constraints	5.3	x		MIP
Trunk constraints	5.4	x		MIP

tenance constraint $l \in L$ in model (BM) "only" states that the hyperarcs associated with x-variables with value one have to feasible w.r.t. l.

Conclusion. We conclude this section by Table 10.1, which summarizes the industrial details of the RSRP that are explained in [25, Chapters 4 & 5]. This shows the modeling power of hypergraphs for rolling stock rotation optimization.

Table 10.1 lists all requirements that we consider in our industrial application. By column "feasibility" ("optimality") we indicate if the corresponding requirement has to be "primarily" considered in order avoid infeasible (suboptimal) rolling stock rotations when solving RSRP instances. But those columns are rather secondary.

The last column of Table 10.1 denotes our argument for the usefulness of the hypergraph, i.e., it denotes if the implementation of the corresponding requirement is mainly accomplished by ROTOR's hypergraph or by ROTOR's MIP model.

We proceed with ROTOR's algorithmic key concept when solving industrial RSRP instances.

10.4 Solve via Coarse-to-Fine

In this section we highlight a coarse-to-fine (C2F) approach to solve certain linear programs by column generation. The problems that we address contain layers corresponding to different levels of detail, i.e., coarse layers as well as fine layers. These layers are utilized to design efficient pricing rules. In a nutshell, the method shifts the pricing of a fine linear program to a coarse counterpart. In this way, major decisions are taken in the coarse layer, while minor details are tackled within the fine layer. We elucidate this methodology by applying it to the RSRP.

In fact, the C2F approach is motivated by the RSRP as we introduced it in the previous section. In its industrial application, the RSRP consists of several "layers" that address different levels of detail. The major decisions of the RSRP deal with covering timetabled trips by rolling stock rotations. This is a coarse layer of the problem. At the same time minor decisions, for example, about the detailed arrival of a multi-traction vehicle composition at some station, must be considered for technical reasons. This defines a fine layer.

The paradigm of the approach is to work with a version of the fine model that is restricted to a small subset of variables. This restricted model is iteratively extended by using information from the coarse model. In other words, the coarse model is used to identify the relevant decisions (i.e., variables) of the fine model by (hopefully) focusing the attention exactly to where it is needed.

The variable selection process is handled by column generation, see Luebbecke and Desrosiers [20] for an introduction. The idea is to work with two LP models, one for the coarse and one for the fine layer. A coarse LP model is constructed by aggregating suitable rows of the fine LP model and, hopefully, turns out to be a combinatorial optimization problem of low complexity, e.g., a network flow problem. Variables for the fine LP model are generated using the coarse LP model until convergence. This method aims at a rapid solution progress and at a complete elimination of stalling and tailing-off effects that are due to the fine layer. A comprehensive literature review of related ideas can be found in [25, Section 2.2]. We proceed with the description of the generic C2F idea in the next section and illustrate its application to the RSRP afterwards.

10.4.1 C2F Column Generation for Linear Programs

In this section, we present the C2F idea for general LP models. The C2F idea is embedded in the column generation algorithm (CGA) that we briefly recall:

Column Generation. Given index sets $I = \{1,\ldots,m\}$ and $J = \{1,\ldots,n\}$, a matrix $A \in \mathbb{Q}^{I\times J}$, and vectors $b \in \mathbb{Q}^I$ and $c \in \mathbb{Q}^J$, consider a linear program

$$\min c^T x \quad \text{s.t.} \quad Ax = b, \quad x \in \mathbb{Q}^J_+ \tag{MP}$$

and its dual

$$\max b^T \pi \quad \text{s.t.} \quad A^T \pi \leq c, \quad \pi \in \mathbb{Q}^I.$$

We call program (MP) the master problem. If $|J|$ is very large, the CGA is the method of choice to solve the master problem (MP). By using the CGA one restricts J to a subset $J' \subseteq J$ of columns to solve the restricted master problem (RMP), i.e.,

$$\min c'^T x \quad \text{s.t.} \quad A'x = b, \quad x \in \mathbb{Q}^{J'}_+. \tag{RMP}$$

Note that c' and A' are also restricted to the subset J'. We assume x_j to be zero for $j \in J \setminus J'$ in a vector $x \in \mathbb{Q}^J$, which makes x compatible for both the MP and RMP.

In each iteration of the CGA we try to *price columns* (i.e., to find at least one column that is added to program (RMP)) $j \in J \setminus J'$ by solving the *pricing problem*. The pricing problem is to solve

$$\bar{c} := \min \left\{ c_j - \pi^T a_j \mid j \in J \right\}$$

where $a_j \in \mathbb{Q}^m$ is the column vector of A for column $j \in J$ and $c_j \in \mathbb{Q}$ is the objective coefficient of column j. If $\bar{c} \geq 0$, we have a proof that an optimal solution $x^\star$ for program (RMP) is also an optimal solution for program (MP). Otherwise, we select a set of columns $J^\star \subseteq J$ such that at least one $j \in J^\star$ has negative *reduced cost* $d_j := c_j - \pi^T a_j$, add the columns associated with $J^\star$ to program (RMP), and continue with re-optimizing program (RMP).

Column Selection by Layers. We are free in selecting columns for the set $J^\star$ by a *column selection rule* as long as at least one element of $J^\star$ has negative reduced cost. But, it is obvious that a better column selection rule improves the efficiency of the CGA. In particular, it can be beneficial to add also columns with positive reduced cost as we will see. We address applications where J is enumerated to check every $j \in J$ whether d_j is negative, e.g., the simplex method. We call this enumeration *pricing loop*.

The main idea is to introduce *layers* (precise definition follows) that are utilized to improve two aspects of the CGA. The first one is to speed-up the pricing loop in each iteration of the CGA. The second one is to refine the column selection rule. The latter, aims at reducing the total number of iterations performed by the CGA and to reduce the total number of columns generated.

We restrict our considerations for general LP models to two layers, namely the *coarse layer* and the *fine layer*. The *fine layer* is equal to program (MP). The coarse layer appears by the following considerations.

Let $[\cdot] : I \mapsto [I]$ be a *coarsening projection* that maps the index set I of the equations of program (MP) to a smaller *coarse index set* $[I]$ of size $|[I]| \leq |I|$. We use this notation because the projection $[\cdot]$ induces an equivalence relation on the row indices I, namely, $i \sim j \iff [i] = [j]$. Let $v \in \mathbb{Q}^I$ be a column vector with index set I and let v_i be the element of v with index $i \in I$. We define $\tau(v, i)$ to be the number of non-zero coefficients in v supported by rows equivalent to row i, i.e.,

$$\tau(v, i) := \left| \{ v_k \neq 0 \mid [k] = [i] \} \right|.$$

We denote the *coarse vector* or *coarsening* of v by $[v] \in \mathbb{Q}^{[I]}$. The coarse vector $[v]$ is composed of the following *coarse coefficients*

$$\begin{aligned} [v]_{[i]} &:= \left([v]_{[i]1}, [v]_{[i]2} \right) := \\ &\quad \left(\min \{ v_k \neq 0 \mid k \in I : [k] = [i] \}, \max \{ v_k \neq 0 \mid k \in I : [k] = [i] \} \right) \cdot \tau(v, i) \end{aligned}$$

where we define $[v]_{[i]} := (0, 0)$ if $\{ v_k \neq 0 \mid k \in I : [k] = [i] \} = \emptyset$. Note that $[v]_{[i]}$ is a pair of numbers, namely, the minimal and the maximal non-zero coefficient in the set of rows equivalent to row i, multiplied by the number of non-zero entries. Let $([A_{\cdot j}])_{j=1,\ldots,|J|}$ be the bimatrix of coarse column vectors of A. Typically, this bimatrix contains identical columns caused by the coarsening projection, which is a highly desirable effect. Later, in an implementation, we choose exactly one representative for a set of identical columns but we denote the resulting bimatrix by $[A]$ with columns $[J]$ and assume that $[A]$ has as many columns as A has in order to not leave the vector space of the columns of A. Further, we define the coarse objective coefficient $[c_j] := \min_{i\in J}\{c_i \mid [i]=[j]\}$ for column $j \in J$.

Let $\pi \in \mathbb{Q}^I$ be a dual solution vector of program (MP). We define[2]

$$[\pi]_{[i]} := (\min\{\pi_k \mid k \in I : [k]=[i]\}, \max\{\pi_k \mid k \in I : [k]=[i]\}).$$

Further, let a_j, $j \in J$, be a column vector with objective coefficient c_j. For the ease of notation, the *coarse reduced cost* $[d]$ is defined via coefficients $[d_j] := [c_j] - [\pi]^T \cdot [a_j]$, $j \in J$, where we define the multiplication of pairs as $(a_1,b_1)\cdot(a_2,b_2) := \max\{a_1a_2, a_1b_2, b_1a_2, b_1b_2\}$ for two pairs $(a_1,b_1) \in \mathbb{Q}^2$ and $(a_2,b_2) \in \mathbb{Q}^2$. Note that the coarse reduced cost is *not* the coarsening of the reduced cost vector d. The *coarse reduction* (R) of the master problem (MP) is

$$\min\ [d]^T x \quad \text{s.t.}\ [A]x\,[=]\,[b], x \in \mathbb{Q}_+^{[J]}, \tag{R}$$

where we define

$$[A]x\,[=]\,[b] \quad :\Leftrightarrow \quad [b]_{[i]1} \le \sum_{j\in J} [A_{\cdot j}]_{[i]2} x_j, \quad \sum_{j\in J} [A_{\cdot j}]_{[i]1} x_j \le [b]_{[i]2} \quad \forall [i] \in [I].$$

That is, the coarse reduction (R) approximates equations of the MP by two extreme case constraints arising from the minimum and maximum non-zero coefficients in equivalent rows. Note that the objective function of the coarse reduction is to minimize $[d]$ (and not c); the reason for this will become clear in the sequel. We also address the coarse reduction as *coarse LP model* and the MP as *fine LP model.*

The polytopes associated with programs (MP) and (R) are denoted by $P_{(\mathrm{MP})}$ and $P_{(\mathrm{R})}$, respectively. Coarsening has the following simple but important properties.

Lemma 10.1. *The coarse polytope associated with program* (R) *includes the fine polytope associated with program* (MP), *i.e.,* $P_{(\mathrm{R})} \supseteq P_{(\mathrm{MP})}$.

Proof. Let $\sum_{i\in I:[i]=[k]} \sum_{j\in J} A_{ij}x_j = \sum_{i\in I:[i]=[k]} b_i$ for all $[k] \in [I]$ be the system of sums of equivalent rows of program (MP). Every row of (R) is a relaxation of this (already relaxed) system because each coefficient is respectively over- or underestimated.

Lemma 10.2. *The coarse reduced cost always underestimate the (original) reduced cost, i.e.,*

$$[d_j] = [c_j] - [\pi]^T \cdot [a_j] \le c_j - \pi^T \cdot a_j = d_j \quad \forall j \in J.$$

Proof. By definition we have $[c_j] \le c_j$ and each summand in $\pi^T \cdot a_j$ is overestimated by a summand of $[\pi]^T \cdot [a_j]$.

---

[2] Note that the definition is different to the definition of the coarse coefficients. It is a correction of the definition in [6, 25].

Lemma 10.1 shows that the coarse reduction (R) provides an approximation of the fine MP which has fewer rows and, thus, is probably easier to solve. We want to take advantage of this approximation in a CGA for the fine LP model by shifting the pricing to the coarse reduction. A naive way to do this is to solve the coarse reduction w.r.t. the original objective function c (and not $[d]$ for that it is defined) in a first step, producing a set of columns

$$J^\star := \left\{ j \in J \,\middle|\, [x^\star]_{[j]} > 0 \right\} \subseteq J,$$

and then to solve the fine master LP model in a second step, starting from program (MP) restricted to the columns $J^\star$. However, this simplistic procedure is unlikely to work well because of a lack of information exchange between the coarse and the fine LP models.

C2F Column Generation Algorithm. Lemma 10.2 proposes an alternative (w.r.t. the simple approach mentioned at the end of the last section), i.e., to use the coarse reduced cost as objective function for the coarse reduction and, in addition, to prune the pricing loop by the coarse reduced cost. These generic ideas are formalized in Algorithm 10.1 that illustrates one single iteration within a CGA.

```
C2FColumnGeneration()
{
   π* := SOLVE((RMP)); // π* ∈ ℚ^m is an optimal dual solution of (RMP)

   [π*] := COMPUTE(π*); // coarsen dual solution according to [·]

   [d*] := COMPUTE([π*]); // compute coarse reduced cost

   [x*] := SOLVE((R)); // [x*] ∈ ℚ^[J] solves the coarse reduction (R) w.r.t. [d]

   J* := { j ∈ J | [x*]_[j] > 0 }; // select new columns for (RMP) by [x*]

   if( { j ∈ J* | d_j < 0 } = ∅ )
   {
      [J*] := { [j] ∈ [J] | [d_j] < 0 }; // "pricing loop" in coarse layer

      Q := { j ∈ J | [j] ∈ [J*], d_j < 0 }; // "pricing loop" in fine layer

      if( Q ≠ ∅ ) // Are there still columns with negative reduced cost?
      {
         J* := J* ∪ { CHOOSE(Q) }; // add at least one more column
      }
   }

   return J*;
}
```

Algorithm 10.1: C2F column generation for linear programs (only a single iteration is outlined)

The C2F CGA, as outlined by Algorithm 10.1, is based on an optimal dual solution of the fine RMP, see line 3. Then, the fine dual solution is coarsened according to the coarse projection in line 5 and the coarse reduced cost are computed in line 7.

The C2F CGA solves the fine MP by using the coarse reduction as a primary column selection rule, see lines 9 and 11 of Algorithm 10.1. The intention behind this rule is to compute a reasonable combination of (hopefully) improving columns by solving the coarse reduction. Using the coarse reduced cost as an objective aims at a "good combination" of improving columns of negative reduced cost and further columns of positive reduced cost that are "necessary" to complete the construction of the solution. Column selection by the solution of the coarse reduction is crucial for the performance of the C2F approach. Note that a similar idea (without coarsening and in the context of Lagrangian relaxation) has been introduced by Löebel [19, Section 7.1.2] published under the name "Lagrangian pricing".

After columns have been selected through the coarse reduction, the algorithm iterates through a coarse pricing loop, see line 15. Note that this is only necessary (for the proof of convergence) if it turns out that no columns with negative reduced cost have been selected through the coarse reduction. This is indicated by the if-statement in line 13 of Algorithm 10.1. On the basis of the set $[J^\star]$ of coarse columns with negative coarse reduced cost it is straightforward to check if there are fine columns with negative reduced cost. Note that this check is pruned by the results of the coarse pricing loop. By Lemma 10.2 we can not miss any columns in the fine layer with negative reduced cost. That shows that the preselection by $[J^\star]$ is exact. Finally, if the set Q is not empty it is sufficient to select one element of Q in order to ensure global convergence.

Algorithm 10.1 is iteratively called until convergence. It works particularly well if the coarse reduction appears (better to say is arranged) as a simple combinatorial optimization problem such as a standard assignment problem (AP) as it is the case in the RSRP application.

10.4.2 Layers for Rolling Stock Rotation Optimization

Coarsening Rows by Coarsening Nodes. In the previous section, we proposed to coarsen rows of an LP model such that the coarsening projection is "meaningful" w.r.t. the concrete underlying optimization problem. Even if this is the basic idea, we follow an equivalent but more indirect approach when we come to the RSRP-specific approach in this section where we illustrate the implementation of the C2F idea for the RSRP in ROTOR:

We define the coarsening projections in terms of the objects of ROTOR's hypergraph that are associated with ROTOR's MIP model. This is simple to denote and easy to motivate. In addition, this was the application for that the C2F idea was developed originally. To this end, we start with:

A Compromise. The final goal of this section is to solve the LP relaxation of ROTOR's overall MIP formulation for the RSRP. This model is presented in [25, Section 5.1]. The C2F approach is motivated by complex industrial vehicle composition requirements that are extensively described in [25, Chapter 4] in detail.

In order to keep the presentation here compact and understandable at the same time, we apply the following compromise. We explain in a way by that the reader is not required to know all details of ROTOR's overall model, even if we sometimes refer to it. Just keep in mind that this model is a large MIP formulation for that we introduce two additional coarse layers in the following.

In fact, it is enough to remember how the basis model (BM) for the non-maintenance relaxation (also without capacity constraints) of the RSRP as introduced on page 221 works in order to follow the considerations of this section. We briefly define only the required combinatorial details of the model, namely vehicle composition aspects, that we need here.

Vehicle Composition in a Nutshell. We derive the coarse layers for the RSRP from detailed combinatorial aspects of vehicle composition. Therefore, we explain what is meant by *fleet*, *vehicle orientation*, *position*, *vehicle composition*, and *vehicle configuration* in the following.

A *fleet* is a basic type of railway vehicles (e.g., the approximately 40 vehicles of the ICE's first generation owned by DBF form a fleet). The set F denotes the set of fleets.

An *orientation* is an element of the set $O = \{\text{Tick}, \text{Tack}\}$. The orientation describes the two options of how vehicles can be placed on a railway track. This is distinguished at DBF by the position of the first class carriage of the vehicle w.r.t. the driving direction. Tick (Tack) means that the first class carriage is located at the head (tail) of the vehicle w.r.t. the driving direction.

A *(vehicle) composition* c of size $n \in \mathbb{N}_+$ is an n-tuple of the form

$$c = ((f_1, o_1), (f_2, o_2), \ldots, (f_n, o_n)) \in (F \times O)^n,$$

i.e., a vehicle composition defines detailed positions, orientations, and fleets of railway vehicles that are coupled together.

A *(vehicle) configuration* is a multiset of fleets and can be seen as a main characteristic of a vehicle composition. We say that the vehicle configuration k is *realized* by the vehicle composition $c = ((f_1, o_1), (f_2, o_2), \ldots, (f_n, o_n))$ if $k = \{f_1, \ldots, f_n\}$, i.e., if the multiset of fleets used in the composition c is equal to the vehicle configuration k.

In fact, the relation and, moreover, the difference between vehicle composition and vehicle configuration is important for rolling stock rotation optimization and of major interest for the C2F implementation.

Vehicle Composition by ROTOR's Hypergraph. Each node $v \in V$ of ROTOR's hypergraph (V, A, H) (the set of service nodes S is not necessary to be considered in the following) is constructed in such a way that it defines the fleet, the vehicle orientation, the position, and the vehicle configuration (not to be confused with vehicle composition) of a railway vehicle that traverses the node v. Each hyperarc $h \in H$ of ROTOR's hypergraph is created in order to model the movement of a vehicle composition (not to be confused with vehicle configuration). The vehicle composition derives from the nodes that h connects by its standard arcs.

In fact, the main structure of model (BM) is implied by the underlying hypergraph (V,A,H). Therefore, it is natural to first define projections for (V,A,H) (precise definition follows) and to apply the C2F CGA afterwards. In addition, our description gains a lot of convenience from the discussion in terms of nodes, arcs, and hyperarcs. We do not need to discuss every constraint of model (BM) separately (even if this is necessary within ROTOR). Instead, all details (in particular the coarse coefficient matrix) as required in an implementation derive from the hypergraph projections in a straightforward way.

The starting point for a hypergraph projection is the definition of how the nodes of V are projected onto coarser nodes. Based on this definition we expose further details in the next three subsections. To this end, we frequently consult Fig. 10.2. This illustration serves as a running example and shows the relation of three layers for the RSRP plus an additional figure that illustrates the vehicle configurations that are allowed to be chosen for the trips of the input timetable at the very top.

The layers, namely, a *composition layer* $G = (V,A,H)$, a *configuration layer* $[G] = ([V],[A],[H])$, and a *vehicle layer* $[[G]] = ([[V]],[[A]],[[H]])$ induce hypergraphs themselves. They are strongly motivated by the vehicle composition requirements of the RSRP.

In the application at DBF the RSRP must be solved for the composition layer, but many technical rules only apply to the configuration layer, which is much smaller w.r.t. the size of the set of hyperarcs. In addition, we define the vehicle layer to set up a super-coarse RSRP that provides a reasonable description of the major problem characteristics (e.g., the total number of railway vehicles used in a solution) and for that we design the coarse reduction such that it is combinatorially solvable in polynomial time.

We proceed with the discussion of ROTOR's three layers, namely, the composition layer, the configuration layer, and the vehicle layer.

Composition Layer. We define the *composition layer* simply as the hypergraph (V,A,H) of the RSRP. It is illustrated on the very bottom of Fig. 10.2. The composition layer represents the most fine layer on that the RSRP and, in particular, the LP relaxation of model (BM) needs to be solved.

The hypergraph (V,A,H) is constructed in such a way that each hyperarc $h \in H$ identifies a vehicle composition. In order to allow for this construction the nodes V are particularly assembled. Since this assembly is essential for the coarsening projections we describe it in the following.

We define an *event* as a triple (e,t,p) with $e \in \{d,a\}$ defining the departure $(e = d)$ or the arrival $(e = a)$ of an individual railway vehicle at position $p \in \mathbb{Z}_+$ in a vehicle composition while operating the timetabled trip $t \in T$.

A node $v \in V$ of the composition layer is of the form of a four-tuple:

$$v = ((e,t,p),k,f,o) \tag{10.9}$$

where we assume the following notation:

- (e,t,p) identifies an event,
- k denotes the vehicle configuration (see Reuther [25, Section 4.1]),

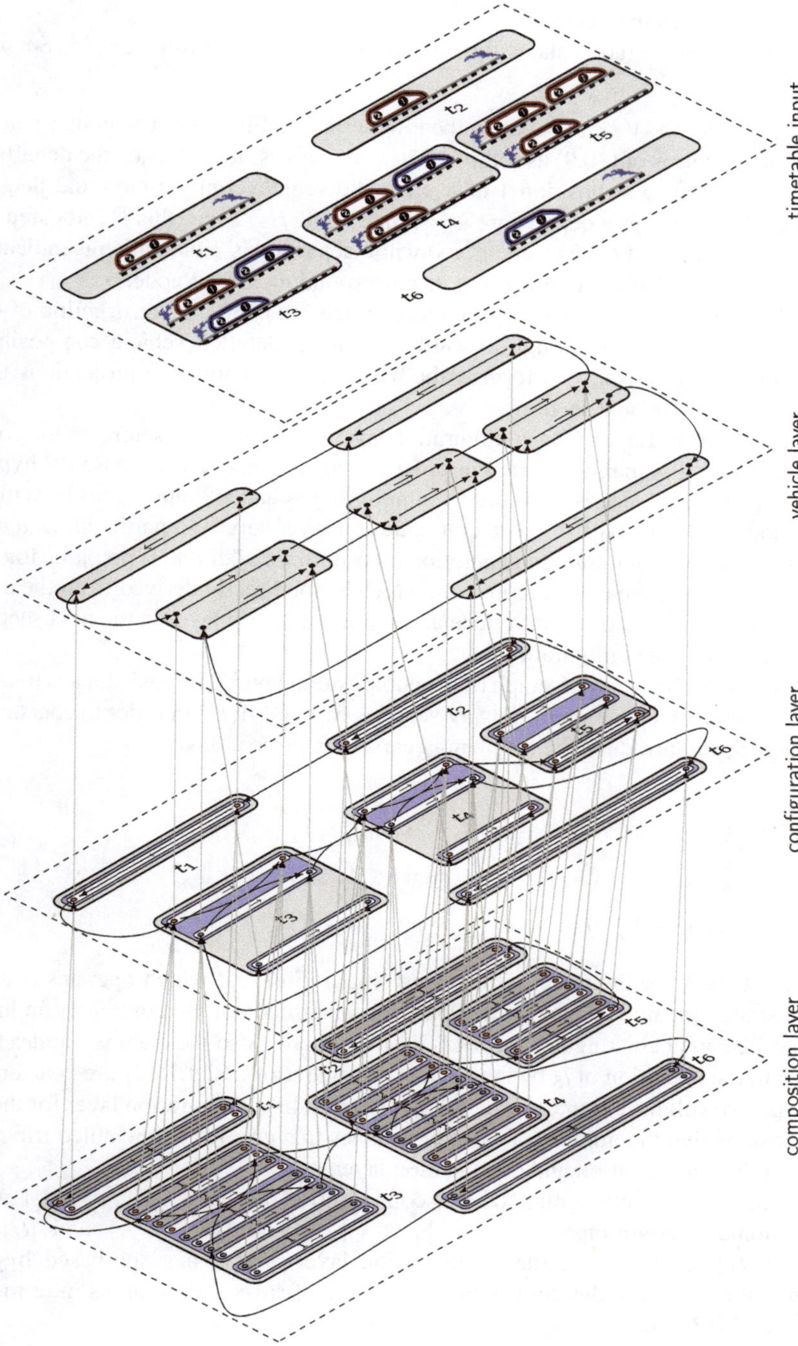

Fig. 10.2: Layers for the RSRP

- $f \in F$ declares the fleet, and
- $o \in O = \{\text{Tick}, \text{Tack}\}$ states the vehicle orientation (see Reuther [25, Section 4.6]).

For example, let $v \in V$ be the node in the lowest line[3] in Fig. 10.2. The node v can be interpreted in terms of (10.9) as follows. The event of v is $(d, t_3, 1)$, i.e., the departure of a railway vehicle at position 1 of trip t_3. The vehicle configuration and fleet in that a vehicle through v departs are $k = \{\text{Blue}\}$ and $f = \text{Blue}$ (this is indicated by the blue color). Finally, the orientation o while departing is Tick, which is indicated by the white box that surrounds v and its corresponding arrival node.

By having nodes that are composed as denoted in (10.9) the construction of the hyperarcs of (V, A, H) is straightforward even under detailed vehicle composition requirements. Moreover, this form is the basis for the coarsening projections that we introduce in the next sections.

Configuration Layer. The configuration layer is a further coarsening of the composition layer. It is mainly created in order to reduce the size for the RSRP hypergraph. The reduced size is achieved by simplifying vehicle compositions to vehicle configurations as we will see. The configuration layer serves in particular as a tool for pruning during our column generation algorithm. In addition, it prepares for the implementation of the coarse reduction, which is not directly derived from the configuration layer. Instead, we introduce an even more coarse layer in the next section that is based on the configuration layer.

Let (V, A, H) be the hypergraph of the composition layer. The basic idea is to omit the orientation and position of railway vehicles from (V, A, H) in order to coarsen it. To this end, we introduce the following coarsening projections:

$$\begin{aligned}
[v] &:= ((e,t,p),k) && \text{for } v = ((e,t,p),k,f,o) \in V, \\
[a] &:= ([v],[w]) && \text{for } \quad a = (v,w) \in A, \text{ and} \\
[h] &:= \{[a] \mid a \in h\} && \text{for } \qquad h \in H.
\end{aligned}$$

Again, let $v \in V$ be the node at the lowest line in Fig. 10.2 which operates $t_3 \in T$. As illustrated the node v is projected onto a coarser node of the configuration layer for that we can not distinguish its orientation anymore. Also the eight red nodes that belong to the operation of t_5 by the vehicle configuration $\{\text{Red}, \text{Red}\}$ are coarsened. Note that we still have four corresponding nodes in the configuration layer for them. This ensures that the number of railway vehicles that cover the timetabled trip t_5 is equal in the configuration and composition layer.

We denote the hypergraph of the configuration layer as $([V], [A], [H])$ with the canonical definitions: $[V] := \{[v] \mid v \in V\}$, $[A] := \{[a] \mid a \in A\}$ and $[H] := \{[h] \mid h \in H\}$. In this way, the configuration layer is also a graph-based hypergraph that completely determines the coarsening of rows and columns in terms of ROTOR's MIP model.

[3] In order to decrease confusion: The blue circle with the lowest y-coordinate in a natural Cartesian system of coordinates for Fig. 10.2 is meant.

Since the projections of the composition layer onto the configuration layer omit the orientation and the position of railway vehicles, the hyperarcs of $[H]$ can be interpreted as connections of timetabled trips with vehicle configurations (not to be confused with vehicle compositions). Indeed, this is the original motivation for defining the configuration layer. Vehicle configuration plays a much more essential role in the application than vehicle composition. Most of the time-dependent constraints, e.g., the minimal time needed for cleaning or refueling, refer "only" to the configuration layer, i.e., they are independent of the concrete vehicle composition that is realized. Of course, vehicle composition is important and necessary but it mainly causes large hypergraphs. Vehicle configuration is much more responsible for the hardness of the RSRP.

A comparison of the size of the composition and configuration layer, shows that the cardinality of the set of hyperarcs in the composition layer (V,A,H) is exponential in the size of the set of hyperarcs in the configuration layer $([V],[A],[H])$, see Reuther [25, Section 2.4.2].

Vehicle Layer. The vehicle layer is the coarsest layer that is created in ROTOR for the RSRP. Its purpose is to derive an implementation of the coarse reduction. Recall, that the coarse reduction is mainly solved to identify the "right" columns while generating and we are, fortunately, not strictly bound to the generic procedure proposed in Sect. 10.4.1. Therefore, the implementation of the coarse reduction in ROTOR is strongly related but formally not completely equal to the generic LP version. More precisely, it is based on the coarse reduced cost of the configuration layer and the underlying graph of the vehicle layer is a further coarsening of the configuration layer $([V],[A],[H])$.

The main motivation for the concrete implementation comes from the experience in solving RSRP instances. In these instances, a main characteristic is the number of railway vehicles that are used in a solution. Our observation is that this number is often already very well approximated if we solve a very simplified version of the original problem, namely a AP. In the following we describe how we setup this AP on the basis of the configuration layer.

Let $([V],[A],[H])$ be the hypergraph of the configuration layer and consider the following further coarsening projections:

$$[[v]] := (e,t,p) \quad \text{for } [v] = ((e,t,p),k) \in [V], \text{ and}$$
$$[[a]] := ([[v]],[[w]]) \quad \text{for } [a] = ([v],[w]) \in [A].$$

Given a configuration layer with $([V],[A],[H])$, we define the *vehicle layer* as the (graph-based hyper-) graph $([[V]],[[A]],[[H]])$ with $[[V]] := \{[[v]] \mid [v] \in [V]\}$, $[[A]] := \{[[a]] \mid [a] \in [A]\}$, and $[[H]] := \{\{[[a]]\} \mid [[a]] \in [[A]]\}$.

The set $[[H]]$ is only defined for the sake of an uniform notation. It is not a set of proper hyperarcs (anymore). In fact, we disassemble the hyperarcs to standard arcs in the vehicle layer in order to gain a coarse reduction which can be solved very quickly.

Since we want to price on the basis of the standard arcs of an AP, we have to migrate the coarse reduced cost, which are still associated with hyperarcs, to them.

We do this via the *partial coarse reduced cost* $[[c]] : [[A]] \mapsto \mathbb{Q}$ as:

$$[[c]]([[a]]) := \min\left\{ \frac{[d_h]}{|h|} \,\middle|\, [a] \in [h] \in [H] \right\}$$

for each arc $[[a]] \in [[A]]$. Note that this formula is based on the coarse reduced cost of the configuration layer, but $[[c]]([[a]])$ for $[[a]] \in [[A]]$ does formally not denote reduced cost in terms of LP. Its purpose, namely to transfer the coarse reduced cost associated with hyperarcs of the configuration layer to the standard arcs of the vehicle layer, is heuristic. See Reuther [25, Section 2.4.3] for a motivation of this definition.

As denoted, to further coarsen the node v from $[v]$ to $[[v]]$ means to also omit the vehicle configuration k from v. This has the following essential effect. In the vehicle layer $([[V]], [[A]], [[H]])$ there is only a single option of how to cover a timetabled trip. This is illustrated in Fig. 10.2. On the contrary, in the composition layer there are many options for the operation of the timetabled trip t_3, i.e., $|H(t_3)|$ is large. This is reduced to two possible vehicle configurations in the configuration layer and further simplified to two departure and arrival pairs in the vehicle layer.

Finally, we are ready to denote the *assignment pricing problem* (APP) that serves as coarse reduction in our specialized C2F CGA for the solution of the LP relaxation of ROTOR's MIP model for the RSRP. It reads:

$$\begin{aligned}
\min \sum_{[[a]]\in[[A]]} [[c]]([[a]])\, x_{[[a]]}, & & \qquad\qquad \text{(APP)}\\
\sum_{[[a]]\in[[A]]([[v]])^{\text{in}}} x_{[[a]]} = 1 & \qquad \forall\, [[v]] \in [[V]], \\
\sum_{[[a]]\in[[A]]([[v]])^{\text{out}}} x_{[[a]]} = 1 & \qquad \forall\, [[v]] \in [[V]], \\
x_{[[a]]} \in \{0,1\} & \qquad \forall\, [[a]] \in [[A]].
\end{aligned}$$

The decision variables in program (APP) correspond to the standard arcs of the vehicle layer. A solution of the assignment pricing problem (APP) is a set of cycles that cover all the departure and arrival pairs of the vehicle layer.

The assignment pricing problem (APP) is an AP, which we solve by the Hungarian method [18] in ROTOR's implementation. The pricing strategy behind model (APP) is simply: Whenever a hyperarc of the composition layer projects to an arc of the AP's solution, we add it to the RMP, i.e., price it.

For more details about ROTOR's C2F implementation, especially computational results, see Reuther [25, Chapter 2]. ROTOR's C2F method is the algorithmic key technology in solving large and complex industrial RSRP instances. One (re-optimization) example is presented in the next section. The associated RSRP instance would not have been solved without the C2F procedure.

10.5 Apply via Re-optimization the Köln-Rhein-Main Construction Site

We conclude this chapter with **bad news** that are actually very good news:

> Bauarbeiten: ICE-Strecke von Frankfurt nach Köln zeitweise gesperrt [4]

This news appeared in the "FAZ" (a prominent German newspaper). It announced that the railway tracks for the high-speed ICE lines between Frankfurt (Main) and Cologne would be completely closed on four succeeding weekends in May 2015. The reason was that Deutsche Bahn had to rebuild 17 km long rails. In this section, we call the corresponding occasion the Köln-Rhein-Main construction site (KRM).

In general, the KRM was bad news: It caused an additional driving time of approximately 60 min between Frankfurt and Cologne because the railway vehicles (i.e., ICEs) were redirected through rails via the Rhine. The rails of the redirection are not appropriate for high-speed operations. Therefore, the KRM directly affected the timetable of eight ICE lines that usually run over the involved rails via Montabaur with high speed and a high frequency, see Fig. 10.3.

However, the KRM is good news for mathematical optimization in the following sense: Before the news were published, ROTOR was used by DBF in order to compute appropriate rolling stock rotations for the KRM. This application is the subject of this section.

Redirection via the Rhine. The KRM had drastic consequences for the operation of the ICE vehicles of DBF compared to other construction sites. The main reason is that an increased driving time for eight (out of overall 27) ICE lines causes unusually many "standard" rolling stock rotations to become infeasible. In fact, it is the increase by 60 min that matters here because a driving time extension of, e.g., half an hour can sometimes be handled more easily: Many ICE lines operate with a period of 1 h and, therefore, an often successful approach is to shift connections between timetabled trips (i.e., turns) by one period. For the KRM it was not obvious how to implement this approach, i.e., an easy solution was not at hand.

Scenario Isolation. During the planning of a construction site by DBF, rotation planners try to make minimally invasive changes to the existing timetable and rolling stock rotations. To this end, it is desirable to isolate the planning scenario as much as possible from the remaining part of the ICE network. This is done in order to minimize negative side effects (e.g., the propagation of delays and a decreased volume of passenger of a construction site) on other parts of the ICE network.

For the KRM the isolation procedure was obvious: The timetable changes[4] as illustrated in Fig. 10.3 suggest that the operation (in particular the rolling stock rotations) of all ICE vehicles of the category ICE-W[5] of DBF are affected by the KRM.

---

[4] We do not provide the detailed changes of departures and arrivals here. The important aspect is that the eight lines illustrated in Fig. 10.3 were directly affected by the 60 min increased driving time.

[5] The "W" indicates that the ICE-W vehicles are equipped with "Wirbelstrombremsen", i.e., eddy current brakes.

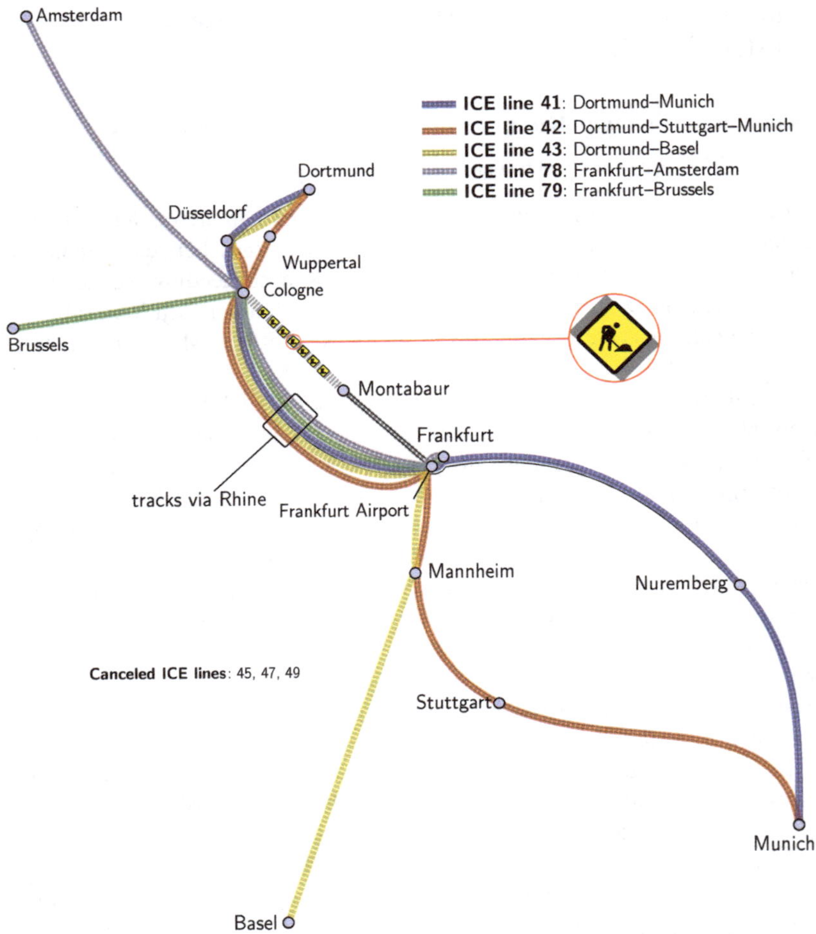

Fig. 10.3: Redirection for the Köln-Rhein-Main construction site (KRM): All ICE lines between Montabaur and Cologne were closed in order to rebuild rails (as indicated by the sleepers without rails) on four succeeding weekends in May 2015. Note that the figure only illustrates the situation at the weekend when the KRM is performed, i.e., the ICE lines 41, 42, 43, 45, 47, 49, 78, 79, and 82 operate as usual on the high-speed tracks between Frankfurt and Cologne from Monday to Friday. Note also that the placement of the locations is true to scale

The corresponding ICE lines are the lines 41, 42, 43, 45, 47, 49, 78, 79, and 82. This might not really appear as a minimally invasive isolation at first glance. In fact, in case of the KRM it was the minimal set of ICE lines to consider!

It turned out that line 82, which connects Frankfurt to Paris, could be treated separately. The ICE vehicles for line 82 have special equipment for France and are usually not mixed with the other ICE vehicles on other ICE lines.

RSRP Setup. For the remaining eight ICE lines an instance of the RSRP dedicated to the KRM was set up by rotation planners of DBF as follows. The original timetable for the seven remaining ICE lines for Monday to Friday was taken without modifications. The timetabled trips of the ICE lines 41, 42, 43, 78, and 79 were modified on the weekend in order to reflect the increased driving time caused by the redirection via the Rhine. The trips of the ICE lines 45, 47, and 49 that operate on the weekend were canceled. Then, allowed vehicle configuration requirements were specified for the "KRM timetable". Not surprisingly, rotation planners of DBF reported that the setup phase took a long time because a lot of data had to be manually modified for the KRM.

The standard rolling stock rotations were declared as reference rotations. Thus, the KRM RSRP was a proper re-optimization scenario. It contained all industrial requirements described in [25, Chapters 4 & 5] except for infrastructure capacity constraints which were not actually needed.

Optimization Rounds. It is far too much to expect that ROTOR immediately computed an outstanding result for such a complex industrial instance. Beside ROTOR's pure computation time it is almost always certain that the first computations will reveal data issues. Naturally, such a complex RSRP instance, which was manually set up "under fire", will not immediately model what is desired in the first shot. Therefore, we are not surprised that approximately four rounds were necessary to obtain the first usable rolling stock rotations. And even if ROTOR had produced those rotations in the first place, further rounds were needed to tune the timetable and rotations.

Keep in mind, that rolling stock rotations often give rise to further timetable changes which involve further optimization rounds. What we learn from this is that optimization (equally whether automatic or manual) can easily run into a loop that will never stop. This is a serious issue that all optimization approaches share and which can be minimized by, e.g., software tools dedicated to the preparation of ROTOR's input data and also for further processing ROTOR's output data. Fortunately, also those software tools are at hand at DBF. Thus, the optimization rounds, indeed, came to an end.

The KRM RSRP. Most of the final results were obtained from ROTOR by computing a solution to a certain RSRP instance at DBF. We call this instance KRM RSRP in this section. In fact, it was contributed by DBF to our test set at Zuse Institute Berlin (ZIB) and it is called `RSRP_111R` in [25, Section 7.3]. The timetable of the KRM RSRP consists of 169 trains, 1025 timetabled trips, and 838 timetabled trip sequences. Seven reference rotations are to be re-optimized and also seven fleets are given. The rotation templates for these reference rotations lead to a refinement of 15 original vehicle configurations to 30 refined configurations. ROTOR's hypergraph for this instance has 69,751,046 hyperarcs and one maintenance constraint (i.e., the regular inspection) is contained.

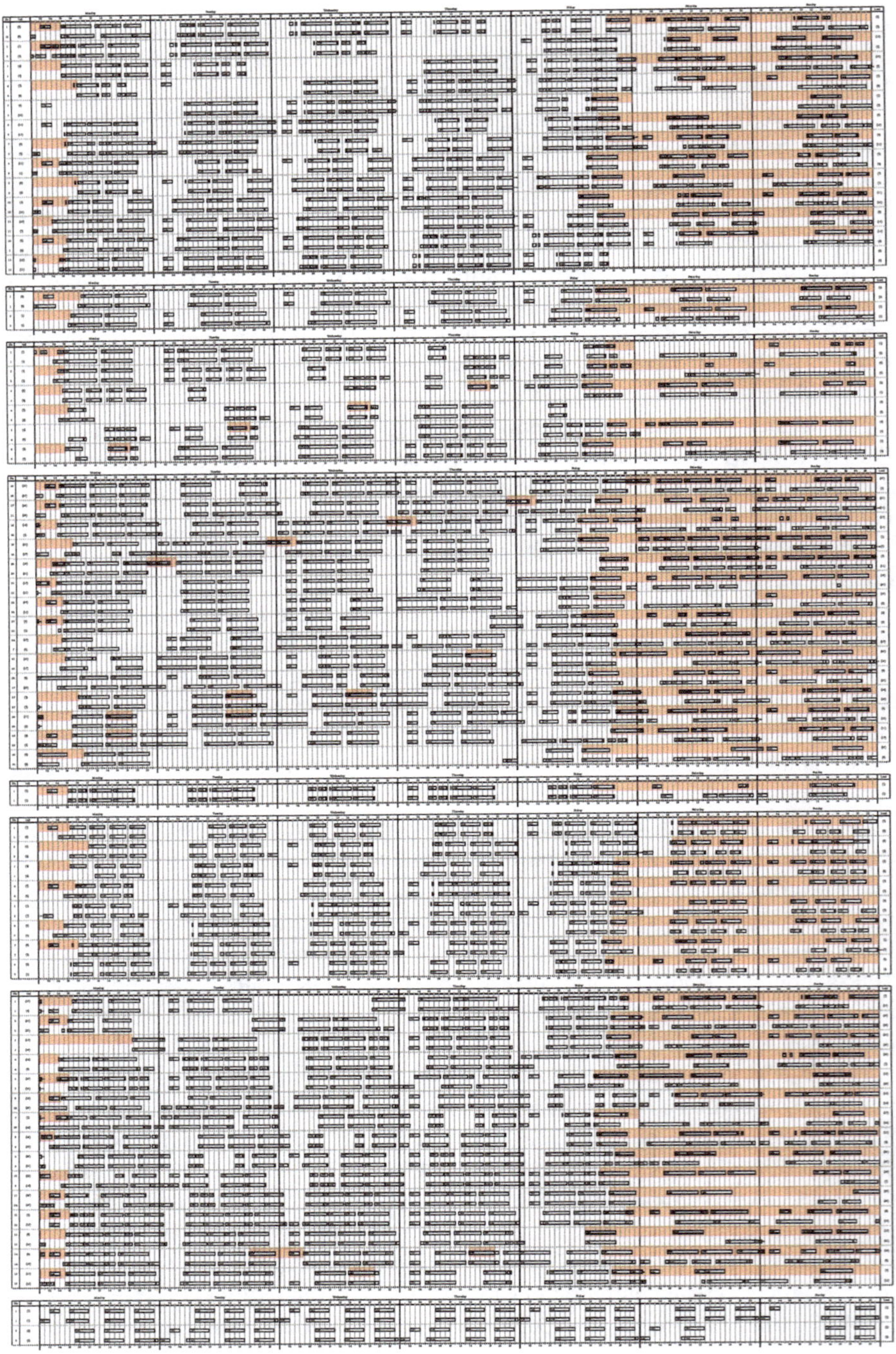

Fig. 10.4: KRM rotations in comparison to the reference rotations

At the end of the year 2014, ROTOR version 2.3 computed a solution to the KRM RSRP within approximately 1 day at DBF with an integrality gap below 1%, see Reuther [25, Section 7.5].

KRM Rotations. The eight rolling stock rotations that ROTOR produced for the KRM RSRP, namely the KRM rotations, are illustrated in Fig. 10.4. The rows of the figure alternate between the reference rotations and the KRM rotations. The red blocks indicate those parts of the reference rotations (which would have been operated if the KRM did not exist) that became infeasible on the occasion of the timetable changes for the KRM.

Even if the whole page with anonymous rollings stock rotations appears as slightly over-ambitious to the reader, the figure clearly shows that the KRM rotations are very (very) similar to the reference rotations between Monday noontime and Friday noontime. This provides the opportunity to easily bridge between the reference rotations and the KRM rotations on all the eight boundaries of the four KRM weekends. These rotations were made “camera-ready” by rotation planners of DBF in order to operate them in May 2015.

References

1. Ahuja RK, Liu J, Orlin JB, Sharma D, Shughart LA (2005) Solving real-life locomotive-scheduling problems. Transp Sci 39:503–517. ISSN: 1526-5447. https://doi.org/10.1287/trsc.1050.0115 (cited on page 216)
2. Ahuja RK, Möhring RH, Zaroliagis CD (eds) (2009) Robust and online large-scale optimization: models and techniques for transportation systems. Lecture notes in computer science, vol 5868. Springer, Berlin, Heidelberg. ISBN: 978-3-642-05464-8. https://doi.org/10.1007/978-3-642-05465-5 (cited on page 217)
3. Anderegg L, Eidenbenz S, Gantenbein M, Stamm C et al (2003) Train routing algorithms: concepts, design choices, and practical considerations. In: Proceedings of the 5th workshop on algorithm engineering and experiments (ALENEX). SIAM, Philadelphia, PA, pp 106–118 (cited on page 216)
4. Bauarbeiten: ICE-Strecke von Frankfurt nach Köln zeitweise gesperrt. German. Frankfurter Allgemeine Zeitung (FAZ) (2015). faz.net/-gzg-7zfb7 (visited on 09/11/2017) (cited on page 235)
5. Behrendt S (2008) Dienstreihenfolgeplanung mit ganzzahliger Optimierung. German. Diplomarbeit, Technische Universität Berlin (cited on page 216)
6. Borndörfer R, Reuther M, Schlechte T (2014) A coarse-to-fine approach to the railway rolling stock rotation problem. In: Funke S, Mihalák M (eds) 14th workshop on algorithmic approaches for transportation modelling, optimization, and systems (ATMOS 2014). OpenAccess series in informatics (OASIcs), vol 42. Schloss Dagstuhl-Leibniz-Zentrum fúr Informatik, pp 79–91. ISBN: 978-3-939897-75-0. https://doi.org/10.4230/OASIcs.ATMOS.2014.79 (cited on page 226)

7. Cacchiani V, Caprara A, Toth P (2012) A fast heuristic algorithm for the train unit assignment problem. In: Delling D, Liberti L (eds) 12th workshop on algorithmic approaches for transportation modelling, optimization, and systems, ATMOS 2012. OpenAccess series in informatics (OASIcs), vol 25. Schloss Dagstuhl, Wadern, pp 1–9. ISBN: 978-3-939897-45-3. https://doi.org/10.4230/OASIcs.ATMOS.2012.1 (cited on page 217)
8. Cordeau J-F, Soumis F, Desrosiers J (2001) Simultaneous assignment of locomotives and cars to passenger trains. Oper Res 49:531–548. ISSN: 0030-364X. https://doi.org/10.1287/opre.49.4.531.11226 (cited on page 216)
9. DB Fernverkehr AG (2015) German intercity-express (ICE) lines. http:bahn.de (visited on 09/01/2015) (Not cited.)
10. Euler R, Gamrath G (2017) Website of hydraw – hy(pergraph)draw(ing). http:hydraw.zib.de (cited on page 215)
11. Fioole P-J, Kroon L, Maróti G, Schrijver A (2006) A rolling stock circulation model for combining and splitting of passenger trains. Eur J Oper Res 174(2):1281–1297. ISSN: 0377-2217. https://doi.org/10.1016/j.ejor.2005.03.032 (cited on page 216)
12. Giacco GL (2014) Rolling stock rostering and maintenance scheduling optimization. PhD thesis, Roma Tre University, HDL: 2307/4429 (cited on page 217)
13. Giacco GL, Carillo D, D'Ariano A, Pacciarelli D, Marín A (2014) Short-term rail rolling stock rostering and maintenance scheduling. In: Transportation research procedia. 17th EURO working group on transportation, EWGT2014, Sevilla, vol 3, pp 651–659. ISSN: 2352–1465. https://doi.org/10.1016/j.trpro.2014.10.044 (cited on page 217)
14. Giacco GL, D'Ariano A, Pacciarelli D (2014) Rolling stock rostering optimization under maintenance constraints. J Intell Transp Syst 18(1):95–105. https://doi.org/10.1080/15472450.2013.801712 (cited on page 217)
15. Haahr J (2015) Reactive robustness and integrated approaches for railway optimization problems. PhD thesis, DTU Management Engineering (cited on page 217)
16. Haahr JT, Lusby RM, Larsen J, Pisinger D (2014) A branch-and-price framework for railway rolling stock rescheduling during disruptions. Technical report, DTU Management Engineering (cited on page 217)
17. Haahr, J, Wagenaar J, Veelenturf L, Kroon L (2015) A comparison of two exact methods for passenger railway rolling stock (re)scheduling. Technical report ERS-2015-007-LIS, ERIM report series research in management, HDL: 1765/78317 (cited on page 217)
18. Kuhn HW (1955) The Hungarian method for the assignment problem. Nav Res Logist Q 2(1–2):83–97. ISSN: 1931–9193. https://doi.org/10.1002/nav.3800020109 (cited on page 234)
19. Löbel A (1997) Optimal vehicle scheduling in public transit. PhD thesis, TU Berlin. URN: urn:nbn:de:0297 - zib - 10169 (cited on pages 216, 228)

20. Lübbecke ME, Desrosiers J (2005) Selected topics in column generation. Oper Res 53(6):1007–1023. https://doi.org/10.1287/opre.1050.0234 (cited on page 224)
21. Maráti G, Kroon L (2005) Maintenance routing for train units: the transition model. Transp Sci 39:518–525. ISSN: 1526–5447. https://doi.org/10.1287/trsc.1050.0116 (cited on page 216)
22. Maróti G, Kroon LG (2007) Maintenance routing for train units: the interchange model. Comput Oper Res 34(4):1121–1140. https://doi.org/10.1016/j.cor.2005.05.026 (cited on page 216)
23. Mellouli T, Suhl L (2007) Rotation planning of locomotive and carriage groups with shared capacities. In: Geraets F, Kroon L, Schoebel A, Wagner D, Zaroliagis CD (eds) Algorithmic methods for railway optimization. Lecture notes in computer science, vol 4359. Springer, Berlin, Heidelberg, pp 276–294. ISBN: 978-3-540-74245-6. https://doi.org/10.1007/978-3-540-74247-0_15 (cited on page 216)
24. Nielsen LK (2011) Rolling stock rescheduling in passenger railways: applications in short-term planning and in disruption management. PhD thesis, Erasmus University Rotterdam, HDL: 1765/22444 (cited on page 217)
25. Reuther M (2017) Mathematical optimization of rolling stock rotations. PhD thesis, TU Berlin. https://doi.org/10.14279/depositonce-5865 (cited on pages 216, 218, 219, 220, 221, 222, 223, 224, 226, 228, 230, 232, 233, 234, 237, 239)
26. Schlechte T (2012) Railway track allocation - models and algorithms. PhD thesis, TU Berlin. https://doi.org/10.14279/depositonce-3124 (cited on page 214)
27. Thorlacius P, Larsen J, Laumanns M (2015) An integrated rolling stock planning model for the Copenhagen suburban passenger railway. J Rail Transp Plann Manag 5(4):240–262. ISSN: 2210-9706. https://doi.org/10.1016/j.jrtpm.2015.11.001 (cited on page 217)
28. Toth P, Vigo D (eds) (2014) Vehicle routing. Society for Industrial and Applied Mathematics, Philadelphia, PA. https://doi.org/10.1137/1.9781611973594 (cited on page 217)
29. Wagenaar J, Kroon L, Schmidt M (2016) Maintenance appointments in railway rolling stock rescheduling. Technical report ERS-2016-001-LIS, ERIM report series research in management Erasmus Research Institute of Management, HDL: 1765/79441 (cited on page 217)
30. Ziarati K, Soumis F, Desrosiers J, Gélinas S, Saintonge A (1997) Locomotive assignment with heterogeneous consists at CN North America. English. Eur J Oper Res 97(2):281–292. https://doi.org/10.1016/S0377-2217(96)00198-1 (cited on page 216)

Chapter 11
Railway Crew Management

Erwin Abbink, Dennis Huisman, and Leo Kroon

Abstract This chapter deals with railway crew management, thereby focusing on the situation at Netherlands Railways (NS). NS is the main operator of passenger trains in the Netherlands. In this context, railway crew management is the process of guaranteeing in the most efficient way that the timetabled trains are supplied with a

E. Abbink (✉)
Netherlands Railways, Process quality and Innovation, P.O. Box 2025, NL-3500 HA Utrecht, The Netherlands
e-mail: erwin.abbink@ns.nl

D. Huisman
Econometric Institute and ECOPT, Erasmus University Rotterdam, P.O. Box 1738, NL-3000 DR Rotterdam, The Netherlands

Netherlands Railways, Process quality and Innovation, P.O. Box 2025, NL-3500 HA Utrecht, The Netherlands
e-mail: huisman@ese.eur.nl

L. Kroon
Rotterdam School of Management, Erasmus University, P.O. Box 1738, NL-3000 DR Rotterdam, The Netherlands

Netherlands Railways, Process quality and Innovation, P.O. Box 2025, NL-3500 HA Utrecht, The Netherlands

R. Borndörfer et al. (eds.), *Handbook of Optimization in the Railway Industry*, International Series in Operations Research & Management Science 268, https://doi.org/10.1007/978-3-319-72153-8_11

train driver and a sufficient number of conductors, thereby satisfying all the relevant practical constraints. Crew management has long term capacity planning aspects as well as short term scheduling, rostering, and dispatching aspects. In this chapter we describe these planning and dispatching processes, thereby focusing on practical issues. We also describe the planning support tools that are used by NS to support these processes.

11.1 Introduction

Netherlands Railways (NS) currently (2015) operates about 4700 commercial passenger trains on each working day. Here a train is not a physical object, but a timetabled train service with a unique train number. This terminology is used in the remainder of this chapter. Each train needs at least a train driver and a number of conductors. The latter number depends on the length and the type of the corresponding rolling stock composition.

To that end, NS employs about 3300 drivers and 3500 conductors, some of which are employed part-time. In order to be able to carry out their work, these crew members possess certain qualifications. For drivers these include route knowledge and rolling stock knowledge. Having route knowledge means that a driver is familiar with the routes on which he[1] has to operate trains. He knows where to expect signals and where to adjust the speed of the trains to specific circumstances in the railway infrastructure. Rolling stock knowledge relates to the specifics of the different types of rolling stock. The conductors are also supposed to have a certain rolling stock knowledge, but this is less detailed than the drivers'.

Carrying out the work on the trains in the most efficient way with the available crew members is the result of an extensive planning and rescheduling process. This chapter discusses several aspects of this process, its organization, and the tools that are used within it. We focus on practical aspects that have not been described so much in literature yet.

The remainder of this chapter is structured as follows. In Sect. 11.2 we describe the main concepts in the crew management process in more detail. In Sects. 11.3 and 11.4 strategic and operational planning processes are presented. Section 11.5 describes the real-time rescheduling process. In Sect. 11.6 we describe two extended versions of the well-known set covering model that are used as the kernel of the planning support systems for the crew (re)scheduling processes at NS. Section 11.7 presents an overview of the developments that lead to the currently available planning support tools within NS. This chapter is finished with Sect. 11.8 where we sketch some further developments that are relevant for crew management.

[1] In order to simplify the formulations in this chapter, we assume that the crew members are male, despite the fact that the number of women among the crew members is increasing. However, they are still a minority.

11.2 Main Concepts in Crew Management

In this section, we describe the main concepts in crew management in more detail. First, we describe the demand side of crew management: the tasks that need to be scheduled to operate the trains. Then we describe the concept of a duty, a sequence of tasks to be carried out by a single crew member. Finally, we describe the depots with their crew members (the supply side of crew management) and rosters, which allocate the duties to the crew members.

11.2.1 Tasks

Each trip of a train from one station to the next defines a number of tasks: one task for a driver and one or more tasks for conductors. A task means that a crew member of the right type and with the right qualifications must be present at the corresponding train, and either drive the train or provide service to the passengers. Each task is defined by a start station and start time, and an end station and end time, as specified by the timetable. Another characteristic of a task is the type of the involved rolling stock.

Note that stations where the crew of a train cannot be changed are not really relevant in the context of crew management. Therefore one can aggregate consecutive tasks of the same type (driver or conductor) on the same train that start or end in a station where the crew of a train cannot be changed. As a consequence, one may assume that each aggregated task starts and ends in a station where a train starts or ends, or where the crew of a train can be changed. In the following, we will use the term "task" for "aggregated task".

Apart from the trains in the commercial operation, also a number of trains are operated for repositioning rolling stock units. These trains are often also timetabled, and must also be supplied with a driver. Usually these trains do not need a conductor, although sometimes such trains are also used for repositioning crew members.

The distribution of the total workload over the weekdays is about 15% per workday, and 13% and 12% on Saturdays and Sundays, respectively. On a normal workday, the total numbers of tasks for drivers and conductors are approximately equal to 10,000 and 12,000, respectively. The workload of the conductors shows two peaks on workdays due to the higher passenger demand during peak hours, resulting in more trains and in trains with more capacity (longer, double-deck).

11.2.2 Duties

In the planning process, the tasks to be carried out are organized into a number of duties. Here each duty is a sequence of consecutive tasks to be carried out by a single crew member and in most cases on a single day. Within each duty, two consecutive tasks end and start at the same station, and the second task starts later than the first task has ended. The fact that duties are supposed to be carried out on a single day is the result of the fact that NS operates only a small number of night trains. Moreover, one of the rules used in the crew scheduling process of NS is that each duty ends at the same location as where it started. This is in contrast with the situation in larger countries, like, e.g., Germany, where an outbound duty may be followed by an inbound duty with an overnight outside rest in between.

On a normal workday, the number of duties for drivers and conductors are approximately equal to 1000 and 1100, respectively.

Example

Figure 11.1 shows nine duties for drivers. The horizontal direction represents time. The green horizontal lines represent the tasks to be carried out. The numbers above the tasks denote the corresponding train numbers. The brown horizontal lines below the tasks indicate that two consecutive tasks are carried out on the same rolling stock composition. If this is not the case, then there must be a transfer time of at least 20 min between two tasks.

For example, the first duty is a duty for depot Amsterdam (Asd), where the duty starts around 6:00. Then it proceeds along The Hague (Gvc), Utrecht (Ut), Eindhoven (Ehv), Sittard (Std), Maastricht (Mt), and Heerlen (Hrl), to return around

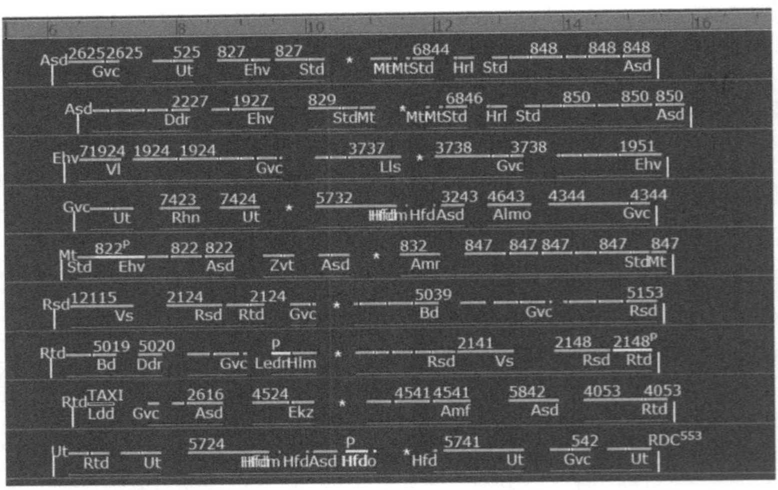

Fig. 11.1: Nine duties for train drivers as shown in the CREWS system

15:30 in Amsterdam. This duty transfers four times from one rolling stock composition to another, as is shown by the brown horizontal lines. The * (star) in the middle of the duty indicates the meal break.

The yellow horizontal lines in the fifth, seventh, and ninth duty represent crew repositioning tasks, also called passenger tasks, indicated by a "P". These tasks are not used for driving a train from one station to another but for repositioning a driver from one station to another. In the other duties (not shown in the figure) there are other tasks for drivers on the involved trains, which are used for actually driving these trains.

11.2.3 Depots

The drivers and conductors of NS currently operate from 28 depots: each crew member belongs to one crew depot, where each of his duties starts and ends. The depots are typically located at the main stations in the network (like Amsterdam, Rotterdam, Utrecht, and Eindhoven), but also at the stations closer to the borders of the country, either the natural borders of the country (like Den Helder and Vlissingen) or the borders with the neighbouring countries (like Groningen, Enschede, and Maastricht). The former depots are usually large, since they are visited by many trains, and, as a consequence, they can easily cover a large number of tasks. The latter depots are usually smaller. A main reason for the existence of these depots is that their crew members can cover the tasks in the early morning from these stations as well as the tasks in the late evening towards these stations.

11.2.4 Rosters

The rosters describe the allocation of the duties to the crew members. Within NS, the crew members of each crew depot have been allocated to a number of roster groups. The crew members within each roster group have more or less the same qualifications.

The duties of each depot are allocated to the different roster groups: each roster group has its own duties. In this allocation the contents of the duties and the qualifications of the members of the roster groups are taken into account. The allocation of the duties to the crew members of the roster groups is based on a *periodic* roster. As a consequence, if the roster is carried out completely as planned, then the crew members already know a long time in advance on which days they will be on or off duty. However, in practice there are usually deviations from the periodic roster, due to vacations or illness of crew members.

Currently in many companies the allocation of duties to crew members is handled by an *individual rostering* system. Here crew members can specify their preferences for days and times on which they want to be on or off duty. This provides much more flexibility than a periodic roster. Although some experiments with such an individual rostering system have been carried out, NS has not implemented such a system yet.

11.3 Strategic Planning

With respect to the capacity and the organization of the workforce, decisions have to be made to make sure that the capacity of the workforce matches as well as possible with the work to be carried out, both quantitatively and qualitatively. The latter refers to the required qualifications of the crew members.

Strategic decisions have to be taken with respect to the number, the locations and the capacities of the depots, both on the relatively short term and on the longer term. Changing the number or the locations of the depots is not usual. It is more common to change the capacities of the depots by hiring new crew members. Firing crew members is very uncommon, but of course in the long run crew members quit, since they retire or change jobs. Furthermore, transferring or delegating crew members from one depot to another is incidentally possible.

It takes one (for conductors) to three (for drivers) years to fully train a new crew member. This includes both theoretical training and practical on-the-job training. As a consequence, the long term capacity management process has to anticipate on long term changes in the workload. These are mainly determined by long term changes in the line system, the timetable, the rolling stock capacity, and the labor rules. Conversely, the decisions taken within the capacity management process, e.g. with respect to hiring new crew members, have long lasting effects on the capacity of the workforce.

A problem that is often faced in practice is that the capacity of the workforce as a whole is sufficient to cover the total workload, but that part of the capacity of the workforce is available at the wrong locations. This may have a negative influence on the efficiency of the crew scheduling process, since in such situations more repositioning tasks will have to be scheduled in order to move the crew members to the locations of the tasks to be carried out. According to the collective labour agreements, such repositioning tasks are also considered as part of the work to be carried out, and are paid as such.

Despite the strategic impact and relevance of the long term capacity management process, models and tools for supporting this process are hardly available. The studies usually do not go beyond "what-if" analyses based on a forecasted timetable and rolling stock circulation. As a consequence, in the remainder of this chapter we focus on the operational and real-time crew management processes, given the number, locations, and capacities of the crew depots.

11.4 Operational Planning

The operational crew planning process can be split in several steps, as depicted in Fig. 11.2. The major part of the operational planning process is carried out within

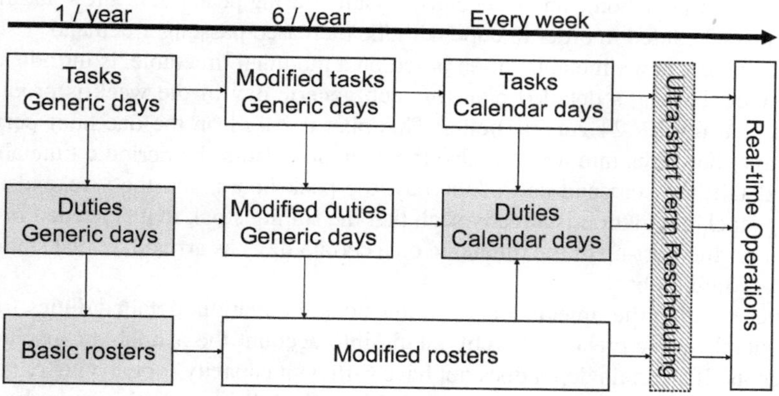

Fig. 11.2: Planning phases

the NS headquarters in Utrecht. However, the process of planning the rosters for the crew depots, based on the planned crew duties, is carried out in the depots themselves.

In each stage of the operational planning process as well as in the real-time rescheduling process, the crew (re)scheduling process follows the timetabling and rolling stock (re)scheduling process. This is caused by the fact that the timetable and the rolling stock circulation determine the tasks to be carried out by the crew, as explained before.

The tasks are input for scheduling the duties. During the duty scheduling process, the duties are *anonymous*. That means that the duties have not yet been allocated to individual crew members. Thus the qualifications of the individual crew members can not be taken into account in this stage of the planning process.

The finalized duties are used for creating the rosters. About six times a year, the generic tasks, duties and rosters are modified. Each week, the schedules are adapted to the specific exceptions on a calendar day.

When we expect severe weather conditions, we can adapt the schedules and reduce the number of trains to be operated in advance. We call this Ultra-short rescheduling, on the day before operation. We will discuss the most important steps, the planning of generic duties, basic rosters, calendar duties and the Ultra-short term rescheduling, in the following sections.

11.4.1 Planning for Generic Duties

In principle, the timetable that is operated by NS is a periodic timetable with a period of 1 h. Most of the lines in the system are even operated twice per hour with an interval of exactly 30 min. However, there are slight differences between peak hours and off-peak hours and weekends. Usually during peak hours a few additional trains are scheduled in order to cope with the increased passenger demand.

Each year a new timetable, often based on a modified timetable, is introduced in December. That is, a detailed plan for each generic day of the week is produced. It is a plan for 7×24 generic hours. This plan is based on the one hour periodic timetable, but takes into account also the deviations from the periodic timetable in more detail: the start and end of the day, the peak hours, and the weekends. The rolling stock circulation is already such that the rolling stock that is needed in each station for the start-up of the timetable on a certain day has arrived on that station in the preceding night.

After creating the timetable and rolling stock circulation, detailed duties for the crew members are created, thereby taking into account the available capacities of the depots. If a certain depot does not have sufficient capacity for carrying out all its tasks, then additional capacity is provided by scheduling repositioning tasks from other depots. In each depot also a number of stand-by duties is scheduled. Crew scheduling is the process of creating the total set of crew duties from scratch.

The generic set of duties for the complete week is used as input for the crew rosters in the depots. The generic plan is modified only a few times per year. When modifying the generic plan, the existing crew rosters in the depots are taken into account. The crew duties are modified such that they fit as well as possible within the existing crew rosters.

11.4.2 Crew Rostering

As it was mentioned in Sect. 11.2.4, NS currently uses a periodic crew rostering system for allocating the anonymous duties to the crew members. Within NS this process is carried out completely manually, although some experiments with automated rostering support have been carried out, see Hartog et al. [7] and Abbink [1].

As described in Sect. 11.2.4, each crew member belongs to exactly one roster group in a crew depot.

The rostering process is carried out once per year in the crew depots, usually in the autumn when the duties for the generic days in the next yearly timetable have been completed and communicated with the crew depots.

Usually, a first step in the creation of a *detailed roster* for a roster group is the creation of a *pattern roster*, describing the pattern of early, late, night and rest duties. Herein also some additional duties must be scheduled, such as stand-by duties and compensation duties. A compensation duty is an extra day off due to a certain amount of irregular working times. Given the many aspects to be taken into account, creating a new pattern roster by hand is extremely difficult. Therefore, the previous roster is often used as a basis for the new one, with only minor adjustments.

The second step in the rostering process is the allocation of the duties to the different roster groups. This is accomplished by a negotiation process between the representatives of the roster groups. In this process the characteristics of the duties and the qualifications of the crew members of the roster groups are taken into account. Also a fair allocation of the attractive and less attractive duties among the roster groups is aimed at. See also Sect. 11.6.2 for a further description of this aspect.

Finally, the duties that have been allocated to a roster group are attached to its pattern roster. If at some point in this rostering process some roster groups cannot find a feasible roster, then the allocation of the duties to the roster groups must be revised or the pattern roster must be adjusted.

Examples of constraints that have to be considered in the rostering process are the following: the rest time between two duties is at least 12 h, except when the first duty finished after 2:00 a.m., then it is at least 14 h. The minimum rest time between two duties is 14 h, if the first duty finishes after 2:00 a.m., the day following the day this first duty starts. The maximum working time per week is 45 h, and over a period of 13 consecutive weeks the average working time per week is at most 40 h. At least once every 3 weeks there is a Red Weekend: a rest period of at least 60 h, starting not later than Saturday 0:00 a.m. and ending not earlier than Monday 4:00 a.m.

From the crew members perspective, the quality of a roster is determined by the sequencing of the roster days and the variety in the duties. Some examples: a series of duties with the same type directly after each other is preferred over sequences of different types of work. Here the type of a duty is either an early, late or night duty. Two or more adjacent days off are preferred. Similar duties should be spread over the roster. A lot of variety in the work is preferred (routes, rolling stock).

For more details and for optimization models that can be used in the rostering process, we refer to [7] and [1]. Here the constraints to be satisfied are modeled in an explicit way. Caprara et al. [6] describe models for crew rostering based on set covering models that are very similar to the standard models for solving crew scheduling problems, where the constraints to be satisfied are modeled more implicitly.

11.4.3 Planning for Calendar Duties

In the following stage, a detailed plan for the calendar days is generated. Usually, each calendar day requires a modified timetable. The latter may be due to the fact that certain trains have to be cancelled because of maintenance or extension of the railway infrastructure, or due to the fact that extra trains have been scheduled for a special event, such as an exposition or a sports event. Especially during the weekends often a lot of maintenance of the infrastructure is carried out.

The planning process for the calendar days is an ongoing process that is carried out on a week-by-week basis. Thus this planning process covers $52 \times 7 \times 24$ calendar hours. Again the crew duties are modified such that they fit as well as possible within the existing rosters in the depots. As far as possible, the same constraints are taken into account as in the duties for the generic days.

The planning process for the calendar days is a *rescheduling* process , in which one of the aims is to leave as many as possible of the duties the same as they are in the plan for the generic days. However, sometimes it is inevitable to modify a large number of duties.

11.4.4 Ultra-Short Term Rescheduling

In order to be able to better cope with snow during winters and other extreme weather conditions, NS decided to develop an ultra-short term rescheduling process. The aim of this ultra-short term rescheduling process is to reduce the number of trains in case of expected snow or other extreme weather conditions. The reduced number of trains is realized by decreasing the frequencies of the trains, in particular on the main corridors. Due to the reduced number of trains, the remaining trains can be controlled better by traffic control, and thus the probability of a snow-ball effect of delays and cancellations is reduced. Therefore, the predictability of the railway system remains higher than without these measures.

The decision to adapt the timetable, the rolling stock circulation and the crew duties is based on the weather forecast on the day before the adapted schedules are to be operated. The final go/no go decision for the adaptations is taken at 15:20 and communicated at 16:00 on that day. Thus the timetable, the rolling stock circulation and the crew duties must be adapted during the upcoming evening and night, although some pre-processing can be done already in advance. In this ultra-short term crew rescheduling process, the availability of fast and automated rescheduling tools has proven to be indispensable.

A consequence of the application of the ultra-short term rescheduling process was that the railway system remained better in control during bad weather conditions. On the other hand, the reduced frequencies of the trains obviously also lead to a reduced transport capacity. Therefore, after the first applications of this process, this reduction in transport capacity was compensated by increasing the capacities of the remaining trains where possible.

11.5 Real-Time Operations

The proof of the pudding is in the eating: the duties have to be carried out during the real-time operations. On the day-of-operations, the allocation of duties to crew members is determined by the crew rosters. However, it may happen that this is not possible due to illness or vacation of crew members. In that case, first the crew members that have been allocated to a stand-by duty may be used to replace the absent crew members. If these crew members are active already, then an ad hoc solution must be found. The crew dispatchers in the dispatching centres are responsible for covering as many tasks and duties by a crew member as possible.

Given the allocation of duties to crew members, rescheduling the duties is not necessary as long as the timetable and the rolling stock circulation are carried out as planned. However, if there are delays or disruptions, the duties may have to be rescheduled. If a train has a serious delay and the driver has to transfer to another train at a next station, then the other train may get a delay as well if the duties are not rescheduled. In that case, the crew dispatchers will try to find an alternative crew member who can carry out the tasks that are delayed otherwise. Also the delayed driver will have to get an alternative continuation of his duty.

In case of a serious disruption, for example due to malfunctioning infrastructure or rolling stock or due to an accident, usually a number of trains (or parts thereof) are cancelled. In such a situation, the trains that are cancelled are selected based on a certain *incident scenario* for the actual disruption. As a consequence, at least the duties covering the tasks corresponding to the cancelled trains have to be rescheduled, since the involved crew members may not be able to reach the next tasks in their duties in time. However, in order to get some more flexibility in the rescheduling process, also some additional duties may be rescheduled. The crew dispatchers usually aim at minimizing the number of rescheduled duties.

Examples of aspects that have to be taken into account when rescheduling the crew duties are the following: The rescheduled duties should still as much as possible satisfy the general constraints for the crew duties. Thus there is preferably a meal break at an appropriate time and place if the duty length exceeds a certain minimum length. Furthermore, the length of the duty may not be extended too much, and the duty should end at its own depot. Due to the many constraints to be taken into account, and due to the fact that the crew duties do not follow a periodic pattern (as is more or less the case for the rolling stock), crew rescheduling is usually considered as the most difficult part of the disruption management process.

The real-time rescheduling process is carried out in five regional dispatching centers, directed by the central Operational Control Center Rail located in Utrecht. The fact that this process is carried out from *five* dispatching centers requires a lot of coordination between these centers, which significantly complicates the real-time crew rescheduling process.

11.6 Optimization Models

In this section, models and solution techniques for the crew scheduling (see Sect. 11.4.1) and crew rescheduling (see Sect. 11.5) are described. We start with a description of a number of relevant objectives and constraints.

11.6.1 Objectives

NS considers three important objectives in crew planning: (1) efficiency, (2) robustness, and (3) quality of work. Efficiency means that the total crew costs are as small as possible, both the planned costs and the realized costs. In principle, this means that the workload should be carried out by a minimum number of crew members, since these are the main cost drivers in this planning process. In the real-time rescheduling process additional costs should be avoided by minimally extending the original duty lengths (additional time must be paid), and by minimally using taxis for repositioning crew members, additional to the taxis already included in the planning phases.

The robustness of the crew duties, i.e. preventing propagation of delays via the crew duties, depends on several elements, including the transfer times of the crews when transferring from one rolling stock composition to another. The recoverability of the duties in case of a disruption may depend on the structure of the duties. If the rescheduling process is carried out manually, then duties with a relatively simple structure may be easier to recover than more complex duties. However, it has been demonstrated also that the availability of effective crew rescheduling tools eliminates this advantage of simply structured duties, see Vlugt [14].

The quality of work is the quality of the duties as it is perceived by the crew members. This quality of work is addressed via labor rules and company agreements, for example, on the amount of variation in the duties as specified by the so-called "Sharing-Sweet-and-Sour" rules. These rules are described in the next section.

Since several objectives have to be taken into account, a trade-off between these objectives must be made until a solution has been found that is acceptable with respect to all objectives. The availability of effective planning support tools is indispensable here.

11.6.2 Constraints

The constraints that have to be respected in the crew scheduling process are rules at the duty level and constraints at the depot level. Examples of constraints at the duty level are the following: each duty starts and ends at the home depot, the length of a duty should not exceed a certain maximum duty length, each duty longer than 5:30 h should have a meal break of sufficient length, the transfer time when transferring from one train to another should be at least 20 min, and each duty should have a certain variation. The latter is expressed by an upper bound on the "repetition in duty", which means that a duty cannot cover any part of the railway network more than a certain number of times.

Examples of constraints at the depot level are the following: per depot the number of weekly duties should not exceed a certain number, per depot the weekly average duty length should not exceed 8 h, per depot the weekly percentage of duties longer than 9 h should not exceed a certain upper bound, and per depot the weekly percentage of night duties should not exceed another upper bound.

Also the so-called "Sharing-Sweet-and-Sour" rules are to a large extent constraints at the depot level. These rules aim to *quantify* the fair allocation of the "sweet" and "sour" workloads among the 28 crew depots. "Sweet" mainly represents the variety in routes and lines as well as the work on intercity trains. "Sour" mainly represents the work on lines with a lot of anticipated passenger aggression and the work on relatively old rolling stock. For example, for each depot at least 35% of the work should be on intercity trains. Furthermore, there is also an upper bound on the *standard deviation* of these percentages over the 28 depots in order to guarantee a fair allocation of the different types of work, see Abbink et al. [2].

11.6.3 Crew Scheduling: Model

The crew scheduling problem for generic days, as described in Sect. 11.4.1, is the problem of most efficiently covering a number of tasks by a number of feasible duties from scratch. As is well-known, the crew scheduling problem can be described by an extended set covering model, see Caprara et al. [5]. Modelling the complex constraints directly usually leads to models that are hard to solve.

To that end, let T be the set of tasks to be covered and let D be the set of potential duties. The subset $D_t \subset D$ consists of the set of potential duties covering task $t \in T$. Every duty d has costs c_d. Furthermore, let S be the set of additional constraints at the depot level, and let l_s and u_s be the corresponding lower and upper bound for constraint $s \in S$. Finally, let $w_{s,d}$ be the coefficient of duty $d \in D$ for constraint $s \in S$.

Next, the binary decision variable X_d indicates whether potential duty $d \in D$ is included in the solution or not. Then the extended set-covering model, with additional constraints at the depot level, can be formulated as follows:

$$\min \sum_{d \in D} c_d X_d \tag{11.1}$$

subject to

$$\sum_{d \in D_t} X_d \geq 1 \qquad \text{for all } t \in T \tag{11.2}$$

$$l_s \leq \sum_{d \in D} w_{s,d} X_d \leq u_s \qquad \text{for all } s \in S \tag{11.3}$$

$$X_d \in \{0,1\} \qquad \text{for all } d \in D \tag{11.4}$$

Here (11.1) is the objective function, which states that the sum of the duty costs is to be minimized. Constraints (11.2) guarantee that for each task $t \in T$ at least one duty covering this task is selected. It may sometimes be better to cover a task more than once. If, for example, on a certain day the number of tasks going out of a crew depot differs from the number of tasks going into that crew depot, then over-covering is necessary. Moreover, even if over-covering is not necessary, it may sometimes be more efficient to allow over-covering. By allowing over-covering, other tasks may be covered more easily, resulting in a larger decrease in costs than the additional costs for the over-covered tasks.

Besides the regular tasks that *must* be covered, the formulation also allows to include a number of additional tasks in the model. These are tasks outside the set T, so that they need not be covered according to (11.2). For example, one can add possible taxi trips to reposition crew members from one location to another. These taxi trips have certain costs, but may make the duties feasible and/or more efficient. Tasks related to shunting activities at stations are examples of other additional tasks. They can be included in the duties, but they can also be performed by dedicated shunting crew members locally at the stations, when the crew duties become inefficient by including these additional shunting tasks.

The additional constraints (11.3) are related to the constraints at the depot level, such as the number of duties per crew depot, or the average length of the duties per crew depot. In these constraints, u_s can be considered as the availability of a certain resource, and $w_{s,d}$ as a parameter describing the amount of this resource used by duty d. For example, let K be the set of depots, and let k_d and l_d denote the crew depot and the length of duty d, respectively. Furthermore, let C_k denote the maximum number of duties for depot k (given the capacity of depot k), and let L denote the maximum average length of the duties of each crew depot. Then the following constraints (11.5) and (11.6) guarantee that the maximum number of duties and the average duty length for each crew depot $k \in K$ are not exceeded, respectively. These constraints have the general structure described by constraints (11.3).

$$\sum_{d \in D:\ k_d = k} X_d \leq C_k \qquad \text{for all } k \in K \tag{11.5}$$

$$\sum_{d \in D:\ k_d = k} (l_d - L) X_d \leq 0 \qquad \text{for all } k \in K \tag{11.6}$$

11.6.4 Crew Scheduling: Solution Technique

The solution process for solving the extended set covering model usually consists of a duty-generation module and a duty-selection module. The algorithm first generates a large set of potential feasible duties. A duty is feasible if it satisfies all constraints at the duty level. That is, it takes into account, for example, the maximum duty length, and the location and duration of the meal break. Then the above described extended set covering model is used to select the subset of feasible duties that covers all tasks in the most efficient ways.

However, since the set of feasible duties can be extremely large, enumerating all possible feasible duties a priori is usually not a practical approach. Therefore a standard approach is to use dynamic column generation . This technique does not generate all duties a priori, but it generates them *on-the-fly* during the solution process. Within the solution process, dual information from the current solution is used to determine whether it is useful to extend the number of duties. An appropriate way to determine this dual information is based on Lagrangian Relaxation instead of Linear Programming. This dual information is used in the duty-generation module, which checks for the existence of additional duties that may be useful to improve the solution. The duty-generation problem can be solved as a resource constrained shortest path problem in a network representing the crew tasks, see Caprara et al. [5].

If dynamic column generation is applied, then the solution process iterates between the duty-generation module (also called pricing problem) and the duty-selection module (also called the master problem). The algorithm may delete feasible duties that were generated in earlier stages and whose effectiveness turns out to be low during later stages of the process. The latter is done to keep the number of active duties manageable. Usually the duty-selection module heuristically looks for a solution for the overall model, based on the currently available set of feasible duties. After a number of iterations, the procedure may activate a fixing procedure to select some duties that appear to be particularly efficient, and to fix them as belonging to the final solution. Then the algorithm repeats this process on the tasks that have not yet been covered by the fixed duties.

Solving an instance of NS for a single day may take at least several hours to compute a feasible solution with sufficient quality. Moreover, currently also instances of NS for a complete week can be solved in up to a week of computation time, see Abbink et al. [4]. The advantage of solving instances for a complete week is that several of the constraints are expressed in weekly averages per depot, such as the constraints on the maximum weekly average duty length per depot. Therefore the resulting solution is usually more efficient than a solution that is obtained by requiring such constraints on a day-by-day basis.

11.6.5 Real-Time Crew Rescheduling: Model

The model and solution process as described in Sects. 11.6.3 and 11.6.4 focus on the planning process from scratch. As explained, it may take at least several hours to compute a feasible solution with sufficient quality for an instance involving a single day. Although this is an impressive improvement in comparison with a manual planning process, it is far too long for the real-time disruption management process, as described in Sect. 11.5. There a solution is needed within minutes, and hence a suboptimal solution is better than no solution at all.

Moreover, the real-time crew rescheduling problem is different from the crew scheduling problem (11.1), (11.2), (11.3), and (11.4) in the planning stage. First of all, the real-time crew rescheduling process has to take into account the fact that a number of duties have started already. The parts of these duties that have been carried out already cannot be changed anymore. Thus for each active duty a feasible *completion* must be found. Furthermore, whereas the crew scheduling problem (11.1), (11.2), (11.3), and (11.4) aims at the generation of anonymous duties, one has to deal now with duties that have been allocated to individual crew members, each with their own qualifications. When rescheduling the duties, these have to be respected.

In addition, in the crew scheduling problem (11.1), (11.2), (11.3), and (11.4) all tasks *must* be covered, since the timetable and rolling stock circulation are fixed. In the real-time crew rescheduling process this may not be true anymore. If it turns out that certain tasks cannot be covered by a completion of a duty, these tasks will have to be cancelled. If this is a task for a driver, then this means that the corresponding train will not be operated. This is highly undesirable in itself, but in particular since this also has consequences for the rolling stock circulation. If it seriously influences the rolling stock circulation, then another iteration of timetable rescheduling, rolling stock rescheduling and crew rescheduling will be required. Note that, if the driver agrees, then it is allowed that a train is operated without a conductor, but this is undesirable as well.

An issue that also plays a role in the real-time crew rescheduling process is the fact that the duration of the disruption is usually not known at the start of the disruption, when the initial rescheduling of the duties has to be carried out. A common approach here is to make an educated guess of the duration of the disruption, e.g. based on experiences with similar disruptions in the past. If later on it turns out that the duration of the disruption is different from the initially assumed duration, then this is considered as a second disruption. At that point in time, another iteration of timetable, rolling stock and crew rescheduling will have to be carried out.

Altogether, assuming a certain duration of the disruption, the timetable that will be operated can be determined based on the involved incident scenario. From this timetable and the rescheduled rolling stock circulation, the crew tasks to be covered can be determined.

Now the real-time crew rescheduling problem can be described as follows. As in (11.1), (11.2), (11.3), and (11.4), let T be the set of crew tasks to be covered. Furthermore, let D be the set of original duties at the moment of rescheduling. This set may be equal to the set of originally planned duties, but it may also be different from that set due to earlier disruptions on the same day.

For each original duty $d \in D$, let D_d be the set of feasible completions of duty d. Then, for each original duty d and for each feasible completion $d' \in D_d$, let $X^d_{d'}$ be a binary decision variable, indicating whether or not feasible completion $d' \in D_d$ is used to complete original duty $d \in D$. Furthermore, for each task $t \in T$, let Y_t be a binary decision variable indicating whether or not task t is cancelled. Then the real-time crew rescheduling problem can be described as follows:

$$\min \sum_{d \in D} \sum_{d' \in D_d} c^d_{d'} X^d_{d'} + \sum_{t \in T} \kappa_t Y_t \tag{11.7}$$

subject to

$$\sum_{d \in D} \sum_{d' \in D_d:\, t \in d'} X^d_{d'} + Y_t \geq 1 \qquad \text{for all } t \in T \tag{11.8}$$

$$\sum_{d' \in D_d} X^d_{d'} = 1 \qquad \text{for all } d \in D \tag{11.9}$$

$$X_{d'}^{d} \in \{0,1\} \qquad \text{for all } d \in D;\ d' \in D_d \qquad (11.10)$$

$$Y_t \in \{0,1\} \qquad \text{for all } t \in T$$

The objective function (11.7) does not only focus on minimizing the total costs of the completions of the duties, but also the costs of canceling tasks. Here κ_t represents the costs of canceling task $t \in T$. These canceling costs are very high and dominate the other cost components, since canceled tasks must be avoided as much as possible. Constraints (11.8) specify that each task $t \in T$ is either covered by at least one completion of a duty, or is canceled. Constraints (11.9) require that for each original duty $d \in D$ exactly one completion is selected.

Note that each original duty has the trivial completion consisting of taking a taxi back to the home depot. However, in that case no additional tasks are covered by the completion. If such trivial completions are selected for many original duties, then many trains will be cancelled due to absence of the required crew. Note that in the real-time crew rescheduling process it is hard to deal with constraints at the depot level, such as constraints (11.3).

11.6.6 Real-Time Crew Rescheduling: Solution Technique

In the real-time crew rescheduling process, short computation times are essential. If one needs a solution within minutes, but has to wait for it for hours, then this solution is useless.

In order to deal with this time constraint and with the other details of the disruption management process [12], developed a solution method to be used for real-time disruption management. Their algorithm is based on column generation techniques combined with Lagrangian heuristics. In order to handle the very large number of duties in practical instances, they first define a core problem with a limited number of duties to be rescheduled. The core problem may exist of only the duties that *must* be rescheduled.

If some tasks remain uncovered in the solution of the core problem, they perform a neighborhood search to improve the solution by defining a new core problem for each uncovered task. This new core problem is typically small such that a large part of the solution of the previous step is fixed. Computational experiments with real-life instances show that this method is capable of producing good solutions within a couple of minutes of computation time, in particular in case of a relatively large disruption.

As mentioned before, a complicating issue in the real-time crew rescheduling process is the fact that the duration of a disruption is usually not known at the start of the disruption. As a consequence, at the start of the disruption it is uncertain for a number of tasks whether they have to be carried out or not. In order to deal with this uncertainty [13], developed a method that is to some extent robust against the duration of the disruption. They consider a minimum and a maximum duration of the disruption. Furthermore, in their rescheduling model they focus on rescheduling the duties for the minimum duration of the disruption, but in such a way that, also in the case that the disruption lasts longer, the duties can be modified easily to cope with this. They call this approach *quasi-robust*, since the rescheduled duties are not feasible independently of the duration of the disruption, but the duties can be made feasible relatively easily under all possible durations of the disruption.

Abbink et al. [3] also developed a solver for supporting the real-time crew rescheduling process based on multi-agent techniques. Here crew members are represented by virtual agents. If, due to a disruption, some agents cannot carry out certain tasks within their duties, then these agents may ask their virtual colleagues to support them by taking over some of these tasks. A virtual colleague may be able to take over these tasks *directly* or *indirectly*. The latter means that he may be able to take over these tasks, but only if someone else takes over some of his own tasks. In this way a recursive negotiation process between the agents is carried out, which usually leads to a solution in the end.

Although this approach performed especially well in disturbed situations involving not too many duties and relatively small delays, it performed on average worse than the set covering based approach developed by Potthoff et al. [12]. However, recently this agent-based approach method was enhanced into a local search heuristic which can be used either as a mini-solver in itself or as a supporting tool in the neighborhood search that is part of the set covering based approach of [12].

11.7 Planning Support System CREWS

The crew scheduling process within NS was carried out traditionally by many planners in a manual way, to some extent supported by relatively simple information systems that mainly provided administrative functions. However, within NS it was recognized already in the early 1990s that crew scheduling is one of the most complex planning processes of a railway operator. Therefore, it was decided that automated planning support, also providing duty *generating* functions, was needed. This was the start of a long research and development process.

11.7.1 CREWS: The First Phase

At that time NS chose for the CREWS system developed by Siscog in Portugal. This system provided an extensive and user friendly system interface that was appreciated by the planners. The algorithmic support was provided by an algorithm based on the A^*-algorithm, see Morgado and Martins [11]. Unfortunately, it turned out that this algorithm was not sufficiently powerful for finding solutions that were satisfactory for the planners. In particular, due to the structure of the A^*-algorithm, it could take into account the constraints at the duty level, but it could hardly deal with the constraints at the depot level.

Therefore the algorithmic part of the CREWS system was enhanced by combining it with the system TURNI developed by Double-Click in Italy, see Kroon and Fischetti [9]. TURNI was complementary to CREWS in the sense that it consisted of a powerful algorithm, based on dynamic column generation, and a very simple user-interface. This combination of CREWS and TURNI was used satisfactorily for a couple of years in the early 2000s.

Amongst others it was used to analyze a large number of different constraint sets to be used in the crew scheduling process, in the end leading to the "Sharing-Sweet-and-Sour" rules, see Abbink et al. [2]. Since this combination of CREWS and TURNI was able to deal with these complex rules, which are really hard to deal with manually, it was also appreciated by the planners. The combination of CREWS and TURNI was used for generating the generic crew duties, as described in Sect. 11.4.1.

11.7.2 CREWS: Stand-Alone

A next step in the development was the implementation into CREWS of a dedicated model and solution technique for supporting the crew rescheduling process for calendar days as described in Sect. 11.4.3. This model is very similar to the model (11.7), (11.8), (11.9), and (11.10), since this process aims at *rescheduling* a set of existing duties, instead of on scheduling the duties from scratch. However, also constraints similar to (11.3) must be taken into account to deal with constraints at the depot level. Moreover, this process is still a *planning* process. Therefore solution quality is more important than computation time. The applied solution technique is very similar to the one described in Sect. 11.6.4, see Huisman [8].

In a next step, the solution technique provided by TURNI was replaced by an algorithm similar to TURNI's, see Abbink et al. [4]. As a consequence, CREWS became powerful enough to operate on its own, and thus TURNI became redundant.

One of the main advantages of the application of CREWS in the planning process was that it allowed to carry out "what-if" analyses, by changing the parameters of the system. As a consequence, it became possible to study the trade-off between different objectives, such as efficiency and robustness. The latter is usually impossible in a manual planning process, where one is usually satisfied already after having found one feasible solution. Additional advantages of the use of the algorithmic support provided by CREWS were a more efficient planning process as well as a reduction of the throughput time of the planning process. For more details on these advantages, see Kroon et al. [10] and Abbink [1].

11.7.3 CREWS: Real-Time Dispatcher

The next step in the extension of the CREWS system aimed at providing support for the real-time crew rescheduling process. To that end [12], developed the rescheduling algorithm described in Sect. 11.6.6. This algorithm was implemented in the real-time dispatching module of CREWS, also called CREWS-RTD.

This system is based on the assumption that the real-time crew rescheduling process is carried out in a centralized way by a single solver. At the moment that the solver is started, the status quo of the duties is *frozen*: the duties cannot be modified manually anymore, but only by the solver. Once a feasible and satisfying solution has been obtained, it is checked against the duties in the real-world, and communicated to the relevant stake-holders.

CREWS-RTD was first used in a rescheduling process that was set up in parallel to the manual real-time crew rescheduling process. Currently this system is used regularly in disrupted situations due to delayed infrastructure maintenance or incidents lasting several days. However, implementing and applying algorithmic support in a real-time rescheduling process turns out to be even more complex than applying such support in a planning process. Therefore, some further development and implementation steps will be needed to make the use of CREWS-RTD in the real-time rescheduling process common practice. These will be described in the next section.

11.8 Further Developments

In this chapter we described the status quo in the area of railway crew management in the Netherlands. The operational crew planning process is currently to a large extent and satisfactorily supported by the crew scheduling tool CREWS. Nevertheless, there are still several developments in the crew management area that may lead to further improvements.

In particular, the implementation of the real-time crew dispatching system CREWS-RTD is not yet complete. The system has been used many times in disrupted situations, but its application in real-time rescheduling situations is certainly not yet common practice. To improve this, it may first be useful to extend the system with a tool that validates the duties in the system against what happens in reality, so that, once the solver is used, it starts with a correct initial situation.

Furthermore, the disruption management process is still to be organized in a more effective way. For example, a more centralized approach of the disruption management process would be more effective than the current decentralized approach. Also the application of objective rules describing which rescheduled duties are feasible or not will be helpful in speeding up the disruption management processes. This will enable *silent* digital communication between dispatchers and crew members, and eliminate time-consuming negotiations.

Another improvement would be to change the crew rescheduling rules in the disruption management process so that the crew rescheduling process becomes less dependent on the rolling stock rescheduling process, and can hence be carried out in parallel. This may save a lot of time in this time-critical process. A lower dependence can be achieved, e.g. by stating that in case of disruptions one conductor per train is sufficient.

Furthermore, there used to be not much flexibility in the crew scheduling and rostering process. In principle all duties per crew type are scheduled based on the same rules: *one size fits all*. For crew members it is possible to have a *part-time* job, but then part-time refers to a *limited* weekly number of *full size* duties. However, times are changing, and in order to remain attractive as an employer, NS may have to introduce more flexibility in the duties. For example, a category of *shorter* duties may be especially attractive to certain categories of employees. Note that shorter duties may also be attractive for NS. Indeed, shorter duties enable a more efficient covering of the workload during peak hours. Introducing such part-time duties will have a significant impact on the scheduling and rostering process. Recently some first steps into the direction of more flexibility in the rosters have been made.

Other developments involving the crews, in particular the drivers, are the application of driver advisory systems and energy efficiency. A driver advisory system supports the driver in his activities. The current driver advisory system of NS (called RouteLint) just informs the driver about his environment: what is the position, speed, and delay of the own train, and which trains are present in the direct neighborhood of his own train. The next generation driver advisory system will support the driver actively with speed advices: that is, where to accelerate to which speed and with which acceleration, and where to reduce the speed again? Important criteria for driver advisory systems are punctuality and energy efficiency.

At this moment (2015), automatic train operation (ATO) does not exist yet for heavy rail systems. However, several metro and light rail systems are operated already by ATO. ATO is a challenging topic that receives a lot of research attention nowadays. Although there are still many practical issues to be solved, it may not be impossible that in the far future also trains in heavy rail systems will be operated by ATO. This will simplify the railway crew management process to a process focused on the conductors. Anyway, whether ATO is applied or not, railway crew management will be a challenging topic, both from a practical and from a scientific point of view.

References

1. Abbink EJW (2014) Crew management in passenger rail transport. PhD thesis, Erasmus University Rotterdam, Rotterdam. Erasmus Research Institute of Management. ISBN: 978-90-5892-374-5. HDL: 1765/76927 (cited on pages 250, 251, 261)

2. Abbink EJW, Fischetti M, Kroon LG, Timmer GT, Vromans MJCM (2005) Reinventing crew scheduling at Netherlands railways. Interfaces 35:393–401. https://doi.org/10.1287/inte.1050.0158 (cited on pages 255, 261)
3. Abbink EJW, Mobach DGA, Fioole PJ, Kroon LG, van der Heijden EHT, Wijngaards NJE (2009) Actor-agent application for train driver rescheduling. In: Proceedings of the 8th international conference on autonomous agents and multiagent systems, AAMAS '09, vol 1. International Foundation for Autonomous Agents and Multiagent Systems, pp 513–520. ISBN: 978-0-9817381-6-1 (cited on page 260)
4. Abbink EJW, Albino L, Dollevoet T, Huisman D, Roussado J, Saldanha RL (2011) Solving large scale crew scheduling problems in practice. Public Transp 3:149–164. https://doi.org/10.1007/s12469-011-0045-x (cited on pages 257, 261)
5. Caprara A, Fischetti M, Toth P, Vigo D, Guida PL (1997) Algorithms for railway crew management. Math Program 79:125–141. https://doi.org/10.1007/bf02614314 (cited on pages 255, 257)
6. Caprara A, Toth P, Vigo D, Fischetti M (1998) Modeling and solving the crew rostering problem. Oper Res 46:820–830. https://doi.org/10.1287/opre.46.6.820 (cited on page 251)
7. Hartog A, Huisman D, Abbink EJW, Kroon LG (2009) Decision support for crew rostering at NS. Public Transp 1:121–133. https://doi.org/10.1007/s12469-009-0009-6 (cited on pages 250, 251)
8. Huisman D (2007) A column generation approach for the railway crew rescheduling problem. Eur J Oper Res 53:1007–1023. https://doi.org/10.1016/j.ejor.2006.04.026 (cited on page 261)
9. Kroon L, Fischetti M (2001) Crew scheduling for Netherlands railways: "destination customer". In: Lecture notes in economics and mathematical systems, vol 505. Springer, Berlin, Heidelberg, pp 181–201. https://doi.org/10.1007/978-3-642-56423-9_11 (cited on page 261)
10. Kroon LG, Huisman D, Abbink EJW, Fioole PJ, Fischetti M, Maróti G, Schrijver A, Steenbeek A, Ybema RJ (2009) The new Dutch timetable: the OR revolution. Interfaces 39(1):6–17. https://doi.org/10.1287/inte.1080.0409 (cited on page 261)
11. Morgado EM, Martins JP (1998) CREWS-NS: scheduling train crews in the Netherlands. AI Mag 19:25–38 (cited on page 260)
12. Potthoff D, Huisman D, Desaulniers G (2010) Column generation with dynamic duty selection for railway crew rescheduling. Transp Sci 44:493–505. https://doi.org/10.1287/trsc.1100.0322 (cited on pages 259, 260, 262)
13. Veelenturf LP, Potthoff D, Huisman D, Kroon LG, Maróti G, Wagelmans APM (2014) A quasi-robust optimization approach for crew rescheduling. Transp Sci 50(1):204–215. https://doi.org/10.1287/trsc.2014.0545 (cited on page 259)
14. Vlugt D (2010) Robust railway crew schedules. MSc thesis, The Netherlands: Econometric Institute, Erasmus University Rotterdam, HDL: 2105/7985. http://hdl.handle.net/2105/7985 (cited on page 254)

Chapter 12
Train Dispatching

Leonardo Lamorgese, Carlo Mannino, Dario Pacciarelli, and Johanna Törnquist Krasemann

L. Lamorgese
Optrail, Rome, Italy
e-mail: leonardo.lamorgese@optrail.com

C. Mannino
SINTEF DIGITAL & University of Oslo, Oslo, Norway
e-mail: carlo.mannino@sintef.no

D. Pacciarelli
Università degli Studi Roma Tre, Dipartimento di Ingegneria, Rome, Italy
e-mail: pacciarelli@ing.uniroma3.it

J. T. Krasemann (✉)
BTH - Blekinge Institute of Technology, Karlskrona, Sweden
e-mail: johanna.tornquist.krasemann@bth.se

R. Borndörfer et al. (eds.), *Handbook of Optimization in the Railway Industry*,
International Series in Operations Research & Management Science 268,
https://doi.org/10.1007/978-3-319-72153-8_12

Abstract Train rescheduling problems have received significant attention in the operations research community during the past 20–30 years. These are complex problems with many aspects and constraints to consider. This chapter defines the problem and summarizes the variety of model types and solution approaches developed over the years, in order to address and solve the train dispatching problem from the infrastructure manager perspective. Despite all the research efforts, it is, however, only very recently that the railway industry has made significant attempts to explore the large potential in using optimization-based decision-support to facilitate railway traffic disturbance management. This chapter reviews state-of-practice and provides a discussion about the observed slow progress in the application of optimization-based methods in practice. A few successful implementations have been identified, but their performance as well as the lessons learned from the development and implementation of those system are unfortunately only partly available to the research community, or potential industry users.

12.1 Introduction

12.1.1 Background and Scope

The challenge of managing disturbances and delays in railway traffic systems has most likely existed ever since the launch of the first public railway system in the beginning of the nineteenth century. With enlargements of the railway systems and service networks, as well as introduction of new technology and more complex organizational structures, the potential sources of faults and their knock-on effects increase. Managing railway traffic networks is today not only a technical achievement and challenge, but appears at times to become more of an organizational challenge. An organizational challenge in the sense that there are nowadays often so many subsystems, stakeholders and dependencies which hampers the ability to overview and manage network activities in a proactive way and with a system perspective. Railway system stakeholders thus need to jointly and stepwise start incorporating effective decision-supporting protocols and software in a much larger extent than now and evaluate the strengths and weaknesses in a systematic and transparent manner so that lessons learned reach beyond individual project groups and system suppliers.

The efforts dedicated in industry and in the research community to develop and evaluate principles, methods and software for decision-support in railway traffic management have increased significantly over the past 20 years. Larger European projects such as COMBINE, ARRIVAL and ON-TIME has resulted in relevant results and pinpointed several challenges, but more emphasize on practical applications and evaluations is needed. The results from the large national tenders seen in Europe recent years, and the forthcoming developed large-scale traffic management systems (TMSs), will hopefully contribute to an increased knowhow in the field and shed some light on many important practical and theoretical questions. We will discuss this aspect further later on in this chapter.

In this chapter, we first briefly introduce the research domain and main terminology with emphasize on optimization-based decision-support for railway traffic management during disturbances. In Sect. 12.2, two common alternative modeling approaches and problem formulations are presented and discussed. In Sect. 12.3, common types of algorithmic approaches are presented and discussed, while Sect. 12.4 presents current and planned practical implementations.

12.1.2 Aspects of Disturbance Management in Railway Traffic Systems

A disturbance in a railway network can occur due to a smaller incident such as an over-crowded platform and unexpectedly long boarding times causing minor delays, which the affected train may be able to recover from if there is sufficient buffer in the timetable. Disturbances can also be more significant and occur due to, e.g., rolling-stock breakdowns, power shortages, or signaling system failures.

Larger disturbances are in the context of railway traffic management sometimes referred to as disruptions, although the words generally can be considered synonymous. The Oxford online dictionary defines disruption as *disturbance or problems which interrupt an event, activity, or process*.

The distinction between smaller and larger disturbances has been discussed in e.g., [4]. There, the following definition is used: *... disturbances are relatively small perturbations of the railway system that can be handled by modifying the timetable, but without modifying the duties for rolling stock and crew. Disruptions are relatively large incidents, requiring both the timetable and the duties for rolling-stock and crew to be modified* (ibid). Hence, the distinction primarily is based on what type of actions that may be needed to cope with the incident rather than the initial sources of disruption. In this chapter we adopt the same distinction for sake of clarity.

Disruptions often result in significant knock-on delays and longer period of partial system unavailability. The railway system state transition over time can be illustrated as a bathtub, see, e.g., [14], where the traffic is reduced to function only partially during the disruption. That reduced level of traffic is then maintained until the system goes back to full capacity again via a transition plan.

When a railway traffic network suffers from a disturbance or disruption, which affects the scheduled railway transport services, the timetable needs to be modified. The re-scheduling of the timetable consists of two main parts:

1. Traffic re-scheduling, where focus is on network capacity and the need of the infrastructure manager (IM) to revise the timetable and allocation of track resources for the affected trains to minimize delays;

2. Transport service re-scheduling where focus is on the transport operating companies (TOC) and their need to handle the timetable from a train service point of view explicitly considering train connections and the effects on the rolling-stock and crew schedules.

The latter part includes the delay management problem, where emphasize is on effective policies for managing train connections and passenger flows during disturbances, in order to minimize passenger delays given a predefined set of available train services. In contrast to traffic re-scheduling, the delay management problem does traditionally not consider network capacity issues although the recent trend is to incorporate an increasing level of detail and realism in the models [11].

Although the majority of research so far has focused on the mentioned perspectives and types of re-scheduling problem individually, the interest in integrated approaches is increasing, see, e.g., [9].

Depending on the organizational structure of the railway systems and what authority and control the traffic managers have, the decision-making may be distributed between several different stakeholders. In fully deregulated networks such as the national railway systems of Sweden and Norway, the control of the infrastructure and traffic management lies on a neutral national transport authority, while the trains and associated transport services are operated by several different private companies. The decision-making during disturbances and disruptions is then depending on two, or more, different organizations. The different types of re-scheduling decisions can be divided as follows [15]:

- Re-timing of trains by allocating new arrival and departures times, including modification of speed profiles and halting schedules.
- Re-ordering of trains by adjusting the meet-pass plans[1]
- Local re-routing, by allocating alternative tracks on the line between two stations, or within the stations.
- Global re-routing by allocating alternative paths in the network.
- Cancellations and/or turning trains earlier than expected.

The first three can normally be made by the IM without consulting the TOCs, but the last two requires consultation with affected TOCs. In this chapter, we will continue discussing only real-time railway traffic disturbance management and focus on the infrastructure manager perspective during re-scheduling.

The development and application of such computational real-time re-scheduling support encompasses several challenges related to:

1. Human-computer interaction and requirements engineering concerning how to define and configure the computational support as functionalities.
2. Specification and formulation of the specified re-scheduling problems as well as development of appropriate solution methods.
3. System integration and communication including input data availability.

[1] Meet-pass events are central operations in train dispatching and in particular in single-track lines. In such lines, indeed, trains can only meet or pass each other in a station or sections where you have parallel tracks. When trains are delayed, dispatchers must often identify suitable new meet or pass points.

Here we focus on the second aspect and particularly optimization-based models and solution approaches. The research concerning development and application of decision-support for railway traffic disturbance management has received significant attention, which can be seen by comparison of different surveys and literature reviews over the years, see e.g., [13, 21, 35].

Proposed models and problem formulations can be described and compared with respect to a number of key properties such as infrastructure representation and level of granularity, time representation and objective(s). Railway traffic and the associated train occupation of network resources is often modelled as train events, where the events are assigned to specific time slots for the associated network resources. The problem of deciding (a) which resource to assign to each train event, (b) in what order different train events should be allocated the resources and (c) during which time period, is then the core problem. This problem is often referred to as the Train Dispatching problem (TD), see [18] and can be seen as a Job Shop Scheduling Problem with *blocking and no-wait constraints*, see [27].

The TD problem is formulated in several different ways depending on how capacity limitations of the network are modelled. The models are often classified as macro models or micro models, although there is no exact definition of what details each level includes. Macro models typically disregard the railway signalling system and consider the lines between stations as a set of parallel tracks. Micro models consider these lines as a set of train paths, each defined by a chain of block sections, crossings and signals.

In [24], two alternative MILP formulations to model the capacity restrictions on stations on a macro level are proposed and benchmarked: A *non-compact* formulation which counts the number of pairs of trains that simultaneously meet at a specific station and then ensures this respects the capacity limit, and a *compact* formulation where the station track occupancy by each train is modeled explicitly. Such approaches are often sufficient for single- and double-track lines, where there are less complex junctions and stations. Examples of models which focuses on complex and busy stations and network areas are presented in [6, 30]. There are also hybrid models which combine the use of macro and micro models. For example, Caimi et al. [5] use a macro model for the compensation zones of the network, with simple topology and low traffic, and a more detailed model for condensation zones, where it is necessary to include more detail in order to ensure feasible solutions. A different approach is proposed in [18], in which two-level strategies are adopted to optimally solve the dispatching problem on a large network, where macro models are used for the upper level and micro models are used for the lower level.

The majority of the problem formulations seen in the literature use a continuous time representation, where disjunctive constraints (also referred to as big-M constraints) are used to decide on the pairwise order of trains on allocated track segments. There are also a significant number of researchers who use a discrete time representation, see, e.g., [3, 28]. In [16] results from a comparative analysis of these two alternative ways of representing track occupancy over time are presented.

The objective(s) of the re-scheduling approaches have traditionally been to minimize train delays in different ways considering, e.g., maximum consecutive delay, or train delays with different weights. The focus on minimization of passenger delays and inconvenience rather than train delays has, however, increased, see, e.g., [34]. There is also a significant stream of research that also include aspects of energy-efficient driving. Such approaches consequently include a more fine-grained model of the infrastructure properties, network topology and train speed profiles in order to compute train trajectories instead of approximations of run times.

For a more detailed description of common alternative modelling approaches and problem formulations we refer to [22].

12.2 Alternative Graph and Disjunctive Formulation

This section focuses on microscopic models for train dispatching from the IM perspective, i.e., with the purpose of adjusting the timetable in response to disturbances in order to minimize train delays.

Train traffic is controlled worldwide by signals, interlocking and Automatic Train Protection (ATP) systems, which set up train routes, enforce train speed restrictions and impose a minimum safety separation between trains to avoid train accidents. Fixed block ATP systems ensure safety by allowing at most one train at a time running on a *resource* of the network. Examples of resources are the platforms of a station or the *block sections* of a line, i.e., portions of track between two consecutive signals. The path of a train from its origin to its destination is therefore a sequence of resources, each assigned exclusively to that train for a specific time interval, called *occupation time*. This time takes into account the dwell time on a stop platform, or the minimum traversing time of a block section, which depends on several factors, such as the length of the block section, the train speed and length, and includes the time between the entrance of the train head (its first axle) in the block section and the exit of its tail (the last axle), plus additional time margins to release the occupied block section and to take into account the sighting distance.

Due to the safety systems, a *primary delay* of a train due to an unexpected delay may easily propagate to other trains in the network, causing *secondary delays*. In fact, at least in areas with dense traffic, the amount of secondary delay due to propagation may significantly exceed that of the originating primary delay, see [17]. The task of the dispatcher is then to adjust the timetable to limit as much as possible delay propagation by keeping a feasible plan of operations. Specifically, given a railway network and train circulations in the network in a given time horizon, the TD problem is the real-time problem faced by the dispatchers, which consists in choosing:

1. a *route* for each train, i.e., a sequence of resources from its initial position, or entry point, to its final position, or exit point, such that the train can physically move from each resource to the next in the sequence;
2. a *sequence* of trains, for each resource traversed by multiple trains;
3. a *schedule* for each train, prescribing the time at which the train should start the occupation of each resource along its route from its entry to its exit point. The

schedule is feasible if there is no *conflict*, i.e., if no two trains try to occupy the same resource at the same time. A train keeps a resource occupied from the start of the occupation for at least the prescribed occupation time. However, having reached the end of the resource, the train can keep it occupied for an additional time if the subsequent resource on its route is not available, and blocks the current resource preventing other trains from entering it. This fact may lead to a *deadlock* status when there is a circular precedence among a set of trains, each of them waiting for the access to a resource blocked by resource blocked by another train in the set.

The main objective of the dispatcher is to design a deadlock-free schedule minimizing a function of train delays. Typical objectives proposed in the literature range from the minimization of the average delay, or of the maximum secondary delay [10], to the minimization of convex functions of the train delays [25]. Microscopic models to represent feasible solutions are based on the observation that by viewing the trains as jobs and the resources as machines, the train dispatching problem is similar to the job shop scheduling problem, in which the occupation time of a train on a resource corresponds to the processing time of a job on a machine. However, there are some specific aspects of the TD problem that must be taken into account. For example a train, having reached the end of a resource, cannot enter the subsequent one if the latter is occupied by another train. In the scheduling theory this is known as a *blocking constraint*. Moreover, other constraints can be defined, such as a minimum departure time from specific resources (platform stops), or a maximum travelling time between pair of resources that can be useful to ensure that a train does not take too long to reach the next station. The latter constraint can be modelled as a generalized no-wait constraint. Hence, the TD problem can be viewed as a job shop scheduling problem with blocking and no-wait constraints. Once a routing is fixed for each train, this type of problem can be effectively formulated by a disjunctive formulation based on the so called *alternative graph*, see [27]. The alternative graph generalizes the disjunctive graph model of the job shop scheduling problem by Roy and Sussmann [31].

An alternative graph is a triple $\mathscr{G} = (N, F, A)$, with N being the set of nodes, F the set of *fixed* directed arcs and A the set of pairs of *alternative* directed arcs. Arcs in F and A are weighted, and let w_{ij} be the length of arc (i, j). A *selection* S is a set of arcs obtained from A by choosing at most one arc from each alternative pair. The selection is *complete* if exactly one arc from each pair in A is selected. Given a pair $[(i, j), (h, k)] \in A$ and a selection S, if $(i, j) \in S$ then arc (i, j) is *selected* and arc (h, k) is *forbidden*. The pair is *unselected* if neither (i, j) nor (h, k) is selected in S. Given a selection S, $\mathscr{G}(S)$ denotes the digraph $(N, F \cup S)$. The selection S is *consistent* if the graph $\mathscr{G}(S)$ has no positive length cycles. An *extension* S' of a consistent selection S is a consistent selection such that $S \subset S'$. Given two nodes $i \in N, j \in N$, then $l^S(i, j)$ denotes the length of the longest path from node i to node j in $\mathscr{G}(S)$.

In the disjunctive formulation of the TD problem based on the alternative graph, set N includes two dummy nodes 0 and n, called *start* and *finish* respectively, such that for each node $i \in N$ there is a directed path from 0 to i and a path from i to n in the digraph $\mathscr{G}(\emptyset) = (N, F)$. To each node in $i \in N \setminus \{0, n\}$ is associated the event that a train starts the occupation of a resource. Let τ_i and ρ_i be the train and the resource associated to event i, respectively. Finally, let t_i be a variable associated to the starting time of event $i \in N \setminus \{0\}$, while $t_0 = 0$. Note that setting a value t_i for all $i \in N \setminus \{0, n\}$ corresponds to choosing a schedule for all trains.

Each arc (i, j), either fixed or alternative, represents a precedence relation constraining the start time (t_j) with respect to the start time (t_i). A fixed arc $(i, j) \in F$, where $\tau_i = \tau_j$ and ρ_j is the resource traversed by τ_i immediately after ρ_i, represents the constraint $t_j \geq t_i + w_{ij}$, where w_{ij} is the traversing time of resource ρ_i by τ_i. A fixed arc $(0, i) \in F$ represents the entrance of train τ_i in the network, or the departure time of τ_i from a platform stop ρ_i, with w_{0i} being the entrance time or the departure time of τ_i, respectively. A fixed arc $(i, 0) \in F$, with $w_{i0} < 0$, represents a firm deadline constraint $t_i \leq -w_{i0}$ for the start of occupation of train τ_i on resource ρ_i (recall that $t_0 = 0$). A deadline arc can be used, e.g., to represent the initial position of train τ_i at time 0, in combination with arc $(0, i)$ with $w_{0i} = -w_{i0}$.

A fixed arc $(i, n) \in F$, where i represents the arrival of τ_i at a platform stop ρ_i, is used to compute the delay of τ_i at a station. To this aim, let δ_i represent the arrival time of τ_i at ρ_i scheduled in the published timetable. By setting $w_{in} = -\delta_i$, arc (i, n) represents the constraint $t_n \geq t_i - \delta_i$, i.e., the arrival delay of τ_i at ρ_i. Similarly, if ε_i denotes the originating delay of τ_i, setting $w_{in} = -\delta_i - \varepsilon_i$ corresponds to the consecutive delay of τ_i at ρ_i. Hence, the minimization of the maximum (consecutive) delay can be obtained by minimization of t_n.

Alternative arcs are used to represent sequencing decisions. Given two trains τ_i and τ_h traversing the same resource $\rho_i = \rho_h$, is necessary to decide a sequencing for the two trains. Let ρ_j and ρ_k be the resources traversed by τ_i and τ_h immediately after ρ_i and ρ_h, respectively (see, Figs. 12.1 and 12.2). Then, arcs (i, j) and (h, k) are added to F with weights w_{ij} and w_{hk} equal to the traversing time of ρ_i by τ_i and τ_h. In the example in Fig. 12.2, two trains A and B share resources 1 and 4. Hence, for resource 1, $\tau_i = \tau_j = A$, $\tau_h = \tau_k = B$, $\rho_i = \rho_h = 1$, $\tau_j = 2$, $\tau_k = 3$. The pair $[(j, h), (k, i)] \in A$ represents the two sequencing alternatives. Arc (j, h) corresponds to giving precedence to train $\tau_i = A$, since τ_h can enter $\rho_i = 1$ only after $\tau_i = A$ has moved from resource $\rho_i = 1$ to $\rho_j = 2$, while arc (k, i) corresponds to giving precedence to train $\tau_h = B$. Here, w_{jh} represents the minimum time needed to release ρ_i after the entrance of τ_i in ρ_j and to make it available for τ_h. A similar discussion holds for w_{ki}.

A feasible schedule for the TD problem can be obtained by defining a complete consistent selection S and fixing all variables $t_i = l^S(0, i)$, for all $i \in N$. Then t_n is the maximum (consecutive) delay of the schedule. Therefore, the TD problem is the problem of finding a complete consistent selection S such that the length of the longest path $l^S(0, n)$ is minimum.

To summarize, the formulation based on the alternative graph of the TD problem is a disjunctive program:

$$
\begin{aligned}
&\min t_n \\
&s.t.\ \ t_j - t_i \geq w_{ij} && (i, j) \in F \\
&\qquad (t_h - t_j \geq w_{jh}) \vee (t_i - t_k \geq w_{ki}) && [(j, h), (k, i)] \in A
\end{aligned}
$$

This formulation can be easily converted into a Mixed Integer Linear Program by associating a binary variable x_{jhki} to each disjunction, i.e., to each pair in A, equal to one when (j,h) is selected and equal to zero if (k,i) is selected. Then, the TD problem can be formulated as a MILP as follows:

Fig. 12.1: A simple instance with two trains. The region contains four block sections (from 1 to 4), each can be occupied by at most one train at a time. Block Sects. 12.2 and 12.3 can be viewed as a station, where block Section 12.2 is a siding. Trains A and B are running in the same direction

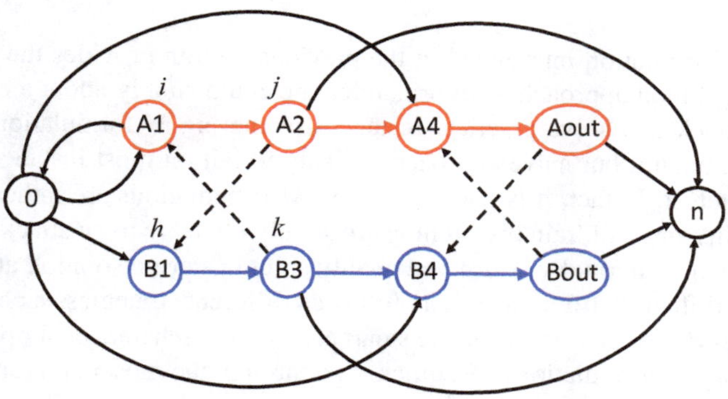

Fig. 12.2: The alternative graph corresponding to the instance of Fig. 12.1 (with solid fixed arcs and dashed alternative arcs)

$$
\begin{array}{lll}
\min t_n & & \\
s.t. & t_j - t_i \geq w_{ij} & (i,j) \in F \\
 & t_h - t_j + M(1 - x_{jhki}) \geq w_{jh} & [(j,h),(k,i)] \in A \\
 & t_i - t_k + M x_{jhki} \geq w_{ki} & [(j,h),(k,i)] \in A \\
 & x \in \{0,1\}^{|A|} &
\end{array}
$$

The MILP formulation can be extended to the more general TD problem in which both a route and a schedule must be defined for each train, i.e., the TD problem with routing flexibility. This can be obtained by enlarging sets F and A to contain all possible arcs for all possible train routes, and by adding for each alternative route variables $y_{uv} \in \{0,1\}$ equal to 1 if route u is chosen for train v, and 0 otherwise.

Let F_u be the set of fixed arcs associated to route u. Alternative pairs are associated to all resources shared by two routes u and $\bar{u}$. Let n_T be the number of trains to be scheduled, R_v be the set of routes of train v, $R = \bigcup_{v=1,\ldots,n_T} R_v$, $A_{u\bar{u}}$ be the set of pairs of alternative arcs associated to the resources shared by routes u and $\bar{u}$, $A = \bigcup_{u,\bar{u}\in R} A_{u\bar{u}}$. Then, the train scheduling and routing formulation is the following:

$$
\begin{array}{ll}
\min \quad t_n & \\
s.t. & \\
t_j - t_i + M(1-y_{uv}) \geq w_{ij} & (i,j) \in F_u; u \in R_v; v = 1,\ldots,n_T \\
t_h - t_j + M(1-x_{jhki}) + M(1-y_{u\tau_h}) + M(1-y_{\bar{u}\tau_j}) \geq w_{jh} & [(j,h),(k,i)] \in A_{u\bar{u}}; u,\bar{u} \in R \\
t_i - t_k + Mx_{jhki} + M(1-y_{u\tau_i}) + M(1-y_{\bar{u}\tau_k}) \geq w_{ki} & [(j,h),(k,i)] \in A_{uv}; u,\bar{u} \in R \\
\sum_{u\in R_v} y_{uv} = 1 & v = 1,\ldots,n_T \\
x \in \{0,1\}^{|A|}, y \in \{0,1\}^{|R|} &
\end{array}
$$

12.3 Algorithmic Aspects

The big-M formulation introduced in the previous section provides the basis for a number of solution approaches. In principle, one could simply adopt a commercial solver to attack and solve practical instances of the big-M formulation. Unfortunately, such natural but naive approach is likely to fail on most instances of some practical interest. In fact, it is known that big-M formulations are rather weak and the typical instances of train dispatching are simply too large to be attacked directly. The solver would normally return bad quality solutions or no solution at all. To get around this difficulty different authors followed different strategies, such as embedding the big-M formulation into some smart algorithmic schema, or simply avoiding the use of big-M formulations. We quickly go through the most common options:

1. Heuristics of various type
2. Branch&Bound, with bound computed with some combinatorial methods
3. Alternative formulations
4. Decomposition methods
5. A hybrid of the above approaches

Concerning point 1, the literature is very vast and varied, so we refer the reader to the mentioned survey papers. As for point 2, the literature reports very few attempts, as, e.g., [25] for mass transit and [10] for main line.

Proceeding with point 3, the classical alternative to the big-M formulation for scheduling problems is the so called *time-indexed* formulation, where the time horizon is discretized, and we have one binary variable y_{it} for each atomic movement (i.e., the occupation of a rail resource by a train) and each period. y_{it} is 1 if and only if the atomic movement i starts in period t. A major drawback of this approach is that it tends to introduce a huge number of binary variables and packing constraints, even for small dispatching instances. Because of the tight computing times enforced by the application—a solution must be returned in a few seconds in order to be assessed by dispatchers in real time. Time-indexed formulations are in general preferred for off-line problems such as train timetabling. Only a few attempts have been made also in train dispatching, such as Caimi et al. [6], Lusby et al. [23], Meng and Zhou [28], and Şahin et al. [32]. Recently, a promising approach with only a few continuous variables and as few binary variables as for the big-M formulation has been presented in [19].

Point 4: An alternative and quite popular technique to tackle complicated mixed integer linear programs is decomposition. The term *decomposition* basically denotes the act to replace the original problem with a sequence of smaller subproblems. The solutions to the smaller problems are then re-combined or extended to the original large problem. If one fails in this phase, some kind of *regret* mechanism must be put in place and the problem solved from the start. The two most common decompositions may be seen as operating in time and space, respectively.

Decomposing in time: rolling horizon. Rolling horizon is a classical decomposition approach (see, e.g., [28, 29, 36, 38]) in which the time horizon is decomposed into smaller intervals and a subproblem is associated with each interval. Then the subproblems are solved in chronological order. At each iteration, the (part of the) solution associated with the previous subproblems is fixed, and only "few" additional variables and constraints are left.

Decomposing in space: the macroscopic/microscopic decomposition principle. A different but still very popular type of decomposition approach is the so-called macroscopic-microscopic decomposition approach [33]. A typical two-stage implementation of this approach goes as follows: in a first stage stations are considered as points with infinite capacity, and trains are scheduled along the line(s). In the second stage, one checks if the schedule so found can be extended to an overall solution, namely reconsidering in detail the actual topology of the stations. If one fails in this phase, some of the decisions taken in the first stage must be reconsidered. Different approaches differ in the way this mechanism is implemented. Most papers resort to heuristic regret approaches [11, 12] or no approaches at all (e.g. [37]). In contrast, an exact regret mechanism is developed in [18] and [20], based on integer programming and logic Benders' decomposition. The basic ideas goes as follows. If we let t be the scheduling vector, and x be the binary vector associated with the disjunctive terms, then the generic big-M formulation introduced in the previous section may be represented as:

$$\min c^T t \tag{12.1}$$

$$\begin{array}{llll} s.t. & & & \\ (i) & A_L x_L + & B_L t & \geq w_L, \\ (ii) & & B_S t + A_S x_S & \geq w_S \\ (iii) & & t \text{ real}, \quad x_L, x_S \text{ binary} & \end{array}$$

Vector x has been written as $x = (x_L, x_S)$ to distinguish between decisions associated with stations (x_S) and decisions associated with tracks between stations (x_L).

Observe that matrices A_L, A_S and vectors w_L, w_S contain in general big-M coefficients.

Next, we may identify blocks (i) and (ii) in Program (12.1). The two blocks only share a small subset of t variables, namely those corresponding to arrivals and departures from stations. Indeed, leaving a station amounts to entering the line between to successive stations, whereas leaving a track between two stations amounts to entering the second station. It follows that block (i) may be seen as corresponding to the macroscopic problem of controlling trains between stations, whereas block (ii) corresponds to the microscopic problem associated with stations.

In Benders' decomposition algorithm, we first solve the restricted problem (*master*) obtained by dropping block (12.1.*ii*). This corresponds to taking dispatching decisions for trains running on the line, neglecting what can happen in stations. Let (x_L^*, t^*) be the optimal solution to the master problem. In order to established if t^* can be extended to a feasible solution to (12.1.*ii*) we need to solve the so called *slave* problem. It should be apparent by now that the slave problem corresponds to solving a microscopic feasibility problem (for all stations), where arrival and departure times for all trains are fixed (by the master). If the slave problem has a solution (t^*, x_S^*) then we are done as (x_L^*, t^*, x_S^*) is an optimal solution to (12.1). Otherwise (x_L^*, t^*) cannot be extended to a feasible solution for the whole problem, and we identify an inequality $q^T x_L + r^T t \leq k$ which is satisfied by all of the feasible solutions to (12.1) but is violated by (x_L^*, t^*). Such inequality is added to the master problem and the process is iterated. A nice feature of the approach is that the slave problem further decomposes into a number of independent sub-problems, one for each station.

A different approach to space decomposition is presented in [8], where entire administrative regions (*areas*) are actually collapsed into single nodes in order to carry out inter-area coordination among trains.

Finally, another type of decomposition approach consists in partitioning trains into groups of one or more elements and then solve the problem associated with each group in sequence, as in [1].

Interestingly, the three decomposition approaches introduced above mimic, in some sense, the actual behavior of human dispatchers, which somehow apply a combination of these methods. Indeed, potential conflicts are typically solved by dispatchers by neglecting the microscopic details (of the stations or of the tracks between stations). Then train movements within stations are then controlled in the necessary detail. Also, a dispatcher typically focuses on one or two trains at a time, and only for the next few movements or conflict.

12.4 State of Practice

Even if the optimization literature has been drilling into train dispatching for over 30 years, to our knowledge there are very few operative TMSs which rely on optimization algorithms to take or suggest dispatching decisions. This gap is also highlighted in recent surveys [4, 7] and papers such as Zhan et al. [38]. There are several reasons for this discrepancy. There is little doubt that the first attempts to automatize the dispatching process resulted in disappointing failures. This was possibly due to the use of inadequate techniques (e.g., rule based systems rather than optimization models and algorithms). The unavoidable consequence was an immediate and long-lasting skepticism of railway infrastructure managers and train operators in the sheer possibility of such automation. Another factor of resistance is related to the particular relation between the two major actors of the TMS market, namely infrastructure managers and (large) system vendors. None of those actually is willing to take the risk of innovation. By one hand, developing optimization based TMSs requires large investments by the vendors, with uncertain outcomes. On the other hand, railway operators, until very recently, did not press the vendors with specific technological requirements, probably because of lack of strong motivations (such as a real competitive market) or incentives. Indeed, in Europe infrastructure managers are typically state-hold companies and they operate in monopolistic markets.

The situation seems now to be on the verge of a rapid change. A growing awareness towards the potential of optimization-based TMSs is tangible. Infrastructure managers and operators around Europe are starting to explicitly request the use of optimization modules within TMSs, as we have observed in recent tenders, e.g., Denmark, Sweden and Norway. This fact actually forces the large vendors to pursue a stronger collaboration with research centers, since developing optimization tools for train dispatching is not an easy task.

Indeed, a few, large system vendors, finally started to claim their TMSs embed some sort of optimization algorithms. This is the case, for instance, of the General Electrics Movement Planner (see GE Movement Planner 2016). Unfortunately, the only available material is a few brochures, where the company states to implement *business objective based optimization*. Other large companies made similar claims with poor supporting documentation. Also, after some personal (but far from exhaustive) investigation, we could not identify any of the OR specialists involved in the development of such softwares, let alone scientific publications. A remarkable exception is represented by Alstom, a large multinational company headquartered in France, specialized in signaling systems and TMSs. A few years ago, in order to prepare for a large tender in Sweden where the use of optimization was explicitly required, Alstom decided to engage three academic groups with strong and long lasting expertise in railway optimization. After a sort of pre-qualification, one of these groups was selected to develop the optimization part of final product and is now collaborating with Alstom. The optimization engine is now integrated in the TMS offered by Alstom, but not yet in operation.

Indeed, according to what reported in the literature and what personally overheard or learned at specialized workshops, meetings, etc., the only few TMSs in operation actively using optimization described in the scientific literature are described in the remaining of this section (see also [2] for more details).

In all existing systems, the optimization routines are typically embedded in a loop. First, the TMS acquires real-time information regarding the status of the network (e.g., train positions, speeds, resource availability etc.). This information is fed into the optimization modules, which, combining it with the required "static" information (e.g., network layout, train connections), return one or more solutions to the current dispatching problem. Dispatchers may also "manually" interact with the systems providing further information (e.g., train delays or cancellations, network disruptions, fixed meeting points). The total time allowed for this loop is rather tight, typically a few seconds, setting a limit on the size and type of instances which can be tackled.

To our knowledge, the first "optimization-based" support system reported in the literature was embedded in a TMS developed by Bombardier Transportation, and operated in some terminal stations of the Milano Underground between 2007 and 2008 [25]. The system was dismissed after a year because Bombardier lost a tender for a complete renewal of the TMSs controlling the lines in Milano underground system. The dispatching algorithm was an exact branch&bound. By contract clause, the system had to prove to perform better than the dispatchers in order to be actually purchased by Azienda Trasporti Milanese, the state-hold company managing Milano underground system. To this end, an intensive 1-week on-field test campaign was set up in a terminal stations, which directly compared dispatchers against algorithm. Each day, four "traffic equivalent" 1 h slots were identified: in two of them the algorithm had full control over the trains, while in the other two the control was assigned to the dispatchers. In spite of the small size of the station, the system performed, on average, 8% better than dispatchers both in terms of deviations from timetable and in regularity. Besides paper [25] a popular science description of this experience is presented in [26].

Again Bombardier Transportation is behind the first main line optimization based dispatching system. This was put in operation in Italy in 2011, on a single-track regional line (Trento-Bassano del Grappa), see [24]. The use of the tool was later extended to other lines in Northern, Southern and Central Italy, such as Milano-Mortara, Piraineto-Trapani-Alcamo, Orte-Terontola-Falconara and others. Table 12.1 gives some hints about the size and geometry of such regional lines.

Table 12.1: Infrastructure details

Line	Stops	Stations	Length (m)	Tracks
Trento–Bassano	22	14	95,711	Single
Piraineto–Trapani	12	12	93,532	Single
Alcamo–Trapani	14	13	116,119	Single
Orte–Terontola–Falconara	54	34	283,839	Mixed

Single stands for single-track, mixed stands for single and double-track

Because of very strict business rules adopted by the Italian Infrastructure Manager (RFI), the optimization algorithm embedded in Bombardiers' TMS is a heuristic. For the same reason, the dispatching loop is semi-automated. The algorithm finds alternative solutions each time it is called and presents them to the dispatcher(s) ranked by cost. Statistics show that in 94% of the cases the first solution proposed by the algorithm is accepted. It is worth noticing that such limiting business rules were introduced to help dispatchers to quickly take difficult decisions. If such rules would be ignored or dropped, then the full power of an exact optimization algorithm could be exploited. Indeed, we have tried to quantify the possible advantages to do this in [20], and it turns out that for the mentioned Orte-Terontola-Falconara line a significant average increase of trains on time (+6.2%) and reduction of trains heavily delayed (8.9%). The results presented in Table 12.2 refer to to a specific week day in year 2013.

Table 12.2: Exact vs heuristic approach: comparisons on train punctuality

Method	On time	Delay$\leq$ 10	Delay$\leq$15	Delay>15
Heuristic	84.7%	86.5%	87.9%	12.1%
Exact method	90.9%	95%	96.8%	3.2%

A new release of the above TMSs has been recently deployed in a few lines in Latvia, namely Daugavpils-Eglaine, Daugavpils-Krustpils, Rezekne-Krustpils, Zilupe- Krustpils, Karzava-Rezekne, for a total of 52 stations, with 10 communication points and 8 station gates. These lines are mainly used for freight transportation and run around 100 trains every 20 h. Again due to local business rules, the optimization algorithm is heuristic. A major innovation is that, in a particular modality, freight trains decision can be directly applied by the system without prior acceptance by dispatchers.

In a recent project involving with the Norwegian infrastructure manager (JBV) and train operating companies (NSB, FlyToget, CargoNet), an automatic dispatching system was put in operation at Stavanger control center in Norway, in February 2014. The system controlled trains on the Stavanger-Moi Line, which is 123 km long, with 16 stations, 7 line stops and 28 block sections. On weekdays, the average number of trains every 12 h is around 100. The system implements an exact, MILP algorithm described in [18, 20]. Like in previous cases, the system presents solutions in real-time to dispatchers which decide whether to accept the solutions. The system was well received by dispatchers and management. The use of the system in Stavanger for real-time dispatching was put on hold for legal reasons towards the end of 2014, because of a competitive tender issued by JBV for the renewal of the entire Norwegian signaling (and centralized traffic control) system.

Finally, concerning large stations rather than lines, the TMSs in Roma Tiburtina station (12 line points, 30 stopping points and 62 interlocking routes) and the multi-station of Monfalcone in Italy have been equipped with optimization algorithms which re-schedule and re-route trains. The optimization modules include both heuristic and exact algorithms. Tiburtina station is the second largest station in Rome, and is considered one of the most important and complex stations in Italy.

References

1. Bettinelli A, Santini A, Vigo D (2017) A real-time conflict solution algorithm for the train rescheduling problem. Transp Res B 106:237–265. (cited on page 276)
2. Borndörfer R, Klug T, Lamorgese L, Mannino C, Reuther M, Schlechte T (2017) Recent success stories on integrated optimization of railway systems. Transp Res C Emerg Technol 74:196–211. https://doi.org/10.1016/j.trc.2016.11.015 (cited on page 277)
3. Cacchiani V, Toth P (2012) Nominal and robust train timetabling problems. Eur J Oper Res 219(3):727–737. https://doi.org/10.1016/j.ejor.2011.11.003 (cited on page 269)
4. Cacchiani V, Huisman D, Kidd M, Kroon L, Toth P, Veelenturf L, Wagenaar J (2014) An overview of recovery models and algorithms for realtime railway rescheduling. English. Transp Res B 63:15–37. https://doi.org/10.1016/j.trb.2014.01.009 (cited on pages 267, 277)
5. Caimi G, Chudak F, Fuchsberger M, Laumanns M, Zenklusen R (2011) A new resource-constrained multicommodity flow model for routing and scheduling. Transp Sci 45(2):212–227. https://doi.org/10.1287/trsc.1100.0349 (cited on page 269)
6. Caimi G, Fuchsberger M, Laumanns M, Lüthi M (2012) A model predictive control approach for discrete-time rescheduling in complex central railway station areas. Comput Oper Res 39(11):2578–2593. https://doi.org/10.1016/j.cor.2012.01.003 (cited on pages 269, 275)
7. Corman F, Meng L (2015) A review of online dynamic models and algorithms for railway traffic management. IEEE Trans Intell Transp Syst 16(3):1274–1284. https://doi.org/10.1109/tits.2014.2358392 (cited on page 277)
8. Corman F, D'Ariano A, Pacciarelli D, Pranzo M (2012) Optimal inter-area coordination of train rescheduling decisions. Transp Res E Log Transp Rev 48(1):71–88. https://doi.org/10.1016/j.sbspro.2011.04.508 (cited on page 276)
9. Corman F, Ariano AD, Marra AD, Pacciarelli D, Samà M (2016) Integrating train scheduling and delay management in real-time railway traffic control. Transp Res E Log Transp Rev. https://doi.org/10.1016/j.tre.2016.04.007 (cited on page 268)
10. D'Ariano A, Pacciarelli D, Pranzo M (2007) A branch and bound algorithm for scheduling trains in a railway network. Eur J Oper Res 183(2):643–657. https://doi.org/10.1016/j.ejor.2006.10.034 (cited on pages 271, 274)
11. Dollevoet T, Corman F, D'Ariano A, Huisman D (2014) An iterative optimization framework for delay management and train scheduling. Flex Serv Manuf J 26(4):490–515. https://doi.org/10.1007/s10696-013-9187-2 (cited on pages 268, 275)
12. Dollevoet T, Huisman D, Kroon L, Schmidt M, Schöbel A (2014) Delay management including capacities of stations. Transp Sci 49(2):185–203. https://doi.org/10.1287/trsc.2013.0506 (cited on page 275)

13. Fang W, Yang S, Yao X (2015) A survey on problem models and solution approaches to rescheduling in railway networks. IEEE Trans Intell Transp Syst 16(6):2997–3016. https://doi.org/10.1109/TITS.2015.2446985 (cited on page 269)
14. Ghaemi N, Goverde RMP (2015) Review of railway disruption management practice and literature. In: 6th international conference on railway operations modelling and analysis-RailTokyo2015. UUID: 9f911a03-155e-4dd3-a64e-3416fd603aa1 (cited on page 267)
15. Hansen IA (2010) State-of-the-art of railway operations research. In: Computers in Railways X. Timetable planning and information quality, Chap A 4. WIT Press, Boston, pp 35–47. https://doi.org/10.2495/CR060561 (cited on page 268)
16. Harrod S, Schlechte T (2013) A direct comparison of physical block occupancy versus timed block occupancy in train timetabling formulations. Transp Res E Log Transp Rev 54:50–66. ISSN: 1366-5545. https://doi.org/10.1016/j.tre.2013.04.003 (cited on page 269)
17. Kecman P, Corman F, D'Ariano A, Goverde RMP (2013) Rescheduling models for railway traffic management in large-scale networks. Public Transp 5(1–2):95–123 (cited on page 270)
18. Lamorgese L, Mannino C (2015) An exact decomposition approach for the real-time train dispatching problem. Oper Res 63:48–64. https://doi.org/10.1287/opre.2014.1327 (cited on pages 269, 275, 279)
19. Lamorgese L, Mannino C (2016) A non-compact formulation for job-shop scheduling problems in transportation. In: Algorithmic methods for optimization in public transport (Dagstuhl Seminar 16171) Dagstuhl Reports, vol 6(4). Schloss Dagstuhl, Wadern. https://doi.org/10.4230/DagRep.6.4.139 (cited on page 275)
20. Lamorgese L, Mannino C, Piacentini M (2016) Optimal train dispatching by benders'-like reformulation. Transp Sci 50(3):910–925. https://doi.org/10.1287/trsc.2015.0605 (cited on pages 275, 279)
21. Lusby RM, Larsen J, Ehrgott M, Ryan D (2011) Railway track allocation: models and methods. OR Spectrum 33(4):843–883. https://doi.org/10.1007/s00291-009-0189-0 (cited on page 269)
22. Lusby R, Larsen J, Ryan D, Ehrgott M (2011) Routing trains through railway junctions: a new set-packing approach. Transp Sci 45(2):228–245. https://doi.org/10.1287/trsc.1100.0362 (cited on page 270)
23. Lusby RM, Larsen J, Ehrgott M, Ryan DM (2013) A set packing inspired method for real-time junction train routing. Comput Oper Res 40(3):713–724. https://doi.org/10.1016/j.cor.2011.12.004 (cited on page 275)
24. Mannino C (2011) Real-time traffic control in railway systems. In: Caprara A, Kontogiannis S (eds) Proceedings of the 11th workshop on algorithmic approaches for transportation modelling, optimization, and systems (ATMOS 2011). OpenAccess Series in Informatics (OASIcs), vol 20. Schloss Dagstuhl, Wadern, pp 1–14. ISBN: 978-3-939897-33-0. https://doi.org/10.4230/OASIcs.ATMOS.2011.1 (cited on pages 269, 278)

25. Mannino C, Mascis A (2009) Optimal real-time traffic control in metro stations. English. Oper Res 57:1026–1039. https://doi.org/10.1287/opre.1080.0642 (cited on pages 271, 274, 278)
26. Mannino C, Mascis A (2010) Fast track to fixing rail delays-award-winning automated rail re-routing system saves time and money. OR MS Today 37(2):28 (cited on page 278)
27. Mascis A, Pacciarelli D (2002) Job shop scheduling with blocking and no-wait constraints. English. Eur J Oper Res 143(3):498–517. https://doi.org/10.1016/s0377-2217(01)00338-1 (cited on pages 269, 271)
28. Meng L, Zhou X (2014) Simultaneous train rerouting and rescheduling on an N-track network: a model reformulation with network-based cumulative flow variables. Transp Res B Methodol 67:208–234. https://doi.org/10.1016/j.trb.2014.05.005 (cited on pages 269, 275)
29. Nielsen LK, Kroon L, Maróti G (2012) A rolling horizon approach for disruption management of railway rolling stock. Eur J Oper Res 220(2):496–509. https://doi.org/10.1016/j.ejor.2012.01.037 (cited on page 275)
30. Pellegrini P, Marlière J, Rodriguez D (2014) Optimal train routing and scheduling for managing traffic perturbations in complex junctions. Transp Res B 59:58–80. https://doi.org/10.1016/j.trb.2013.10.013 (cited on page 269)
31. Roy B, Sussmann B (1964) Les problèmes d'ordonnancement avec contraintes disjonctives, Note D.S. No. 9 bis. French. Tech. rep. SEMA, France (cited on page 271)
32. Şahin G, Ahuja RK, Cunha CB (2010) Integer programming based solution approaches for the train dispatching problem. Tech. rep. Sabanci University (cited on page 275)
33. Schlechte T, Borndörfer R, Erol B, Graffagnino T, Swarat E (2011) Micro-macro transformation of railway networks. English. J Rail Transp Plann Manage 1(1):38–48. https://doi.org/10.1016/j.jrtpm.2011.09.001 (cited on page 275)
34. Toletti A, Weidmann U (2016) Modelling customer inconvenience in train rescheduling. In: 16th Swiss transport research conference (STRC), May 2016. HDL: 20.500.11850/117003 (cited on page 270)
35. Törnquist J (2006) Computer-based decision support for railway traffic scheduling and dispatching: a review of models and algorithms. In: Kroon LG, Möhring RH (eds) Proceedings of the 5th workshop on algorithmic methods and models for optimization of railways (ATMOS'05). OpenAccess series in informatics (OASIcs), vol 2. Schloss Dagstuhl, Wadern. ISBN: 978-3-939897-00-2. https://doi.org/10.4230/OASIcs.ATMOS.2005.659 (cited on page 269)
36. Törnquist J (2007) Railway traffic disturbance management: an experimental analysis of disturbance complexity, management objectives and limitations in planning horizon. Transp Res A Policy Pract 41(3):249–266. https://doi.org/10.1016/j.tra.2006.05.003 (cited on page 275)

37. Zhan S, Kroon LG, Veelenturf LP, Wagenaar JC (2015) Real-time high-speed train rescheduling in case of a complete blockage. Transp Res B Methodol 78:182–201. https://doi.org/10.1016/j.trb.2015.04.001 (cited on page 275)
38. Zhan S, Kroon LG, Zhao J, Peng Q (2016) A rolling horizon approach to the high speed train rescheduling problem in case of a partial segment blockage. Transp Res E Log Transp Rev 95:32–61. https://doi.org/10.1016/j.tre.2016.07.015 (cited on pages 275, 277)

Chapter 13
Delay Propagation and Delay Management in Transportation Networks

Twan Dollevoet, Dennis Huisman, Marie Schmidt, and Anita Schöbel

T. Dollevoet (✉)
Econometric Institute and ECOPT, Erasmus University Rotterdam, P.O. Box 1738, NL-3000 DR Rotterdam, The Netherlands
e-mail: dollevoet@ese.eur.nl

D. Huisman
Econometric Institute and ECOPT, Erasmus University Rotterdam, P.O. Box 1738, NL-3000 DR Rotterdam, The Netherlands

Netherlands Railways, Process quality and Innovation, P.O. Box 2025, NL-3500 HA Utrecht, The Netherlands
e-mail: huisman@ese.eur.nl

M. Schmidt
Rotterdam School of Management, Erasmus University Rotterdam, P.O. Box 1738, NL-3000 DR Rotterdam, The Netherlands
e-mail: schmidt2@rsm.nl

A. Schöbel
Institute for Numerical and Applied Mathematics, Lotzestraße 16-18, 37083 Göttingen, Germany
e-mail: schoebel@math.uni-goettingen.de

R. Borndörfer et al. (eds.), *Handbook of Optimization in the Railway Industry*, International Series in Operations Research & Management Science 268, https://doi.org/10.1007/978-3-319-72153-8_13

Abstract Should connecting trains wait for delayed feeder trains? Or is it better for the passengers if trains depart on time?

Questions of this type are the subject of delay management, which will be treated in this chapter from the point of view of the passengers. We start by discussing how delays are propagated through a public transportation network, and how such propagations can be modeled using event-activity networks.

We then focus on the question of finding an optimal solution to the delay management problem in case some (known) source delays have occurred. We discuss which decisions can be made and how these can be reflected by variables and constraints in integer programming models. In particular, we show how station capacities and the limited capacity of the tracks can be taken into account. Special emphasis will be given to the discussion of passenger-oriented objective functions. We introduce several ways on how to measure the effects that delays have on passengers and explain how to include the resulting objective functions in the models.

Next, we discuss solution approaches for delay management. In particular, we discuss heuristics that decompose the delay management problem and solve it within short computation times. Finally, we give some insights into delay management in practice. We review some simple delay management strategies used today and discuss recent developments that make it possible to implement more advanced methods in practice as well.

13.1 Introduction and Motivation

In highly connected train systems, as common all over Europe, passengers often have to change trains, since it is impossible to give a direct connection between all origin-destination pairs. In order to provide convenient connections, the timetable is often constructed in such a way that train B departs shortly after train A has arrived. However, if train A has a delay during the operations, the question is whether train B should wait or depart on time. *Delay management* deals with this type of decisions. More precisely, in delay management one looks at (small) source delays of a railway system as they occur in the daily operations. In case of such delays, the scheduled timetable is not feasible anymore and has to be updated to a so-called *disposition timetable*. If a connecting train waits for a delayed feeder train, it gets delayed itself. It is hence not clear in advance whether waiting for the delayed train or departing on time is the better decision for the system as a whole. For railway systems, the limited track capacity has to be taken into account in delay management as well: no two trains may occupy the same piece of track at the same time, and safety distances between trains have to be respected. This has to be taken into account when rescheduling trains in case of delays. In many cases, it may be favorable to change the order in which trains pass a specific piece of track. This type of decision is called *precedence decision*. We may have precedence decisions on the tracks outside and inside stations.

In the next section, we mention various models and solution approaches for delay propagation and delay management. The main question in these models is to decide which trains should wait for delayed feeder trains and which trains better depart on time (*wait-depart decisions*). Note that the approaches and results provided in the following are applicable not only for railway networks but also for other transportation networks such as bus or metro systems. Nevertheless we will use the railway-related notion and talk only of trains and stations instead of buses or bus stops.

With the following simple, real-world example we want to illustrate the key aspects of delay management. The railway network used in the example is depicted in Fig. 13.1.

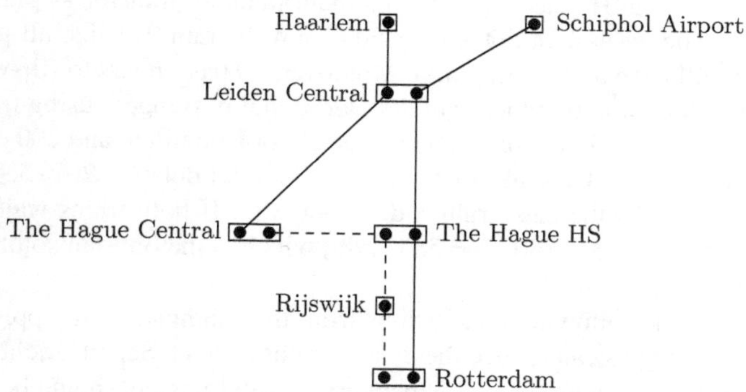

Fig. 13.1: Part of the railway network in the Netherlands. Rectangles represent stations. Dots indicate that trains of a certain line stop at the station. Solid lines represent Intercity trains; dashed lines represent regional trains

In this example, there are three train lines, each with a frequency of 30 min.

Line 1 runs from Haarlem via Leiden Central to The Hague Central. Line 2 connects Schiphol Airport via Leiden Central and The Hague HS with Rotterdam. Both lines are Intercity services, only stopping at the major stations. Line 3 is a local service from The Hague Central via The Hague HS to Rotterdam, stopping among other stations at Rijswijk. The timetable is constructed in such a way that there is a cross-platform transfer of 2 min in Leiden Central between line 1 and 2, vice versa, and in The Hague HS between line 2 and 3, vice versa. For simplicity, assume that there are no running time supplements and there is no slack time in the transfers.

The (relevant) passenger figures can be found in Table 13.1.

Table 13.1: Number of passengers from Haarlem, Schiphol and The Hague Central to The Hague HS, Rijswijk and Rotterdam

	Haarlem	Schiphol	The Hague Central
The Hague HS	100	200	–
Rijswijk	0	70	100
Rotterdam	100	400	200

Suppose that a train of line 1 has a delay of 5 min at arrival in Leiden Central. The main question of delay management is now: Should train 2 wait (for 5 min) or should it depart on time? If train 2 departs on time, 200 passengers (traveling from Haarlem to The Hague HS and Rotterdam), cannot transfer as planned. The total passenger delay is then $200 \cdot 30 = 6000$ min. If train 2 waits, all passengers using train 2 will have a delay of 5 min. Moreover, 70 passengers to Rijswijk might miss their transfer. Note that there are 800 additional passengers using train 2: 600 passengers travel from Schiphol to The Hague HS or Rotterdam and 200 passengers travel from The Hague Central to Rotterdam. The total delay is $200 \cdot 5 + 70 \cdot 30 + 800 \cdot 5 = 7100$ min in the case train 3 does not wait. If both trains wait, the total delay is $200 \cdot 5 + 170 \cdot 5 + 800 \cdot 5 = 5850$ min, which is the optimal solution in this small example.

We can draw the following conclusions from the example. First, applying delay management strategies can reduce the delay for passengers. Second, delay management is non-trivial, because the whole network and all transfers should be taken into account at once.

13.2 Delay Propagation

13.2.1 The Event-Activity Network

The *public transportation network* $PTN = (V,E)$ models the physical railway (or bus) network, i.e., V denotes the set of stations and $E \subseteq V \times V$ the set of direct connections or tracks. Each line in the transportation network is a path through the PTN, and a (passenger) train is a path at a specific time. It is often assumed that all lines in the PTN are served with a common period T, i.e., all departures or arrivals are repeated every T time units.

A *source delay* (also called *primary delay*) of a train is a delay which is caused by an external event, e.g., a signalling problem, a blockage of some track, some construction site that prevented the train from arriving on time, an accident, or simply a crowded station or bad weather conditions. Source delays may spread out into the system. This is due to several propagation rules; the most important ones are listed in Table 13.2, an illustration of the first three is given in Fig. 13.2.

Table 13.2: Delay propagation

(i)	If a train starts with a delay at some station, it is likely to arrive at the next station with some delay (often the train can go faster than planned and reduce its delay)
(ii)	If a train arrives at some station with a delay, it may depart from this station with some delay
(iii)	If a train waits for a delayed feeder train, it may get delayed itself
(iv)	If a train reaches the last station of its line with a delay, it might start its next trip also with a delay
(v)	Delayed trains may block platforms or track sections such that also trains using this infrastructure at a later time get delays

For solving the delay management problem it is important to keep track of these *propagated delays* (also called *secondary delays* or *knock-on delays*). In order to do so within an elegant framework, we introduce event-activity networks. Event-activity networks are widely used in public transportation, in particular for finding a timetable (e.g., [21, 22]), for timetable information (e.g., [15]), and in delay management (e.g., [29]).

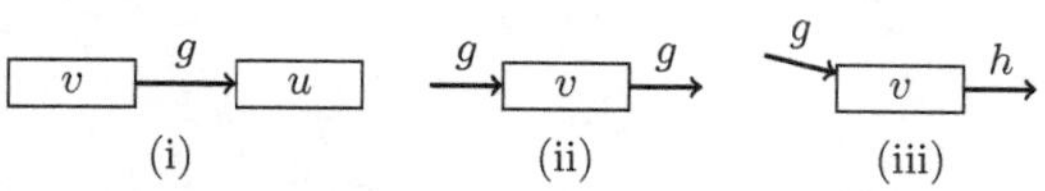

Fig. 13.2: Delay propagation for driving (i), waiting (ii), and changing (iii) activities

Definition 13.1. Let $PTN = (V,E)$ be a public transportation network and $\mathscr{T}$ be a set of trains. We denote by $\mathscr{T}(v)$ the set of all trains that stop at station v. The event-activity network $\mathscr{N} = (\mathscr{E}, \mathscr{A})$ then consists of

1. the set of *events* $\mathscr{E} = \mathscr{E}_{\text{arr}} \cup \mathscr{E}_{\text{dep}}$ where an *arrival event* (v,t,arr) represents the arrival of a train $t \in \mathscr{T}(v)$ at a station $v \in V$ and a *departure event* (v,t,dep) represents the departure of train $t \in \mathscr{T}(v)$ at station $v \in V$,
2. and of a set of directed edges, called *activities* $\mathscr{A} \subset \mathscr{E} \times \mathscr{E}$. The set of activities includes the following subsets:
   - $\mathscr{A}_{\text{wait}}$ (waiting activities) are of type $((v,t,arr),(v,t,dep))$ for some $v \in V, t \in \mathscr{T}(v)$ and represent dwelling of train t in station v,
   - $\mathscr{A}_{\text{drive}}$ (driving activities) are of type $((v_1,t,dep),(v_2,t,arr))$ for some $(v_1,v_2) \in E, t \in \mathscr{T}(v_1) \cap \mathscr{T}(v_2)$ and represent a train t driving from station v_1 to its next station v_2,
   - $\mathscr{A}_{\text{change}}$ (changing activities) are of type $((v,t_1,arr),(v,t_2,dep))$ for some $v \in V, t_1,t_2 \in \mathscr{T}(v)$. They do not represent a *train's* activity, but the possibility for passengers to change from train t_1 to another train t_2 at station v.

A part of an event-activity network is illustrated in Fig. 13.3.

We later add turn-around and headway activities to $\mathscr{A}$. The event-activity network has several important features:

Defining Timetables. First, it is a convenient model for defining *timetables*: A timetable is given by assigning a time to each arrival and each departure of all trains, i.e., it is specified by numbers π_i for all events $i \in \mathscr{E}$. For designing a timetable there are usually lower bounds $L_a > 0$ on the duration of each activity $a \in \mathscr{A}$ given. For a driving activity $a = ((v_1,t,dep),(v_2,t,arr)) \in \mathscr{A}_{\text{drive}}$, the lower bound equals the technically minimal driving time train t needs to go from station v_1 to station v_2. For a waiting activity, the lower bound models the minimum waiting time a train needs to allow passengers to alight and board. Finally, for a changing activity $a = ((v,t_1,arr),(v,t_2,dep)) \in \mathscr{A}_{\text{change}}$, L_a contains the time, a passenger needs for getting off train t_1, changing the platform, and boarding train t_2. Sometimes, also upper bounds are given, but these are usually neglected in delay management.

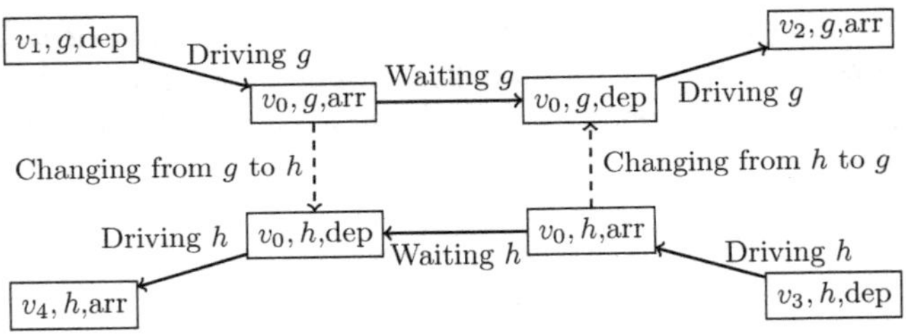

Fig. 13.3: An event-activity network with two vehicles g and h meeting at station v_0

A timetable π is *feasible* if

$$\pi_j - \pi_i \geq L_a \quad \text{for all } a = (i,j) \in \mathscr{A},$$

i.e., if it respects all minimal durations. If $\pi_j - \pi_i > L_a$ the timetable contains some *buffer time*

$$s_a := \pi_j - \pi_i - L_a$$

for activity a. If π is a feasible timetable, all buffer times are greater than or equal to zero.

Order of the Events. Secondly, from the event-activity network one can also read off the order in which events take place: if there is an activity $(i,j) \in \mathscr{A}$ linking event i and event j, then we know that event i takes place before event j. Since this is transitive, we can define an order relation by

$$i \prec j \text{ if there exists a directed path } P \text{ in } \mathscr{N} \text{ from } i \text{ to } j.$$

Note that this is in fact an order relation since $\mathcal{N}$ cannot contain directed cycles because all lower bounds L_a for $a \in \mathcal{A}$ are strictly greater than zero. Hence $\mathcal{N}$ can be ordered topologically by ordering the events $i \in \mathcal{E}$ according to their scheduled times π_i.

Passenger Paths. The third feature of the event-activity network will become important in Sect. 13.3.4. Each directed path between a departure and an arrival event presents a (unique) passenger route, and vice versa.

Delay Propagation. Finally, the fourth (and most relevant) feature of the event-activity network is that it can be used for propagating delays which is discussed in the next section.

13.2.2 Delay Propagation

Here we describe how delays are propagated. We use the event-activity network and proceed in several steps. First, we briefly introduce where delays come from. Then we discuss how the delay of an event i is propagated to its successor j along a single activity $a = (i, j)$ and for a bundle of activities, and finally we show how to compute the delays for a given delay scenario. Let y_i denote the delay of event i. We may assume that $y_i \geq 0$ for all $i \in \mathcal{E}$ since trains are not allowed to depart earlier than planned.

13.2.2.1 Source Delays

There are different possible (external) sources of delay:

- either a source-delayed event (if an event has to be shifted to a later time, e.g., since a driver got sick and his replacement arrives with a delay, or if some repair work is finished at some specific point of time), in this case we have to add the delay d_i to the scheduled time of event i,
- or a source-delayed driving, waiting, or turn-around activity (e.g., the train has to go slowly due to a construction site or due to weather conditions), in this case we add d_a to the minimal duration L_a of the respective activity.

Note that one can transform delayed events into delayed activities easily, but not vice versa. We assume that $d_i = d_a = 0$ for all events and activities without source delays (hence, in particular for all changing and headway activities). A *delay scenario* is then defined as a set of source delays $d_a \geq 0$ for $a \in \mathcal{A}$ and $d_i \geq 0$ for $i \in \mathcal{E}$.

13.2.2.2 Delay Propagation Along a Single Activity

(i) Let a train t have some delay $y_i = d_i$ when it starts from station v_1, i.e., at event (v_1,t,dep). The next activity is a driving activity $a = (i, j)$ from i to

$j = (v_2, t, arr)$. Assume that $d_a = 0$, i.e., a has no additional delay. What is the delay of event j? We have to distinguish two cases:

- $y_i \leq s_a$: Then the delay of the train can be reduced to zero (by driving faster than planned) on its way between stations v_1 and v_2. We obtain a zero delay $y_j = 0$ of event j.
- $y_i > s_a$: Then along activity $a = (i, j)$ the delay y_i can be reduced by s_a but is still propagated to event j. It satisfies $y_j \geq y_i - s_a$.

Together, we obtain

$$y_j \geq [y_i - s_a]^+ \tag{13.1}$$

where $[r]^+ = \max\{r, 0\}$ as usual. Note that we have to write $\geq$ in (13.1) since there could be other delays influencing event j.

(ii) The same observation holds for waiting activities, we obtain $y_j \geq [y_i - s_a]^+$ for a waiting activity $a = (i, j)$.

(iii) Now suppose a train t_1 arrives at some station v with a delay, and there is a changing activity $a = (i, j)$ to another train t_2 with $i = (v, t_1, arr)$ and $j = (v, t_2, dep)$. For changing activities we have another distinction to be made: In case the transfer is maintained, we can use the same formula as for driving and waiting activities and the delay propagates according to $y_j \geq [y_i - s_a]^+$. However, if the transfer is not maintained, the delay from train t_1 will not influence train t_2, i.e., we obtain

$$y_j \geq \begin{cases} [y_i - s_a]^+ & \text{if the transfer is maintained,} \\ 0 & \text{otherwise.} \end{cases}$$

Note that these *wait-depart decisions*, i.e., whether a transfer is maintained or not is in the responsibility of the operator of the public transportation company. In the next section we show how such decisions can be computed in a way which is best for the passengers.

In order to model the other two propagation rules mentioned in Table 13.2 in the same way, we need to introduce additional activities in the event-activity network. These are the following:

(iv) Turn-around activities $\mathscr{A}_{\text{turn}}$ of type $((v_1, t, dep), (v_2, t, arr))$ for $v_1, v_2 \in V, t \in \mathscr{T}(v_1) \cap \mathscr{T}(v_2)$ represent the empty ride of a vehicle from the last station v_1 of its previous trip to the first station v_2 of its next trip. The next trip may also start at the same station, in this case $v_1 = v_2$. The duration L_a of a turn-around activity is set to the minimal time needed before the vehicle can start again, and may include cleaning time and the time for a driver's break. Similarly as for $\mathscr{A}_{\text{wait}}$ and $\mathscr{A}_{\text{drive}}$, also for activities $a \in \mathscr{A}_{\text{turn}}$ delay is propagated according to (13.1).

(v) Headway activities $\mathscr{A}_{\text{head}}$ are of type $((v, t_1, dep), (v, t_2, dep))$ for some $v \in V, t_1, t_2 \in \mathscr{T}(v)$. They represent that train t_2 must wait some time L_a after train t_1 has left before departing from the same station v. Note that for modeling oncoming traffic, it is convenient to allow also headways between departures from different stations. The minimal headway time L_a depends on the speed of the two trains and on the lengths of the blocks on their next track section, see [31] for details. Note that basically we need headway activities between any ordered pair of events i, j using the same infrastructure. Headway activities usually come in pairs (i, j) and (j, i). Both headway activities cannot be satisfied simultaneously. The public transportation company has to make a *precedence decision*, i.e., decide which train goes first. If train t_1 leaves first, headway (i, j) needs to be respected, if t_2 leaves first, (j, i) needs to be respected.

13.2.2.3 Delay Propagation for a Bundle of Activities

For the following let us define

$$\begin{aligned}\bar{\mathscr{A}} = \mathscr{A}_{\text{wait}} \cup \mathscr{A}_{\text{drive}} \cup \mathscr{A}_{\text{turn}} \cup \{a \in \mathscr{A}_{\text{change}} : a \text{ is maintained}\} \\ \cup \{a = (i, j) \in \mathscr{A}_{\text{head}} : i \text{ has precedence over} j\}\end{aligned} \tag{13.2}$$

as the set of activities which are relevant for the delay propagation. These activities have to be chosen by the operator who decides about the wait-depart and about the precedence decisions. Usually, an event j has not only one, but several incoming activities $(i, j) \in \mathscr{A}$. Then event j can only take place after *all* incoming activities are completed.

Example: Let $j = (v, t, dep)$ be the head of a waiting, several changing and headway activities. Then j can take place if the train t is ready to depart and if all passengers from the feeder trains have arrived and the headways are respected.

Hence, the actual delay of j is the maximum of the propagated delays along all incoming activities $a = (i, j) \in \bar{\mathscr{A}}, i \in \mathscr{E}$. We obtain:

$$y_j = \left[\max\{y_i - s_a : a = (i, j) \in \bar{\mathscr{A}}\}\right]^+. \tag{13.3}$$

13.2.2.4 Delay Propagation for a Delay Scenario

We finally can derive a formula for computing the effects of a delay scenario in the network. Hereby, we assume that each event will be scheduled as early as possible in view of the delays. Note that the effects of a delay scenario crucially depend on the decisions of the public transport company, i.e., on the wait-depart decisions in the set $\mathscr{A}_{\text{change}}$, i.e., which trains should wait for delayed feeder trains, and on the precedence decisions to be made (i.e., on the set $\mathscr{A}_{\text{head}}$). Assume that these decisions have been made. Then, in a first step, we sort the events $\mathscr{E}$ topologically. This is possible if and only if the network $(\mathscr{E}, \bar{\mathscr{A}})$ including the wait-depart and precedence decisions does not contain a directed cycle. In the case that the precedence constraints are chosen in the same way as originally planned, the topological order can be found easily by ordering the events according to their scheduled times π_i, $i \in \mathscr{E}$ (independent of the wait-depart decisions).

The delays of the events are then computed iteratively for all events in their topological order, $j = 1, 2, \ldots, |\mathscr{E}|$ as follows.

- If event j does not have any incoming activity, then its delay is set to $y_j = d_j$.
- Else, j has at least one incoming activity i with $(i, j) \in \mathscr{A}$. Since we proceed in topological order we already know the delay y_i for all these events i. We set:

$$y_j = \max\left\{d_j, \max\{y_i - s_a + d_a : a = (i, j) \in \bar{\mathscr{A}}\}\right\}. \tag{13.4}$$

We remark that the delay propagation can also be computed by using the distribution of delays in a periodic event-activity network. The propagation of delays can then be computed via their convolutions. In [18] it is shown that computing the convolutions iteratively leads to a stable process, i.e., it converges in distribution against a stationary delay distribution. Also Kecman et al. [17] consider the prediction of train delays using Bayesian networks.

13.3 Models: Integer Programming Formulations and Objective Functions

In this section, we discuss how delay management problems can be modeled as integer programs (IPs). To this end, we show in Sects. 13.3.1–13.3.3 how delay propagation, precedence decisions on tracks and in stations, and platform re-assignments can be modeled as constraints in an integer program. Based on this, in Sect. 13.3.4 we discuss different objective functions for delay management and explain how they can be modeled in an integer program.

13.3.1 The Basic Model: Wait-Depart Decisions

Let $\mathscr{N} = (\mathscr{E}, \mathscr{A})$ denote the event-activity network. Let us first assume that precedence decisions and platform assignments in stations are fixed, e.g., inherited from the previous timetable. Denote the set of corresponding headway activities as $\bar{\mathscr{A}}_{\text{head}} = \{a = (i, j) \in \mathscr{A}_{\text{head}} : i \text{ has precedence over } j\}$.

Since precedence decisions are fixed, in our first delay management model, we only have to make the wait-depart decisions. To this end, we introduce variables

$$z_a := \begin{cases} 1 & \text{if transfer } a \text{ is not maintained,} \\ 0 & \text{otherwise} \end{cases}$$

for all $a \in \mathscr{A}_{\text{change}}$. As before, the variables y_i describe the delay of event $i \in \mathscr{E}$. Our first delay management model (DM) is as follows:

$$\min \sum_{i \in \mathscr{E}} w_i y_i + \sum_{a \in \mathscr{A}_{\text{change}}} w_a z_a \tag{13.5}$$

such that

$$y_j \geq d_j \quad \forall j \in \mathscr{E}_{\text{arr}} \cup \mathscr{E}_{\text{dep}}, \tag{13.6}$$

$$y_j \geq y_i - s_a + d_a \quad \forall a = (i,j) \in \mathscr{A}_{\text{drive}} \cup \mathscr{A}_{\text{wait}} \cup \bar{\mathscr{A}}_{\text{head}} \cup \mathscr{A}_{\text{turn}}, \tag{13.7}$$

$$y_j \geq y_i - s_a - Mz_a \quad \forall a = (i,j) \in \mathscr{A}_{\text{change}}, \tag{13.8}$$

$$z_a \in \{0,1\} \quad \forall a \in \mathscr{A}_{\text{change}} \tag{13.9}$$

$$y_j \in \mathbb{R} \quad \forall j \in \mathscr{E}_{\text{arr}} \cup \mathscr{E}_{\text{dep}}. \tag{13.10}$$

For each event $j \in \mathscr{E}$, constraints (13.6), (13.7), and (13.8) make sure that

$$y_j \geq \max\{d_j, \max\{y_i - s_a + d_a : a = (i,j) \in \bar{\mathscr{A}}\}\} \tag{13.11}$$

with $\bar{\mathscr{A}}$ as defined in (13.2). Note that if a changing activity $a = (i,j)$ is maintained, $z_a = 0$ and constraint (13.8) simply reads

$$y_j \geq y_i - s_a.$$

On the other hand, if $a = (i,j)$ is not maintained, $z_a = 1$ and constraint (13.8) is always fulfilled for M large enough. It can be shown that $M := \sum_{a \in \mathscr{A}} \max\{d_a - s_a, 0\} + \max\{d_i : i \in \mathscr{E}\}$ is large enough (see [25]).

The objective function (13.5) minimizes the weighted sum of delays at events $i \in \mathscr{E}$ and non-maintained changing activities $a \in \mathscr{A}_{\text{change}}$. Hereby, w_i and w_a are non-negative weights, representing the 'costs' of arriving or departing late, and of not maintaining a transfer. We discuss the choice of these weights in more detail in the first paragraph of Sect. 13.3.4. If the weights w_i are positive for all $i \in \mathscr{E}$, in an optimal solution to (DM) we obtain

$$y_j = \max\{d_j, \max\{y_i - s_a + d_a : a = (i,j) \in \bar{\mathscr{A}}\}\}$$

as required in (13.4).

Note that all models presented in this chapter can be transformed easily to a formulation which uses the actual times $x_i := \pi_i + y_i$ as variables by using the transformation $y_j = x_j - \pi_j$ and $s_a = \pi_j - \pi_i - L_a$ for all $a = (i,j) \in \mathscr{A}$. The times x_j are often called *disposition timetable* since they define the real arrival and departure times in view of the delays.

The above integer programming formulation was first presented in [28] and has been further developed in [4, 30]; see also [29] for an overview of various models. A bi-objective version, which considers the delays of the trains and the number of lost transfers simultaneously has been investigated in [3, 13]. Note that in these early delay management models, the problem was studied without considering capacity restrictions on tracks or stations and turn-around movements, i.e., setting $\bar{\mathscr{A}}_{\text{head}} = \emptyset$ and $\mathscr{A}_{\text{turn}} = \emptyset$.

13.3.2 Adding Precedence Decisions on Tracks

When a train has a delay and precedence decisions on tracks are left unchanged, the following trains will receive a delay, too, unless the buffer times between the trains are large enough. In this section we show how precedence decisions can be integrated in the delay management IP.

In order to avoid delays, it makes sense to allow to change precedence decisions in the delay management step. Remember that in Sect. 13.2 we introduced *two* headway activities whenever two trains use the same piece of track, namely an activity (i,j) and a corresponding activity (j,i) for the events i and j. To impose a precedence constraint for the two trains, we have to select exactly one of these activities. To this end, we introduce variables

$$g_{ij} = \begin{cases} 1 & \text{if } (i,j) \text{ is selected,} \\ 0 & \text{otherwise} \end{cases}$$

for each headway activity $a \in \mathscr{A}_{\text{head}}$. That is, event i is scheduled before event j, if $g_{ij} = 1$. To make sure that for each pair of headway activities exactly one is chosen, we add the constraints

$$g_{ij} + g_{ji} = 1 \quad \forall \text{ pairs of headway activities in } \mathscr{A}_{\text{head}} \tag{13.12}$$

$$g_{ij} \in \{0,1\} \quad \forall a = (i,j) \in \mathscr{A}_{\text{head}}. \tag{13.13}$$

Furthermore, for each headway activity we add the constraint

$$y_j \geq y_i - s_{ij} - M_{head}(1 - g_{ij}) \;\; \forall a = (i,j) \in \mathscr{A}_{\text{head}}, \tag{13.14}$$

where s_{ij} represents the planned buffer time on headway activity (i,j).

Similarly to constraints (13.8), if M_{head} is chosen big enough (see [25]), constraints (13.14) make sure that if and only if event i is scheduled before event j, delay is propagated along activity (i,j).

By adding constraints (13.12), (13.13), and (13.14) to (DM) we obtain the following IP, which we refer to as (DM-prec).

$$\min \sum_{i \in \mathscr{E}} w_i y_i + \sum_{a \in \mathscr{A}_{\text{change}}} w_a z_a \tag{13.15}$$

such that

$$y_j \geq d_j \quad \forall j \in \mathscr{E}_{\text{arr}} \cup \mathscr{E}_{\text{dep}}, \tag{13.16}$$

$$y_j \geq y_i - s_a + d_a \quad \forall a = (i,j) \in \mathscr{A}_{\text{drive}} \cup \mathscr{A}_{\text{wait}} \cup \mathscr{A}_{\text{turn}}, \tag{13.17}$$

$$y_j \geq y_i - s_a - M z_a \quad \forall a = (i,j) \in \mathscr{A}_{\text{change}}, \tag{13.18}$$

$$g_{ij} + g_{ji} = 1 \quad \forall \text{ pairs of headway activities in } \mathscr{A}_{\text{head}} \tag{13.19}$$

$$y_j \geq y_i - s_a - M_{head}(1 - g_{ij}) \quad \forall a = (i,j) \in \mathscr{A}_{\text{head}}, \tag{13.20}$$

$$g_{ij} \in \{0,1\} \quad \forall a = (i,j) \in \mathscr{A}_{\text{head}}, \tag{13.21}$$

$$z_a \in \{0,1\} \quad \forall a \in \mathscr{A}_{\text{change}}, \tag{13.22}$$

$$y_j \in \mathbb{R} \quad \forall j \in \mathscr{E}_{\text{arr}} \cup \mathscr{E}_{\text{dep}}. \tag{13.23}$$

Note that due to Constraints (13.20), it is infeasible to set both g_{ij} and g_{ji} to 1. Namely, if g_{ij} and g_{ji} would both be equal to 1, Constraints (13.20) for (i,j) and (j,i) cannot both be satisfied. Hence, (13.19) can be replaced by

$$g_{ij} + g_{ji} \geq 1 \; \forall \text{ pairs of headway activities in } \mathscr{A}_{\text{head}}. \tag{13.24}$$

Delay management with precedence decisions was first introduced in [31]. Schachtebeck and Schöbel [24, 25] present several heuristic methods to solve the problem. See, e.g., [3] for a model which takes capacity restrictions on the microscopic level into account.

13.3.3 Adding Station Capacities and Platform Re-assignment

Precedence decisions do not only apply to tracks outside stations, but are also important within stations, since two trains cannot occupy the same platform track at the same time. To make sure that this does not happen, even when delays occur, we could use the headways as before. However, it is more efficient to introduce an additional type of headway activities:

(vi) $\mathscr{A}_{\text{station}}$ are of type $(i_1, j_2) = ((v, t_1, dep), (v, t_2, arr))$ for some $v \in V$, $t_1, t_2 \in \mathscr{T}(v)$. They represent that a train t_2 must wait some time L_a to go to a platform after train t_1 has left the platform. To allow precedence changes on platforms, we introduce these activities in pairs, i.e., additionally to (i_1, j_2) we add $(i_2, j_1) = ((v, t_2, dep), (v, t_1, arr))$ and require that exactly one of them is chosen in any feasible solution to the delay management problem, as we did for the headway activities on tracks, too.

We introduce additional variables

$$\bar{g}_{ij} = \begin{cases} 1 \text{ if } (i,j) \text{ is selected,} \\ 0 \text{ otherwise} \end{cases}$$

for each headway activity $a = (i,j) \in \mathscr{A}_{\text{station}}$. That is, event i is scheduled before event j, if $\bar{g}_{ij} = 1$.

Similarly as for the other activities, we denote by $s_a := \pi_j - \pi_i - L_a$ the buffer time of station headway activity $a = (i,j) \in \mathscr{A}_{\text{station}}$. Analogously to the previous paragraph, we can now ensure to obtain a timetable which is feasible in stations, by adding the constraints

$$y_j \geq \; y_i - s_{ij} - M_{stat}(1 - \bar{g}_{ij}) \forall a = (i,j) \in \mathscr{A}_{\text{station}} \tag{13.25}$$

$$\bar{g}_{i_1 j_2} + \bar{g}_{i_2 j_1} = 1 \forall \text{ pairs of station headway activities in } \mathscr{A}_{\text{station}} \tag{13.26}$$

$$\bar{g}_{ij} \in \{0,1\} \forall a = (i,j) \in \mathscr{A}_{\text{station}} \tag{13.27}$$

to (DM-prec), for a large enough M_{stat}. Following the same argumentation as in Sect. 13.3.2, we see that also in (13.26), the equality sign could be replaced with $\geq$ signs.

However, when it is feasible with respect to the station layout, delay may be further reduced by allowing *platform changes*. In the following we present a model from Dollevoet et al. [8] to incorporate platform changes into delay management. To model precedence constraints in this setting, for each arrival event j we define $\mathscr{P}(j)$ to be the set of platforms where arrival event j could take place and introduce a new set of variables which indicate at which platform p arrival event j takes place:

$$c_{jp} := \begin{cases} 1 \text{ if arrival event } j \text{ takes place at platform } p, \\ 0 \text{ otherwise} \end{cases}$$

for all $j \in \mathscr{E}_{\text{arr}}$ and all $p \in \mathscr{P}(j)$. Since a train must depart from the same platform it has arrived at in a station, we do not need to introduce platform assignment variables for departure events.

We extend the set $\mathscr{A}_{\text{station}}$ to contain pairs of activities $(i_1, j_2) = ((v,t_1,dep), (v,t_2,arr))$ and $(i_2, j_1) = ((v,t_2,dep),(v,t_1,arr))$ for all pairs of trains t_1 and t_2 which *could be* scheduled at the same platform at station v. Then we replace constraint (13.26) by

$$\bar{g}_{i_1 j_2} + \bar{g}_{i_2 j_1} \geq c_{j_1 p} + c_{j_2 p} - 1 \quad \forall \text{ pairs of station headway activities in } \mathscr{A}_{\text{station}} \text{ and all } p \in \mathscr{P}(j_1) \tag{13.28}$$

$$c_{jp} \in \{0,1\} \quad \forall j \in \mathscr{E}_{\text{arr}}, \forall p \in \mathscr{P}(j). \tag{13.29}$$

These constraints ensure that $\bar{g}_{i_1 j_2} + \bar{g}_{i_2 j_1} \geq 1$, if $y_{j_1 p} = y_{j_2 p} = 1$, i.e., we impose a precedence constraint on the corresponding trains, if both trains are scheduled on platform p. In the same way as before, if $\bar{g}_{i_1 j_2} + \bar{g}_{i_2 j_1} = 2$ for a pair of station headway activities, Constraints (13.25) cannot be satisfied. This explains why we can use an inequality sign in (13.28).

Furthermore, we require that each arrival event takes place at exactly one platform, that is

$$\sum_{p \in \mathscr{P}(j)} c_{jp} = 1 \quad \forall j \in \mathscr{E}_{\text{arr}}. \tag{13.30}$$

The integer program for delay management with precedence decisions and platform re-assignment (DM-plat) reads as follows:

$$\min \sum_{j \in \mathscr{E}} w_j y_j + \sum_{a \in \mathscr{A}_{\text{change}}} w_a z_a \tag{13.31}$$

such that

$$y_j \geq d_j \quad \forall j \in \mathscr{E}_{\text{arr}} \cup \mathscr{E}_{\text{dep}}, \tag{13.32}$$

$$y_j \geq y_i - s_a + d_a \quad \forall a = (i,j) \in \mathscr{A}_{\text{drive}} \cup \mathscr{A}_{\text{wait}} \cup \mathscr{A}_{\text{turn}}, \tag{13.33}$$

$$y_j \geq y_i - s_a - M z_a \quad \forall a = (i,j) \in \mathscr{A}_{\text{change}}, \tag{13.34}$$

$$y_j \geq y_i - s_{ij} - M_{head}(1 - g_{ij}) \quad \forall a = (i,j) \in \mathscr{A}_{\text{head}}, \tag{13.35}$$

$$y_j \geq y_i - s_{ij} - M_{stat}(1 - \bar{g}_{ij}) \quad \forall a = (i,j) \in \mathscr{A}_{\text{station}}, \tag{13.36}$$

$$g_{ij} + g_{ji} = 1 \quad \forall \text{ pairs of headway activities in } \mathscr{A}_{\text{head}}, \tag{13.37}$$

$$\overline{g}_{i_1 j_2} + \overline{g}_{i_2 j_1} \geq c_{j_1 p} + c_{j_2 p} - 1 \quad \forall \text{ pairs of station headway activities in } \mathscr{A}_{\text{station}}, \text{ and all } p \in \mathscr{P}(j_1) \tag{13.38}$$

$$\sum_{p \in \mathscr{P}(j)} c_{jp} = 1 \quad \forall j \in \mathscr{E}_{\text{arr}}, \tag{13.39}$$

$$z_a \in \{0,1\} \quad \forall a \in \mathscr{A}_{\text{change}}, \tag{13.40}$$

$$g_a \in \{0,1\} \quad \forall a = (i,j) \in \mathscr{A}_{\text{head}}, \tag{13.41}$$

$$\overline{g}_a \in \{0,1\} \quad \forall a = (i,j) \in \mathscr{A}_{\text{station}}, \tag{13.42}$$

$$c_{ip} \in \{0,1\} \quad \forall i \in \mathscr{E}_{\text{arr}}, \forall p \in \mathscr{P}(i), \tag{13.43}$$

$$y_j \in \mathbb{R} \quad \forall j \in \mathscr{E}_{\text{arr}} \cup \mathscr{E}_{\text{dep}}. \tag{13.44}$$

Besides delays, also platform changes are inconvenient for passengers, in particular for boarding passengers if changes are announced last-minute, and if passengers have to walk far to the newly announced platform. In order to take this into account, a term counting (weighted) platform changes could be added to the objective function, or the problem could be considered as a bi-objective problem with passenger delays and number of platform changes as the two objective functions. See [11] for further details.

13.3.4 Delay Management Objectives

In the preceding sections we discussed how to incorporate wait-depart decisions, precedence decisions, and platform re-assignments as variables in an IP model for delay management. Hereby, our objective function was a weighted sum of delays at events and non-maintained transfers.

However, from a passenger's perspective, delays at intermediate departures and arrivals during the journey are less important: in the end a passenger would mostly be interested in his delay upon arrival at his destination. This delay may be due to the delay of the train with which he arrives at his destination, or due to the fact that he missed a transfer because of a delay during the trip. In order to measure *passenger* delays, we need data about the passengers' travel plans as additional input.

We now discuss how *passenger delays* can be accounted for in delay management. In the next paragraphs we describe different ways to include passenger delays in the objective function. For the sake of a more compact representation, we do *not* include precedence decisions or platform assignment in the models presented here. However, the corresponding constraints and variables can simply be added to the integer programs described here if these constraints should be included.

13.3.4.1 Delay Management with Constant Weights

For the simplest delay management formulation with passenger delay as objective, introduced in [30], we assume that we know the *planned route* r for each passenger, which is specified as a path in the event-activity network. Let $\mathscr{R}$ denote the set of all such routes.

There are two different effects that a train delay can have on a passenger:

1. In the first case, the passenger is able to travel on route r as planned (i.e., all transfers on r are maintained). In that case, the delay of the passenger will be the delay of the last train the passenger takes at arrival at his destination. I.e., the passenger is delayed by $y_{i(r)}$ for $i(r)$ being the last event on route r.
2. In the second case, the passenger misses a transfer due to the delay of a feeder train he is in. In that case, the passenger will have to take the next train to his destination. For the sake of simplicity, we assume that in this case the passenger has to wait one time period T until he can continue his journey as planned. Hence, he is delayed by T minutes.

Note that for the second case to be correct we need to assume that the underlying timetable is periodic with period T, that there are no source delays in the next time period, that current delays do not propagate to the next period, and that the passenger indeed waits one time period T, because he has no re-routing options. Delay management with re-routing is considered in a later paragraph.

By setting the factors in the objective function (13.5) to

- $w_i :=$ number of passengers with i as destination event and
- $w_a := T$ multiplied by the number of passengers who use changing activity a,

we obtain an objective function which reflects the passenger delay. We refer to (DM) with the above-defined weights in the objective function as (DM-const).

However, even under the above-listed assumptions, we make a mistake by estimating the total passenger delay with (13.5): In case that a passenger route r contains n missed transfers, this route will contribute a penalty of nT to the objective function, while the delay of the corresponding passenger is only T. Furthermore, if a passenger misses a transfer, and there is a delay at the last event on his initially planned route, both are counted in the objective, while only the penalty for the missed transfer should be counted. Hence, (13.5) with weights as described above *overestimates* the actual passenger delay.

In the next paragraph we see how passenger delay can be modeled more accurately, and give a condition under which the passenger delay estimated by (13.5) is indeed accurate.

13.3.4.2 Delay Management with Fixed Routes

We now modify the delay management formulation from the preceding paragraph, so that passenger delays are computed exactly. This model is based upon the same assumptions as (DM-const). That is, we assume that the underlying timetable is periodic with period T, that there are no source delays in the next time period and current delays do not propagate to the next period, and that the passenger indeed waits one time period T in case of a missed transfer. This model was first introduced in [28].

Again, let $\mathscr{R}$ denote the set of planned passenger routes. Denote by w_r the number of passengers on planned route r. To model passenger delay in an exact way, we introduce additional variables

$$z_r := \begin{cases} 1 & \text{if route } r \text{ is not maintained,} \\ 0 & \text{otherwise} \end{cases}$$

for all routes $r \in \mathscr{R}$.

We then consider the following integer program (DM-route):

$$\min \sum_{r \in \mathscr{R}} w_r (y_{i(r)}(1 - z_r) + T z_r) \tag{13.45}$$

such that

$$y_j \geq d_j \quad \forall j \in \mathscr{E}_{\text{arr}} \cup \mathscr{E}_{\text{dep}}, \tag{13.46}$$

$$y_j \geq y_i - s_a + d_a \quad \forall a = (i, j) \in \mathscr{A}_{\text{drive}} \cup \mathscr{A}_{\text{wait}} \cup \bar{\mathscr{A}}_{\text{head}} \cup \mathscr{A}_{\text{turn}}, \tag{13.47}$$

$$y_j \geq y_i - s_a - M z_r \quad \forall r \in \mathscr{R}, a = (i, j) \in \mathscr{A}_{\text{change}} \cap r, \tag{13.48}$$

$$z_r \in \{0, 1\} \quad \forall a \in \mathscr{A}_{\text{change}}, \tag{13.49}$$

$$y_j \in \mathbb{R} \quad \forall j \in \mathscr{E}_{\text{arr}} \cup \mathscr{E}_{\text{dep}}. \tag{13.50}$$

Constraints (13.46)–(13.47) are the same as in (DM), they ensure that source delays are taken into account and that delays are propagated along driving, waiting, headway, and turn-around activities. Constraints (13.48)–(13.49) replace constraints (13.8)–(13.9) from (DM). They ensure that if a changing activity a is not maintained, all variables z_r for routes r which contain transfer a are set to $z_r = 1$, i.e., the route is marked as not maintained. The other way around, whenever for a route r we have $z_r = 0$, delay is propagated along all changing activities on this route because in this case (13.48) reads

$$y_j \geq y_i - s_a.$$

The objective function (13.45) sums up the delay for each route, which, as specified above, is the delay $y_{i(r)}$ at the end event $i(r)$ of the route r, if r is maintained, and T otherwise.

Since delay variable $y_{i(r)}$ and path variable z_r are multiplied in (13.45), the objective function of (DM-route) is not linear. However, it can easily be linearized by setting $u_r := y_{i(r)}(1 - z_r)$, i.e., u_r is the delay in the last event on route r if the route can be traveled on by the passenger, and 0 otherwise.

Adding the constraints

$$u_r \geq y_{i(r)} - M z_r \ \forall r \in \mathscr{R}, \tag{13.51}$$

$$u_r \geq 0 \ \forall r \in \mathscr{R}, \tag{13.52}$$

and replacing the objective of (13.45) by

$$\min \sum_{r\in\mathscr{R}} w_r(u_r + Tz_r), \tag{13.53}$$

we obtain a linear formulation (DM-route-linear) of (DM-route). As before, $M := \sum_{a\in\mathscr{A}} \max\{d_a - s_a, 0\} + \max\{d_i : i \in \mathscr{E}\}$ is sufficiently large.

As explained beforehand, (DM-const) *overestimates* the objective value of (DM-route). However, for some classes of delay management instances it can be shown that a solution found by (DM-const) is an optimal solution to (DM-route) as well, and that the objective value for both formulations is the same. The most remarkable case are instances where the so-called *never-meet property* for delays holds. The *never-meet property* ensures that each passenger route can be affected by at most one source delay. To formalize this condition for the case of delay management without precedence decisions, denote by $\mathscr{E}_{\text{relevant}}$ all nodes which are delayed when *all* transfers are maintained and by $\mathscr{N}_{\text{relevant}}$ the subgraph induced by these events. Furthermore denote by $\mathscr{E}(i)$ the set of all nodes which can be reached from event i if all transfers are maintained, and by $\mathscr{N}(i)$ the induced subgraph. We say that the never-meet property holds if

1. for each source-delayed event i, $\mathscr{N}(i) \cap \mathscr{N}_{\text{relevant}}$ is a tree, and
2. for each pair of source-delayed events $i \neq j$ $\mathscr{E}(i) \cap \mathscr{E}(j) \cap \mathscr{E}_{\text{relevant}} = \emptyset$.

See [30] for a formal proof of the statement that in case of the never-meet property (DM-const) and (DM-route) are equivalent.

Note that in event-activity networks containing pairs of headway activities, the never-meet property is (almost) never fulfilled since each pair of headway activities forms a cycle. Schachtebeck and Schöbel [25] describe a network reduction technique and show that it is sufficient for the never-meet property to hold in the reduced network to obtain correspondence between the optimal solutions to (DM-const) and (DM-route) and equality of the objective values.

13.3.4.3 Delay Management with Re-routing

The delay management objective functions in the preceding paragraphs were based on the assumption that a passenger has to wait a full time period T if he misses a transfer. However, in many cases there are different ways to reach the destination if a transfer is missed. The problem of making wait-depart decisions to minimize the sum of travel times under the assumption that passengers can re-route is called *delay management with re-routing* (DMwRR).

To introduce the integer programming model for (DMwRR), we first assume that passengers learn about *all* relevant delays and delay management decisions the moment they arrive at their origin station. This could be the case whenever delay management models are used to make a disposition timetable for foreseeable delays, e.g., in case of construction works on the tracks. After presenting the model, we discuss how it can be used in situations where passengers have already left their origin station when delays occur.

Instead of requiring information about all passenger routes $r \in \mathscr{R}$, in (DMwRR) we assume that we know each passenger's origin station, destination station, and arrival time at the origin station. We call these triples OD-pairs, the set of OD-pairs is denoted $\mathscr{OD}$, and the number of passengers belonging to an OD-pair $m \in \mathscr{OD}$ is w^m.

We extend the event-activity network $\mathscr{N}$ by adding an *origin* $i = \text{Org}(m)$ and a *destination event* $j = \text{Dest}(m)$ for each OD-pair $m \in \mathscr{OD}$. Each origin event i is connected by *origin activities* $\mathscr{A}_{\text{org}}(i)$ to all departures from the origin station which take place no earlier than $\text{time}(m)$, the arrival time of the passenger at the origin station. Each destination event j is connected to all arrivals at the destination station by *destination activities* $\mathscr{A}_{\text{dest}}(j)$. We denote the origin and destination events by $\mathscr{E}_{\text{org}}$ and $\mathscr{E}_{\text{dest}}$, respectively, and by $\mathscr{A}_{\text{org}}$ and $\mathscr{A}_{\text{dest}}$ the sets of all origin and destination activities. Note that by computing upper and lower bounds on the passenger travel time, some origin and destination activities may be removed because they would not be used in an optimal solution, see [9] for details.

Every path from origin event to destination event in $(\mathscr{E} \cup \mathscr{E}_{\text{org}} \cup \mathscr{E}_{\text{dest}}, \mathscr{A}_{\text{drive}} \cup \mathscr{A}_{\text{wait}} \cup \mathscr{A}_{\text{change}} \cup \mathscr{A}_{\text{org}} \cup \mathscr{A}_{\text{dest}})$ represents a possible route for the corresponding passenger. To model the passengers' route choice in the re-routing step, we introduce additional binary variables

$$q_a^m = \begin{cases} 1 & \text{OD-pair } m \text{ travels on activity } a, \\ 0 & \text{otherwise} \end{cases}$$

for each activity $a \in \mathscr{A}_{\text{drive}} \cup \mathscr{A}_{\text{wait}} \cup \mathscr{A}_{\text{change}} \cup \mathscr{A}_{\text{org}}(\text{Org}(m)) \cup \mathscr{A}_{\text{dest}}(\text{Dest}(m))$ and each OD-pair $m \in \mathscr{OD}$.

Furthermore, for each OD-pair $m \in \mathscr{OD}$ we introduce a real-valued variable t^m which represents its arrival time at the destination.

Our integer programming formulation of (DMwRR) reads as follows:

$$\min \sum_{m \in \mathscr{OD}} w^m t^m \tag{13.54}$$

such that

$$y_j \geq d_j \quad \forall j \in \mathscr{E}_{\text{arr}} \cup \mathscr{E}_{\text{dep}}, \tag{13.55}$$

$$y_j \geq y_i - s_a + d_a \quad \forall a = (i,j) \in \mathscr{A}_{\text{drive}} \cup \mathscr{A}_{\text{wait}} \cup \bar{\mathscr{A}}_{\text{head}} \cup \mathscr{A}_{\text{turn}}, \tag{13.56}$$

$$y_j \geq y_i - s_a - M z_a \quad \forall a = (i,j) \in \mathscr{A}_{\text{change}}, \tag{13.57}$$

$$q_a^m \leq 1 - z_a \quad \forall m \in \mathscr{OD}, a \in \mathscr{A}_{\text{change}}, \tag{13.58}$$

$$\sum_{a \in \delta^{\text{out}}(i)} q_a^m = 1 \quad \forall m \in \mathscr{OD}, i = \text{Org}(m) \in \mathscr{E}_{\text{org}}, \tag{13.59}$$

$$\sum_{a \in \delta^{\text{out}}(i)} q_a^m = \sum_{a \in \delta^{\text{in}}(i)} q_a^m \quad \forall m \in \mathscr{OD}, i \in \mathscr{E}_{\text{arr}} \cup \mathscr{E}_{\text{dep}}, \tag{13.60}$$

$$1 = \sum_{a \in \delta^{\text{in}}(j)} q_a^m \quad \forall m \in \mathscr{OD}, j = \text{Dest}(m) \in \mathscr{E}_{\text{dest}}, \tag{13.61}$$

$$t^m \geq \pi_i + y_i - M_2(1 - q_a^m) \quad \forall j = \text{Dest}(m) \in \mathscr{E}_{\text{dest}}, a = (i, j) \in \delta^{\text{in}}(j), \tag{13.62}$$

$$z_a \in \{0, 1\} \quad \forall a \in \mathscr{A}_{\text{change}}, \tag{13.63}$$

$$q_a^m \in \{0, 1\} \quad \forall m \in \mathscr{OD}, a \in \mathscr{A}, \tag{13.64}$$

$$y_j \in \mathbb{R} \quad \forall j \in \mathscr{E}_{\text{arr}} \cup \mathscr{E}_{\text{dep}}, \tag{13.65}$$

$$t^m \in \mathbb{R} \quad \forall m \in \mathscr{OD}. \tag{13.66}$$

Hereby $\delta^{\text{out}}(i)$ and $\delta^{\text{in}}(i)$ denote the outgoing and ingoing driving, waiting, transfer, origin, and destination activities of an event i.

Constraints (13.55), (13.56), (13.57), (13.63) and (13.65) are adopted from (DM) and ensure correct delay propagation. For each OD-pair $m \in \mathscr{OD}$, constraints (13.59), (13.60), (13.61) are flow conservation constraints which ensure that each passenger leaves the origin station and arrives at the destination station. In these constraints, we implicitly exclude variables q_a^m that are not defined from the summation.

Constraints (13.62) set the arrival time of a passenger m to the arrival time on the route that this passenger has chosen. Note that for all activities $a \in \delta^{\text{in}}(i)$ but the one chosen by the passenger, $q_a^m = 0$, hence for M_2 big enough, the constraint (13.62) is always fulfilled. However, for the last activity $a' = (i', \text{Dest}(m))$ of the chosen route, $q_{a'}^m = 1$, and hence constraint (13.62) reads $t^m \geq \pi_{i'} + y_{i'}$. This means that the arrival time of passenger m is at least the arrival time on the last arrival event i' on his route. The objective function (13.54) minimizes the sum over the passenger arrival times t^m and hence ensures that t^m is set to $\pi_{i'} + y_{i'}$ in (13.62). Note that minimizing arrival times is equivalent to minimizing passenger delays.

The DMwRR model can also be used to compute optimal wait-depart decisions in case that passengers have already left their origin station. In this case, we simply define the station which passengers reach next after learning about the delays and being presented with re-routing options as $\text{Org}(m)$ and the time they could board trains at these stations as $\text{time}(m)$.

DMwRR was introduced in [7, 9]. In [26] it was shown that the problem is NP-hard, even if there is only one OD-pair. Different heuristic methods to solve the problem using the basic model (DM) are proposed in [6, 27]. Dollevoet et al. [10] describe an iterative method to solve the problem taking into account capacity constraints on a microscopic level.

So far in this section, we have presented several integer programming formulations for the delay management problem. Some of these models incorporate the limited capacity of the infrastructure or the re-routing of passengers in case of delays. Although these models represent reality better, a drawback of these models is their size in terms of binary variables and constraints. In Table 13.3, we report the number of binary variables for some of the models from this section for a number of instances in the Netherlands. For more information on the instances, we refer to Dollevoet et al. [9, 11] and Dollevoet and Huisman [6].
We consider four instances. All instances consider the Mid-Western part of the

Netherlands, where the railway network is very dense. Instances 1 and 2 include a smaller part of the railway network than Instances 3 and 4. Instances 1 and 3 consider only long-distance trains, whereas Instances 2 and 4 also include regional trains. The table lists some characteristics of the instance, the corresponding event-activity network and the sizes of the integer programs for various formulations. It should be noted that (DM-prec) and (DM-plat) consider the same network and timetable, but solve the problem for a time horizon that is a little shorter. Furthermore, for (DMwRR), no number of binary variables is listed for Instance 4. This model was too large even to be loaded into computer memory.

Table 13.3: Characteristics of the event-activity networks and sizes of the integer programs for various delay management models

	Characteristics of the instance					Number of binary variables			
Instance	Stations	Trains	Transfers	OD-pairs	$\lvert\mathcal{E}_{dep}\rvert$	(DM)	(DM-prec)	(DM-plat)	(DMwRR)
1	10	117	1074	355	219	1074	1998	9597	6219
2	34	284	8068	3940	1022	8068	8045	37,972	461,494
3	16	168	7123	914	349	7123	3345	15,913	21,255
4	82	404	13,812	22,256	2053	13,812	12,877	48,841	–

13.4 Heuristics

In the previous section, several models and objective functions for the delay management problem have been discussed. These models can in principle be solved by a commercial IP solver. However, in real-time operations, fast decisions have to be made. In order to make good decisions, the whole network has to be considered. As a consequence, central decision making is required. The central decision maker should communicate each wait-depart decision to a dispatcher responsible for the signals at a specific station. By keeping the signal on "red", the train cannot depart. By changing it to "green", the train might leave. Such a decision process, including the communication between different officers, takes usually a couple of minutes. The computation time of solving a practical instance can therefore be at most a few minutes in a real-world application. Solving the IP models from the previous section to optimality might take too much time. As a consequence, in such cases, heuristics are required. Therefore, in this section, we discuss heuristic algorithms for several delay management models. We first focus on the wait-depart decisions only. Then, we discuss delay management with precedence decisions, delay management with platform re-assignments, and delay management with passenger re-routing.

13.4.1 The Basic Decision: Wait or Depart?

Consider first the model (DM) without any capacity considerations, i.e., with $\mathscr{A}_{\text{head}} = \mathscr{A}_{\text{station}} = \emptyset$. In that case, we only have to determine the wait-depart decisions. Given the wait-depart decisions, the delays and the disposition timetable can easily be computed by using (13.1) for the events in topological order.

In many cases, (DM) can be solved by commercial solvers in a short time. This is demonstrated, for example, in [6, 9]. However, if instances would be too large to be solved by a commercial solver, one could resort to rules-of-thumb to decide which trains should wait and which trains better depart on time. In [1, 19] many rules-of-thumb are described and compared in an online setting. We now describe the most common policies.

A very simple rule-of-thumb is the *no-wait rule*. Here, connecting trains never wait for delayed feeder trains. On the contrary, when using the *all-wait rule*, all connecting trains wait for delayed feeder trains. In between these extremes is the *waiting time rule*. When applying the waiting time rule, a waiting threshold τ_a is determined for every changing activity $a = (i, j) \in \mathscr{A}_{\text{change}}$ in the system. If the arrival of the feeder train, i, is delayed, one first determines how long the connecting train, j, would have to wait in order to maintain the transfer. If this waiting time is smaller than the threshold τ_a, the train waits. Otherwise, it departs on time. As an example, consider a changing activity $a = (i, j) \in \mathscr{A}_{\text{change}}$ and assume that the arrival of the feeder train, i, is delayed. The transfer is maintained if

$$y_i - s_a \leq \tau_a.$$

When determining the waiting thresholds τ_a, one can distinguish between long-distance and regional trains (see [19]), between peak and off-peak hours (see [5]), or use machine learning (see [1]).

Other rules-of-thumb incorporate the number of passengers that want to transfer. For example, Schachtebeck and Schöbel [25] fix all changing activities with a weight w_a above a certain threshold. Kliewer and Suhl [19] compare the number of passengers who want to transfer to the number of passengers already in the feeder train or arriving at the next stop. A similar approach is used in [23], but based on dynamically updating passengers' routes: For every critical transfer, the passengers' routes are computed for the case that it is maintained or it is not maintained (thereby assuming fixed waiting time rules for all other transfers). Based on eight different evaluation criteria it is then decided if such a critical transfer should be maintained or not. The resulting software is currently tested in cooperation with Deutsche Bahn.

For all the rules-of-thumb described so far, the wait-depart decisions can be made instantly.

13.4.2 Adding Precedence Decisions: Which Train Goes First?

If the delay management model includes capacity considerations, one has to decide both on the wait-depart decisions and on the order of the trains. If solving (DM-prec) by a commercial solver takes too much time, one can decide on the order of the trains heuristically. Recall from Sect. 13.3, that the order of the trains is encoded by the variables g_{ij} for $(i,j) \in \mathscr{A}_{\text{head}}$. The model (DM-prec) reduces to a model of type (DM) if the precedence decisions are fixed. Given the precedence decisions, or, equivalently, the set $\bar{\mathscr{A}}_{\text{head}}$, the wait-depart decisions can be obtained by solving (DM) with a commercial solver or by using any of the rules-of-thumb described in the previous section. A wide variety of rules-of-thumb can be applied to determine the precedence decisions. In what follows, we discuss three rules-of-thumb that are discussed in [25].

The heuristics from Schachtebeck and Schöbel [25] fix the precedence decisions based on either the original timetable, or on a preliminary disposition timetable. In the first case, the heuristic is referred to as *First scheduled, first served* (FSFS). Here, the order of the trains is fixed to the order in the original timetable:

$$\bar{\mathscr{A}}_{\text{head}} = \{(i,j) \in \mathscr{A}_{\text{head}} : \pi_i \leq \pi_j\}.$$

In the second case, the heuristic is referred to as *First rescheduled, first served* (FRFS). Here, in the first step, (DM) is solved without any headway constraints ($\mathscr{A}_{\text{head}} = \emptyset$). We obtain a preliminary disposition timetable $x_i := \pi_i + y_i$ for all $i \in \mathscr{E}$. Then, the precedence decisions are fixed according to this preliminary disposition timetable, i.e., as

$$\bar{\mathscr{A}}_{\text{head}} = \{(i,j) \in \mathscr{A}_{\text{head}} : x_i \leq x_j\}.$$

In both cases, the model (DM) is solved with $\bar{\mathscr{A}}_{\text{head}}$ as defined above. This means for the second heuristic, that model (DM) must be solved twice. If this takes too much time, one can also fix both the priority decisions and the wait-depart decisions that are obtained when solving (DM). In that case, the disposition timetable can be found by using (13.1) iteratively in topological order. This heuristic is referred to as (FRFS-FIX).

The heuristics for (DM-prec) are compared on real-world instances from Germany in [25]. It is shown that (FRFS) is slightly better than (FRFS-FIX). Both find close-to-optimal solutions more often than (FSFS). However, there are a few scenarios for which the quality of (FRFS) and (FRFS-FIX) deviates much from the optimum. Combining (FSFS) with (FRFS), by applying both and choosing the best solution, one obtains the benefit of both heuristics. All heuristics can be executed within seconds.

13.4.3 Adding Decisions in Stations: Which Platform to Use?

The model (DM-plat) from Sect. 13.3 includes all precedence decisions and allows to reschedule the platform assignment. Recall that a precedence decision must be made for all pairs of trains that might use a common element in the railway infrastructure. This means that a decision variable $\bar{g}_{ij}$ must be included for all pairs of trains that stop at a common station. As a consequence, real-world instances of this problem become huge and cannot be solved to optimality in a real-time setting. In this section, we discuss an iterative algorithm from Dollevoet et al. [11] to solve (DM-plat). The algorithm optimizes the platform assignment independently from the precedence and wait-depart decisions. By doing so, the platform assignments can be solved individually for each station.

Algorithm 13.1: Delay management with platform reassignment

Data: The parameters w_a for $a \in \mathscr{A}_{\text{change}}$ and w_i for $i \in \mathscr{E}$
Result: Variables z_a for $a \in \mathscr{A}_{\text{change}}$, y_i for $i \in \mathscr{E}$
Init: Fix the platform assignment to the planned platform assignment;
while *Improvements are found & Maximum number of iterations not reached* **do**
  Solve model (DM-prec) with the current platform assignment;
  forall the *Stations v* **do**
    Determine a new platform assignment for station v;
  end
end

The iterative algorithm is presented in Algorithm 13.1. It first fixes the platform assignment to the scheduled one. Then, it optimizes the precedence and wait-depart decisions for fixed variables c_{ip}. Using the disposition timetable, it identifies which delays might be reduced by changing the platform assignment. It then maximizes the *potential* for reducing these delays by optimizing the platform assignment. These steps are repeated as long as the platform assignment changes. Note that, for fixed values c_{ip}, problem (DM-plat) reduces to a problem of type (DM-prec), where only precedence and wait-depart decisions have to be taken. Solving it, we obtain a solution y_i for (DM-prec).

The most interesting step in the iterative algorithm is the optimization of the platform assignment. In this step, we decide on the variables c_{ip}. Recall that $c_{ip} = 1$ if event $i \in \mathscr{E}_{\text{arr}}$ is scheduled at platform p and 0 otherwise. In order to explain the procedure to optimize the platform assignment, we introduce some additional notation. First, we consider all events $i \in \mathscr{E}$ that take place at a given station $v \in V$. For each waiting activity $(i, j) \in \mathscr{A}_{\text{wait}}$, y_i and y_j represent the delays of the train when arriving and departing from station v, respectively. We define the platform occupancy for this waiting activity as the interval $[h_i, h_j]$, where $h_i = \pi_i + y_i - l_i$ and $h_j = \pi_j + y_j + l_j$ take into account the times needed to enter and leave the station, respectively. A platform assignment schedules every waiting activity at v at a certain platform in v, in such a way that the intervals $[h_i, h_j]$ and $[h_{i'}, h_{j'}]$ are disjoint if waiting activities $a = (i, j)$ and $a' = (i', j')$ are scheduled at the same platform. We are now ready to explain how to find a platform assignment that maximizes the potential for reducing delays.

Some of the delays in the disposition timetable might be caused by the platform being occupied when a train is about to enter a station. In order to determine which delays can be reduced by changing the platform assignment, we determine for each arrival event $i \in \mathscr{E}_{\text{arr}}$ at station v, a *target time* τ_i. This target time represents the time when the train would start entering the system if there would be no other trains in the system. Let $a = (k, i) \in \mathscr{A}_{\text{drive}}$ be the unique driving activity whose head is i. If there is a train at the platform before event i takes place, we define the target time by

$$\tau_i = \max\{\pi_i + d_i, \pi_k + y_k + d_a - s_a\} - l_i.$$

Here, again, l_i is the time needed to enter the station. If there is no train at the platform just before event i takes place, or if $y_i = 0$ in the disposition timetable, we define $\tau_i = h_i$. In the platform assignment step, we now look for values $\tau_i \leq q_i \leq h_i$ and a platform assignment, such that the intervals $[q_i, h_j]$ are mutually disjoint for all waiting activities that are scheduled at the same platform. The objective is to minimize

$$\sum_i w_i q_i.$$

Here, w_i is a weight that measures the importance of arrival event i. If $q_i < h_i$, there is a potential for reducing the delay of arrival event i. It is proven in [11], that this problem can be modeled as a flow problem and solved in polynomial time.

This implies that the platform assignment can be solved easily. Note that we only consider the platform occupancy and assume that all trains can use all platforms when determining the platform assignment. In particular, we do not take the routing inside the stations into account. Even in this case, if trains could only use a subset of the platforms, the platform assignment is NP-hard; see [20].

It then remains to solve the delay management model (DM-prec). This can be done, for example, by using the IP formulation from the previous section with a commercial solver or by using any of the rules-of-thumb described above. Numerical experiments in [11] on a medium-size network show that running times can be reduced by 70% by using the iterative approach. Here, the model (DM-prec) is solved by a commercial solver. This speed up comes at the cost of an increase in total delay of at most 1%. In particular, for a medium-size instance including both regional and long distance trains, the running times decrease from around 10 min to less than 3 min. For large-size instances with both regional and long distance trains, the computation time is still too long. For these instances, model (DM-prec) should be solved heuristically as well.

13.4.4 Adding Decisions on Passengers Paths: Which Route to Take?

An integer programming formulation for delay management with passenger re-routing (DMwRR) is given by (13.54)–(13.66). A key feature of this model is its objective function. In contrast to other delay management models, it computes the delay for passengers by finding an alternative travel option whenever a transfer is not maintained. By doing so, the delays for passengers can be computed more realistically.

Recall from the previous section that (DMwRR) integrates the problems of deciding on the wait-depart decisions on the one hand, and on routing the passengers through the railway network on the other. One can find optimal solutions to (DMwRR) by solving its integer programming formulation with a commercial solver. However, it is proven in [9] that (DMwRR) without any capacity considerations is NP-hard even in special cases. In particular, it is shown to be NP-hard in [26] in case there is only *one* OD-pair. As a consequence, solving large-scale instances of (DMwRR) might require a considerable amount of computation time.

13.4.4.1 Passenger Re-routing with Constant Penalties

The heuristics for (DMwRR) decouple the optimization of the wait-depart decisions from the passenger routing. We first decide on the wait-depart decisions by solving any model from Sect. 13.3. These models decide on the wait-depart decisions and determine the delays y_i for all $i \in \mathscr{E}$. Given these values, the routes that minimize the arrival delay can be computed for all OD-pairs individually.

As described in Sect. 13.3.4, the parameters w_i and w_a can be related to the number of passengers that arrive with event $i \in \mathscr{E}$ and make use of changing activity $a \in \mathscr{A}_{\text{change}}$, respectively. To incorporate the delay for passengers who miss a transfer (corresponding to a changing activity $a \in \mathscr{A}_{\text{change}}$), the objective includes a penalty that represents the delay caused by missing the transfer. In Sect. 13.3.4, it is proposed to multiply the number of passengers w_a with the period time T: The penalty is then given by w_aT. In our heuristic, we replace T by a parameter to be determined and denote it by D. Using the set of OD-pairs $\mathscr{OD}$, we then rewrite the objective function as follows.

$$\min_{y,z} \sum_{i\in\mathscr{E}_{\text{arr}}} \sum_{m\in\mathscr{OD}(i)} w_m d_i + \sum_{a\in\mathscr{A}_{\text{change}}} \sum_{m\in\mathscr{OD}(a)} w_m z_a D.$$

Here, the set $\mathscr{OD}(i)$ contains all OD-pairs $m \in \mathscr{OD}$ that arrive with event $i \in \mathscr{E}_{\text{arr}}$ according to the *planned* timetable. Similarly, the set $\mathscr{OD}(a)$ contains all OD-pairs $m \in \mathscr{OD}$ that planned to make use of changing activity $a \in \mathscr{A}_{\text{change}}$. The delay for passengers who miss a transfer is assumed to be equal to D. By varying the value of D, one can evaluate different delay management strategies. It turns out that D should be smaller than the period of the timetable, if there are multiple train lines along each edge $e \in E$ in the physical railway network (PTN). Intuitively, this makes perfect sense: If there are more train lines along an edge in the network, there are multiple travel options along that edge in each period. Therefore, passengers who miss a transfer do not have to wait a complete time period, but can take any other train going along the edge.

In Sect. 13.4.1, we introduced the *no-wait* and *always-wait* policies. By choosing the value of the parameter D equal to zero, we obtain a no-wait policy. On the contrary, if the penalty is set to a very high value, we obtain an always-wait policy.

13.4.4.2 Passenger Re-routing with OD-Dependent Penalties

The heuristic described in the previous paragraph includes a penalty D for all passengers who miss a transfer. The penalty represents the additional delay these passengers incur as a consequence of not being able to transfer from one train to another. In reality, this additional delay will vary among the different OD-pairs. For example, a passenger who misses a transfer will generally wait for the next train to its destination at the transfer station. The additional delay for that passenger then depends on the number of train lines that are operated between the transfer station and the passenger's destination. In order to improve the quality of the heuristic, we now include in the objective a penalty D_m that depends on the OD-pair m. The value of this parameter is updated in an iterative algorithm. We now first give an overview of our approach in Algorithm 13.2 and then discuss its details.

The algorithm initializes the penalties $D_m = 0$ for all OD-pairs $m \in \mathcal{OD}$. It then repeats the following procedure for at most $maxIter$ iterations. It first solves a model without passenger re-routing using the penalties D_m. The output of this model are the wait-depart decisions and the disposition timetable. Given this disposition timetable, the actual routes for the passengers can be computed. For each OD-pair $m \in \mathcal{OD}$, we thus find the arrival delay d_m. From this arrival delay, we can compute the additional delay $D_m = d_m - y_i$ caused by missing the transfer, where i is the event OD-pair m planned to arrive with. We keep track of whether the values D_m have changed in the current iteration. If all values D_m remain unchanged, the algorithm terminates. Otherwise, we obtained new penalties D_m and repeat the procedure.

This iterative heuristic has been implemented and tested on a real-world railway network in [6]. Here, (DMwRR) is used without any headway activities. For these instances, solving (DMwRR) to optimality takes up to 6 min for medium-sized instances. For large instances, no solution can be obtained at all by using the IP. In the heuristic, we implemented (DM-const), also without headway activities. It is shown that the iterative algorithm generally converges within few iterations. The computation times are in the order of 10 s and thus allow for practical application. In terms of quality, solutions obtained by the iterative heuristic are only slightly worse than those obtained by using an exact algorithm. This suggests that the re-routing aspect of delay management can be dealt with by iteratively solving a sequence of delay management models that do not include passenger re-routing.

Algorithm 13.2: DMwRR with OD-dependent penalties D_m

```
Data: The parameters w_a for a ∈ A_change and w_i for i ∈ E, maxIter
Result: Variables z_a for a ∈ A_change, y_i for i ∈ E
Init: iter = 0;
Init: D_m = 0 for all m ∈ OD ;
Init: changed = true;
while changed & iter < maxIter do
    changed = false;
    Solve any model without re-routing with penalties D_m for m ∈ OD;
    forall the OD-pairs m ∈ OD do
        Compute the arrival delay d_m for OD-pair m;
        Let i be the event OD-pair m planned to arrive with;
        if Transfer missed & D_m ≠ d_m − y_i then
            D_m := d_m − y_i;
            changed = true;
        end
    end
    iter = iter + 1;
end
```

13.5 Practical Considerations and Conclusions

In practice, often simple waiting time rules are implemented. Examples of such rules are: "Wait at most y minutes starting from the original timetable" (see Sect. 13.4.1 and, e.g., [1]). Often these rules are static, which means they are updated at most once a year, and are the same for each day and for each period of the day, although passenger demand might significantly differ between peak and off-peak hours, during different days of the week, and during the year. As an example, in the Netherlands, two so-called waiting time rules are applied. One rule is applied for all trains except the last train on a day, and the other rule is applied for a transfer to the last train on a particular route. The first rule states that a train should wait for a feeder train if this feeder train has less than x minutes delay. The delay of the feeder train is measured at a certain specific location, typically a few kilometers before the transfer station. The value of x is determined by the slack in the timetable, the number of transferring passengers and the frequency of the trains. In the Netherlands, the most common value for x is 0, i.e., most often a no-wait policy is applied.

Unfortunately, the use of exact delay management models and advanced heuristics such as presented in this chapter, is still limited in practice. We think that there are five main reasons for this:

1. Although punctuality has always been an important performance indicator in the railway sector (both for operators and for infrastructure managers), this performance is usually measured in terms of train punctuality.[1] As a result, the focus was on running trains on time instead of transporting passengers on time.

[1] Train punctuality is often computed as the percentage of trains that arrive within x minutes of their scheduled arrival time. Usually, this is not measured at all stations but only at a limited subset.

2. The lack of accurate passenger data which are necessary to compute an optimal solution.
3. The computational power required to compute optimal solutions of (advanced) delay management models in real-time.
4. Delays are uncertain in practice while delay management models are deterministic.
5. In some countries, there are several railway operators which on the one hand compete with each other but on the other hand should cooperate, because passengers might have to transfer from one operator to the other.

Although the last reason above is still a topic of ongoing political debate, we believe that the first four reasons can be dealt with and thus the interest in exact optimization models will change in the coming years.

Firstly, railway operators tend to focus much more on improving the service to their passengers and attracting new passengers. As a consequence, they change their focus from running trains on time to transporting passengers on time. In several countries, including Switzerland, Denmark and the Netherlands, the national railway operator reports passenger punctuality as key performance indicator. Passenger punctuality is then measured as an approximation of the percentage of passengers arriving within x minutes of their promised arrival time.

Secondly, smart cards and/or electronic ticketing are introduced in many countries. When using smart cards, a passenger has to check-in at his origin and check-out at his destination station. As an effect, accurate passenger figures become available, even in real-time. The first and second development strengthen each other. A nice example is the introduction of improved performance indicators on passenger punctuality in the contract between the Dutch government on one side and the main Dutch railway operator, Netherlands Railways, and the Dutch rail infrastructure manager, ProRail, on the other side (see [16]). In these indicators, the passenger punctuality gives an indication of the percentage of passengers with a delay of less than 5 respectively 15 min. This means that the difference between the actual arrival time of a passenger and the planned arrival time according to the official travel planner is less than 5 (15) min. Here, the promised arrival time is the arrival time of the fastest option from the check-in to the check-out station, which can be made directly after the check-in time in the official travel planner 2 days in advance. The realized arrival time is the arrival time of the last train in the travel advice if all trains were operated and all transfers were realized. In any other case, the realized arrival time is estimated by the passenger's check-out time. In 2015, the passenger punctuality on the main Dutch railway network was 90.0% on 5 min and 97.0% on 15 min. According to the contract, in 2019 these two figures have to be improved to 91.3% and 97.3%, respectively.

Thirdly, the models as discussed in this chapter become more advanced and thus more realistic. In addition, faster computers and algorithmic developments will decrease computation times. In this way, it is likely that advanced delay management models can be solved fast enough to be used in real-time.

Finally, uncertainty can be taken into account by using a rolling horizon approach. This means that every time new information becomes available, a new instance of delay management is solved, see [19]. Some approaches even try to find robust delay management solutions by taking into account further delays that may occur, see [2].

Nevertheless, the use of exact algorithms is a promising tool when developing rules of thumb as the ones developed in [1, 5]. These references show that more advanced rules, which are still easily applicable in practice, give close to optimal solutions. An example of such a rule might be "If train g has a delay of at most x minutes, then train h should wait at station v", where g, h, and x differ from station to station, between periods of the day and between different days of the week. Such rules can be updated much more frequently than is done currently by simulating different wait-depart strategies and comparing the rule to the optimal solutions of the exact models.

In [5], four critical transfers on the Dutch railway network are analyzed. For different source delays for each train, the optimal wait-depart decisions are computed by the exact delay management model with passenger re-routing. In the optimal decisions, x could vary for each train and thus for each time of the day. Since this is difficult to manage for dispatchers, de Lugt used several statistical methods to get generic rules-of-thumb for different time intervals during the day. As a result the same decision is taken in such an interval. For instance, one could get a certain value for the morning-peak, the off-peak hours and the evening peak. By introducing three intervals instead of the same rule for the whole day (as currently done in practice), the total passenger delay could be reduced by 20% in 3 out of the 4 considered transfers.

In [1] values for x for all *changing activities* of trains g and h are determined, i.e., these values may vary from hour to hour. The authors use a machine-learning approach: For a large set of delay scenarios, the delay management problem is solved exactly. From the resulting data, one can identify how long it makes sense to wait for a delayed train (for every changing activity). The resulting numbers are then used as a quick rule of thumb for the wait-depart decision in new scenarios. Using data from the LinTim library (see [12, 14]) it is shown that this rule of thumb performs best among all tested strategies.

References

1. Bauer R, Schöbel A (2014) Rules of thumb – practical online strategies for delay management. Public Transp 6(1):85–105. https://doi.org/10.1007/s12469-013-0082-8 (cited on pages 306, 312, 314)
2. Cicerone S, Di Stefano G, Schachtebeck M, Schöbel A (2012) Multi-stage recovery robustness for optimization problems: a new concept for planning under disturbances. Inf Sci 190:107–126. https://doi.org/10.1016/j.ins.2011.12.010 (cited on page 314)
3. Corman F, D'Ariano A, Pacciarelli D, Pranzo M (2012) Bi-objective conflict detection and resolution in railway traffic management. Transp Res C 20:79–94. https://doi.org/10.1016/j.trc.2010.09.009 (cited on pages 295, 297)

4. De Giovanni L, Heilporn G, Labbé M (2008) Optimization models for the single delay management problem in public transportation. Eur J Oper Res 189(3):762–774. https://doi.org/10.1016/j.ejor.2006.10.065 (cited on page 295)
5. de Lugt N (2013) Delay management: improving rules of thumb concerning wait-depart decisions. MSc thesis, Tilburg University (cited on pages 306, 314)
6. Dollevoet T, Huisman D (2014) Fast heuristics for delay management with passenger rerouting. Public Transp 6(1–2):74–89. https://doi.org/10.1007/s12469-013-0076-6 (cited on pages 304, 306, 311)
7. Dollevoet T, Huisman D, Schmidt M, Schöbel A (2009) Delay management with re-routing of passengers. In: Clausen J, Di Stefano G (eds) ATMOS 2009. Dagstuhl Seminar Proceedings. ISBN: 978-3-939897-11-8. https://doi.org/10.4230/OASIcs.ATMOS.2009.2143 (cited on page 304)
8. Dollevoet T, Schmidt M, Schöbel A (2011) Delay management including capacities of stations. In: Caprara A, Kontogiannis S (eds) 11th workshop on algorithmic approaches for transportation modelling, optimization, and systems. OpenAccess series in informatics (OASIcs), vol 20. Schloss Dagstuhl-Leibniz-Zentrum fuer Informatik, Dagstuhl, pp 88–99. ISBN: 978-3-939897-33-0. https://doi.org/10.4230/OASIcs.ATMOS.2011.88 (cited on page 298)
9. Dollevoet T, Huisman D, Schmidt M, Schöbel A (2012) Delay management with rerouting of passengers. Transp Sci 46(1):74–89. https://doi.org/10.1287/trsc.1110.0375 (cited on pages 303, 304, 306, 310)
10. Dollevoet T, Corman F, D'Ariano A, Huisman D (2014) An iterative optimization framework for delay management and train scheduling. Flex Serv Manuf J 26(4):490–515. https://doi.org/10.1007/s10696-013-9187-2 (cited on page 304)
11. Dollevoet T, Huisman D, Kroon L, Schmidt M, Schöbel A (2015) Delay management including capacities of stations. Transp Sci 49(2):185–203. https://doi.org/10.1287/trsc.2013.0506 (cited on pages 299, 304, 308, 309)
12. Gattermann P, Harbering J, Schiewe A, Schöbel A (2017) LinTim – integrated optimization in public transportation, June 2017. https://lintim.math.uni-goettingen.de (cited on page 314)
13. Ginkel A, Schöbel A (2007) To wait or not to wait? The bicriteria delay management problem in public transportation. Transp Sci 41(4):527–538. https://doi.org/10.1287/trsc.1070.0212 (cited on page 295)
14. Goerigk M, Schachtebeck M, Schöbel A (2013) Evaluating line concepts using travel times and robustness: simulations with the Lintim toolbox. Public Transport 5(3):267–284. https://doi.org/10.1007/s12469-013-0072-x (cited on page 314)
15. Goerigk M, Knoth M, Müller-Hannemann M, Schmidt M, Schöbel A (2014) The price of strict and light robustness in timetable information. Transp Sci 48:225–242. https://doi.org/10.1287/trsc.2013.0470 (cited on page 289)
16. I&M (2016) Vervoerplan NS 2017 en beheerplan ProRail 2017. Letter of the Dutch parliament from "Ministerie van Infrastructuur en Milieu" (December 13, 2016) (cited on page 313)

17. Kecman P, Corman F, Peterson A, Joborn M (2015) Stochastic prediction of train delays in real-time using Bayesian networks. In: Conference on advanced systems in public transport (CASPT) (cited on page 294)
18. Kirchoff F (2015) Verspätungsfortpflanzung in Bahnnetzen, Modellierung und Berechnung mit Verteilungsfamilien. PhD thesis, University of Technology Claustal, Clausthal-Zellerfeld (cited on page 294)
19. Kliewer N Suhl L (2011) A note on the online nature of the railway delay management problem. Networks 57(1):28–37. https://doi.org/10.1002/net.20381 (cited on pages 306, 314)
20. Lamorgese L, Mannino C (2015) An exact decomposition approach for the real-time train dispatching problem. Oper Res 63:48–64. https://doi.org/10.1287/opre.2014.1327 (cited on page 309)
21. Liebchen C, Möhring R (2007) The modeling power of the periodic event scheduling problem: railway timetables – and beyond. In: Geraets F, Kroon L, Schoebel A, Wagner D, Zaroliagis CD (eds) Algorithmic methods for railway optimization. Lecture notes on computer science, vol 4359. Springer, Berlin, pp 3–40. https://doi.org/10.1007/978-3-540-74247-0_1 (cited on page 289)
22. Nachtigall K (1998) Periodic network optimization and fixed interval timetables. Habilitation. University of Hildesheim (cited on page 289)
23. Rückert R, Lemnian M, Blendinger C, Rechner S, Müller-Hannemann M (2017) PANDA: a software tool for improved train dispatching with focus on passenger flows. Public Transp 9(1–2):307–324 https://doi.org/10.1007/s12469-016-0140-0 (cited on page 306)
24. Schachtebeck M, Schöbel A (2008) IP-based techniques for delay management with priority decisions. In: Fischetti M, Widmayer P (eds) Proceedings of the 8th workshop on algorithmic approaches for transportation modeling, optimization, and systems (ATMOS 2008). Dagstuhl seminar proceedings. https://doi.org/10.4230/OASIcs.ATMOS.2008.1586 (cited on page 297)
25. Schachtebeck M, Schöbel A (2010) To wait or not to wait-and who goes first? Delay management with priority decisions. Transp Sci 44(3):307–321. https://doi.org/10.1287/trsc.1100.0318 (cited on pages 295, 296, 297, 302, 306, 307)
26. Schmidt M (2013) Simultaneous optimization of delay management decisions and passenger routes. Public Transp 5(1):125–147. https://doi.org/10.1007/s12469-013-0069-5 (cited on pages 304, 310)
27. Schmidt M (2014) Integrating routing decisions in public transportation problems. Optimization and its applications, vol 89. Springer, New York. https://doi.org/10.1007/978-1-4614-9566-6 (cited on page 304)
28. Schöbel A (2001) A model for the delay management problem based on mixed-integer programming. Electron Notes Theor Comput Sci 50(1):1–10. https://doi.org/10.1016/s1571-0661(04)00160-4 (cited on pages 295, 301)
29. Schöbel A (2006) Optimization in public transportation. Stop location, delay management and tariff planning from a customer-oriented point of view. Optimization and its applications. Springer, New York. https://doi.org/10.1007/978-0-387-36643-2 (cited on pages 289, 295)

30. Schöbel A (2007) Integer programming approaches for solving the delay management problem. In: Geraets F, Kroon L, Schöbel A, Wagner D, Zaroliagis CD (eds) Algorithmic methods for railway optimization. Lecture notes in computer science, vol 4359. Springer, Heidelberg, pp 145–170. https://doi.org/10.1007/978-3-540-74247-0_7 (cited on pages 295, 299, 302)
31. Schöbel A (2009) Capacity constraints in delay management. Public Transp 1(2):135–154. https://doi.org/10.1007/s12469-009-0010-0 (cited on pages 293, 297)

Index

R. Borndörfer et al. (eds.), *Handbook of Optimization in the Railway Industry*,
International Series in Operations Research & Management Science 268,
https://doi.org/10.1007/978-3-319-72153-8

FSC
www.fsc.org
MIX
Papier | Fördert
gute Waldnutzung
FSC® C083411

Kolibri 360 GmbH
Ferdinand-Jühlke-Straße 7
99095 Erfurt, Deutschland
produktsicherheit@kolibri360.de